W9-DIM-773

WITHDRAWN FROM LIBRARY

MATHEMATICAL PROBLEM SOLVING

MATHEMATICAL PROBLEM SOLVING

ALAN H. SCHOENFELD

School of Education
Department of Mathematics
University of California
Berkeley, California

1985

ACADEMIC PRESS, INC.
Harcourt Brace Jovanovich, Publishers
Orlando San Diego New York
Austin Boston London Sydney
Tokyo Toronto

ACADEMIC PRESS, INC.
Orlando, Florida 32887

United Kingdom Edition published by
ACADEMIC PRESS INC. (LONDON) LTD.
24–28 Oval Road, London NW1 7DX

LIBRARY OF CONGRESS CATALOGING IN PUBLICATION DATA

Schoenfeld, Alan H.
Mathematical problem solving.

Bibliography: p.
Includes index.
1. Problem solving. 2. Mathematics—Problems, exercises, etc. I. Title.
QA63.S35 1985 511.3 85-1360
ISBN 0-12-628870-4 (alk. paper)

PRINTED IN THE UNITED STATES OF AMERICA

87 88 9 8 7 6 5 4 3 2

for Jane

Contents

Preface

In the fall of 1974 I ran across George Pólya's little volume, *How to Solve It.* I was a practicing mathematician, a few years out of graduate school and happily producing theorems in topology and measure theory. Pólya wrote about problem solving, more specifically about the strategies used by mathematicians to solve problems. The book was fun to read. I zipped through it, nodding my head in agreement with the author; the strategies he described for solving problems were, with uncanny accuracy, the kinds of things I did when I did mathematics.

My first reaction to the book was sheer pleasure. If, after all, I had discovered for myself the problem-solving strategies described by an eminent mathematician, then I must be an honest-to-goodness mathematician myself! After a while, however, the pleasure gave way to annoyance. These kinds of strategies had not been mentioned at any time during my academic career. Why wasn't I given the book when I was a freshman, to save me the trouble of discovering the strategies on my own?

The next day I spoke to the colleague who trained our department's team for the Putnam exam, a prestigious nationwide mathematics competition. Did he use Pólya's book? "No," he said. "It's worthless." His teams did quite well, so there must have been some truth in what he said—but at the same time, I had the gut feeling that Pólya had identified something significant. The conflict had to be resolved.

Since then the two major questions that have preoccupied me are, What does it mean to "think mathematically"? and, How can we help students to do it? Those are the issues at the core of this book, which summarizes a decade of efforts to understand and teach mathematical problem-solving skills.

What does this book have to offer, and to whom? It is addressed to people with research interests in the nature of mathematical thinking at any level, to people with an interest in "higher-order thinking skills" in any domain, and to all mathematics teachers. The focal point of the book is a framework for the analysis of complex problem-solving behavior. That framework is presented in Part One, which consists of Chapters 1 through 5. It describes four qualitatively different aspects of complex intellectual activity: *cognitive resources,* the body of facts and procedures at one's disposal; *heuristics,* "rules of thumb" for making progress in difficult situations; *control,* having to do with the efficiency with which individuals utilize the knowledge at their disposal; and *belief systems,* one's perspectives regarding the nature of a discipline and how one goes about working in it. Part Two of the book, consisting of Chapters 6 through 10, presents a series of empirical studies that flesh out the analytical framework. These studies document the ways that competent problem solvers make the most of the knowledge at their disposal. They include observations of students, indicating some typical roadblocks to success. Data taken from students before and after a series of intensive problem-solving courses document the kinds of learning that can result from carefully designed instruction. Finally, observations made in typical high school classrooms serve to indicate some of the sources of students' (often counterproductive) mathematical behavior.

As the scope of the categories in the framework suggests, I argue that coming to grips with any discipline—and, in particular, learning to think mathematically—involves a great deal more than having large amounts of subject-matter knowledge at one's fingertips. It includes being flexible and resourceful within the discipline, using one's knowledge efficiently, and understanding and accepting the tacit "rules of the game." The framework represents an attempt to elaborate a spectrum of behaviors that comprise mathematical thinking. My experience and most of my research observations have been at the secondary and college levels, so the discussions deal with mathematics at those levels. The issues, however, are relevant to investigations of mathematical behavior at any level. In fact, most of the issues discussed here apply in much broader contexts than mathematics. Higher-order skills such as monitoring and assessing one's progress "on line" (and thereby avoiding wild-goose chases) are as important in physics or in writing an essay as they are in mathematics. To address these issues in mathematics I have borrowed or adapted methodologies from artificial intelligence and information-processing psychology (more generally, from cognitive science), research on writing, naive physics, and decision theory, to name just a few. In turn, the methodological tools developed and discussed here can be applied (or adapted) to address higher-order thinking skills in those domains.

As noted above, my interests in understanding and teaching mathematical-thinking skills go hand in hand. The research described here indicates that when instruction focuses almost exclusively on mastery of facts and procedures, students are not likely to develop some of the higher-order skills necessary for using mathematics. It also indicates that when teaching focuses on those skills, students can learn them. This book describes much of the "why" and some of the "how to" of teaching problem solving; it describes the results of instruction that does, and does not, focus on problem solving. It suggests ways that we might wish to teach mathematics, so that our students will indeed learn to think mathematically. I hope it will be of interest to teachers of mathematics at all levels.

Acknowledgments

A number of good friends and colleagues have helped to shape the contents of this book. Working with Karel deLeeuw, I learned to do real mathematics and also to pursue issues that seem important whether or not they happen to be fashionable. It was Ruth von Blum who convinced me that issues of mathematical cognition and of mathematics teaching were worth taking seriously as objects of intellectual inquiry. Fred Reif showed me that such issues can be approached with the same care and precision with which one expects to approach difficult questions in mathematics and science. John Seely Brown has been a consistently valuable critic and source of ideas through the years. To these and many other colleagues I owe sincere thanks; the work described here is much better for their help. The mistakes, of course, are all mine.

This book, and a large part of the research on which it is based, were produced with the help of a research grant from the Spencer Foundation; the earlier empirical work was supported by the National Science Foundation. It is a pleasure to acknowledge that generous support. The grant from the Spencer Foundation enabled David Spanagel, Margaret Davidson, and Roger Meike to take part in the project. David Spanagel's help included developing research tools, videotaping classes in local schools, reviewing and proofreading draft versions of the manuscript, and compiling the author index. Margaret Davidson grappled successfully with the transcription of barely audible audiotapes and typed the myriad draft versions of the manuscript. Roger Meike was our resource for computation, both for statistical analyses and simulation models. Jane Schoenfeld compiled the subject index. Herb Ginsburg and Fred Reif read the $(n-1)$st draft version of the manuscript and made valuable suggestions for its improvement. Each of these people made a significant contribution to the book you are now reading.

It was a pleasure to work with Academic Press, whose staff consistently produced high-quality work, on schedule, as the book evolved from manuscript to finished form.

To all of the friends and colleagues who contributed in so many ways to the book, my most grateful thanks.

Introduction and Overview

This book is about doing, understanding, and teaching mathematical problem solving. Most of the problems discussed are appropriate for college freshmen, with a tolerance of about 2 years; that is, they are generally accessible to tenth or eleventh graders, but mathematics majors at the junior level in college will often find them challenging. Few of the problems require a knowledge of mathematics as sophisticated as calculus in order to be solved. Virtually all, however, require a substantial amount of thinking. Understanding the nature of mathematical thinking is the issue at the core of this book, and pursuing it will lead us to a host of related issues. The following hypothetical experiment introduces some of the major ones.

Two groups of people participate in that experiment. The first group consists of a dozen mathematically talented undergraduates, say, the top 12 first-year mathematics majors at a particular university. The second group consists of a dozen members of the mathematics department at the same university. The mathematicians are randomly selected, save for one condition: They have not done any plane geometry for at least 10 years. (Surprisingly, this is not an unusual condition. Randomly selected mathematicians have not, in all likelihood, done any plane geometry since their high school days.) Each of the participants will be asked to solve a series of geometry problems similar to Problems 1.1 and 1.2, given below.

Problem 1.1 You are given two intersecting straight lines and a point P marked on one of them, as in the figure below. Show how to construct, using straightedge and compass, a circle that is tangent to both lines and that has the point P as its point of tangency to one of the lines.

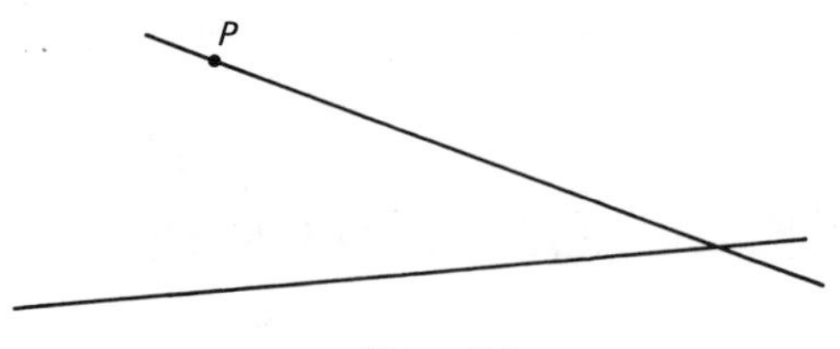

Figure 0.1

Problem 1.2 You are given a fixed triangle T with base B, as in the figure below. Show that it is always possible to construct, with straightedge and compass, a straight line that is parallel to B and that divides T into two parts of equal area. Can you similarly divide T into five parts of equal area?

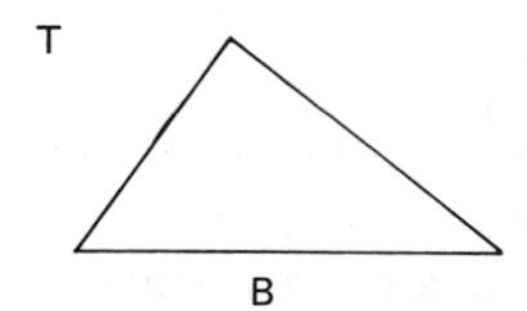

Figure 0.2

Problems like these, which are discussed extensively in the sequel, are "nonstandard" in that they are not typically covered in high school geometry courses. The college students are likely to possess more than enough factual knowledge to solve these problems. However, the nonstandard nature of the problems ensures that the students will not be able to solve them by simply recalling and applying familiar solution patterns (known as problem *schemata*). Nor will the problems be familiar to the faculty. Moreover, since none of the faculty has done geometry for many years, their initial recall of directly relevant facts and procedures is likely to be a good deal more shaky than the students'. We can expect both groups of subjects to find at least some of the problems challenging.

The subjects are trained to solve problems out loud. Their verbal reports as they work are recorded, and the recordings are transcribed. These transcripts (called *protocols*) and the written work produced by the subjects constitute our data. We consider two questions:

1. Which group will do better on the problems, and why?

2. Given the protocol and the written work produced by one of the subjects, would we be able to determine whether that person was a student or a professor?

In answer to the first question most people choose the faculty and then justify their choice by saying that "the faculty are brighter" or "they know more" or "they have more experience." In fact, the faculty do outperform the students, and the contest is not even close. But these justifications are not really explanations. Let us take them one at a time. We can dispatch with "the faculty are brighter" by altering the experiment in the following way. Suppose that we were able to obtain protocols of these faculty members solving the geometry problems when they themselves were students. Would they do better on the problems as students or as professional mathematicians? Provided that the problems were sufficiently difficult, we would probably bet on the faculty members as practicing mathematicians—so simple brainpower is eliminated as an explanation.

What do we mean by "the faculty know more"? The obvious meaning—that they have a better mastery of the facts and procedures required for plane geometry—is false. After a 10-year hiatus, many of the specific facts and procedures that the faculty once knew have faded from memory. Proofs of theorems that were once required, and even knowledge of basic results—for example, that every angle inscribed in the diameter of a circle is a right angle—may well be forgotten. When they begin work on the problems the students will remember more of the basics and thus have the initial advantage over the faculty. (If one is concerned about poor memory on the part of the students, the experiment can be changed again, using in their stead the same college faculty just after they had completed their high school geometry courses. The odds are still with the faculty as professionals, even though they were every bit as bright in high school as they are now and had the advantage in terms of relevant facts and procedures.) So whatever gives the faculty the advantage is more than simple subject matter knowledge. That leaves "experience."

This last answer is not wrong, but it is inadequate. By analogy, consider the answer "aging" to the question, What causes senility? Of course, senility occurs with aging, but saying so really begs the question. The implied and more precise question is, Can we explain the underlying biological and chemical processes that comprise the phenomenon known as senility? Clarifying this question, elaborating on it, and exploring it lead one into deep questions of medical science.

In a similar vein, one can begin to formulate questions about experience with greater precision. When the faculty members sit down to work on the problems, there are fewer resources immediately accessible to them than to the students. Yet the faculty manage, somehow, to see what makes the problem tick, to retrieve or generate the facts they need, to come up with a variety of plausible directions for exploration where the students do not, and so on — all with some sense of purposefulness and efficiency. To specify the nature of these skills and to elaborate the means by which they are achieved is to begin to provide a real explanation of what the faculty's experience means.

Most people would probably be inclined to answer the second question in the affirmative, based on the feeling that the faculty's expertise should somehow be evident in the work they produce. To focus on what constitutes that expertise, let us imagine three conditions under which one might make the discrimination:

1. The solutions are graded by a third party, and we are given the grades earned by the subject on the problems.
2. We are given the final solutions handed in for grading by the subject.
3. We are given the transcripts produced by the subject as he or she worked through the problems, and the scratch work he or she produced.

We would, of course, be least confident of our judgments in the first case and most certain in the third. The point to observe with regard to option (1) is that the scores in themselves are not very rich sources of information; test scores rarely are. They may tell us how well someone performed, but they say virtually nothing about what the person actually did. In option (2) we have much more information, of the type customarily used to judge problem-solving performance. But it, too, is lacking. The difficulty in using such information is that two people may come upon the same solution through entirely different means. For example, one person might stumble upon a construction that "works" and not be able to justify it (the search space is relatively small for some of these problems); another person might derive the same construction logically and coherently. Yet, their descriptions of the construction, which is what they were asked to produce, may be identical. Similarly, there are both good and bad failures. One person may not come to grips with a problem at all, or might waste the allotted time pursuing irrelevant trivia, while another person may make a series of entirely plausible, but ultimately unsuccessful, attempts to solve the problem. In both cases, it is what the person *does,* rather than what the person *produces,* that is the determining factor in making our decision. It is precisely the insights into problem-solving processes that make option (3) so much better a choice.

This brief discussion raises the issues that are central in all that follows. First, it reveals my bias that there is much more to doing and understanding mathematics than simply mastering the subject matter. What one does with the facts at his or her disposal accounts, in large part, for problem-solving success. Second, it points to problem-solving processes as being absolutely central in any discussion of mathematical performance. This book is devoted to the exploration of the mathematical processes, mathematical strategies, and tacit mathematical understandings that constitute thinking mathematically. The focus here is primarily on research, although there is a strong "applied" component to the work as well. Much of the research is inspired, and informed, by classroom aplications.

To sum things up in a phrase, the goal of the research that has generated this book is to make sense of people's mathematical behavior—to explain what goes on in their heads as they engage in mathematical tasks of some complexity. This book is divided into two parts, dealing, respectively, with the theoretical and empirical aspects of understanding and exploring mathematical behavior.

Part I, consisting of Chapters 1 through 5, is an attempt to outline and flesh out a theoretical framework for investigating mathematical thinking. Chapter 1 presents an overview of the framework, and the four chapters that follow elaborate the four major categories in it. Chapter 2 offers a discussion of *resources,* an inventory of the mathematical knowledge possessed by the individual. In brief, What does the individual "know," and how is that knowledge accessed for use? Chapter 3 focuses on *heuristics,* the general mathematical problem-solving strategies or rules of thumb for successful problem solving whose investigation was pioneered by George Pólya. We shall explore the nature of such strategies, and the kinds of knowledge that are actually required to implement them. Chapter 4 considers the issue of *control.* The idea is that mathematical performance depends not only on what one knows but on how one uses that knowledge, and with what efficiency. Competent decision-making can help to ensure success even though one has few resources to begin with, and poor decision-making can guarantee failure despite potential access to a large collection of resources. The nature of such decision-making in mathematical problem solving, and its effects, are explored in detail. Finally, Chapter 5 deals with *belief systems,* the set of understandings about mathematics that establish the psychological context within which individuals do mathematics. We shall see that people's mathematical "world views" determine their orientation toward problems, the tools and techniques that they think are relevant, and even their unconscious access (or lack of access) to potentially related and useful material.

The issues in Part I are quite broad. Contributions to our understanding of those issues come from disciplines as diverse as mathematics education, artificial intelligence (AI), cognitive anthropology, and developmental psychology. While this situation is natural and healthy, there are ways in which it makes for difficulties. Papers within any disciplinary tradition are usually written under the assumption that readers will share the author's paradigms, assumptions, and language. Readers from outside that discipline can find it hard to penetrate the barier of assumptions and language, to uncover relevant results, and to see connections that might otherwise be seen. For that reason my intention in writing Part I is partly tutorial. Where possible there are references to work in other fields that I have found useful.

To put things simply, Part II offers some of the detailed evidence upon which the hypotheses advanced in Part I are based. It offers selections from a series of empirical studies begun in 1975. Most of these studies were dual-purpose, seeking (1) to develop a methodology for investigating a particular aspect of mathematical behavior and (2) to use that methodology to explore and elucidate the framework described in Part I. While this research was being conducted, I taught a series of problem-solving courses based on (and contributing to) the ideas in the framework. These courses served as the laboratories for some of the research. A number of the chapters in Part II characterize students' behavior before, during, and after my problem-solving courses. Part II proceeds in chronological order.

When the research for Chapter 6 was undertaken, it was an open question as to whether students could master heuristic strategies—in any circumstances. Chapter 6 describes a small-scale laboratory study that explores that issue in relatively ideal conditions. The results were positive, and the research moved from the laboratory to the real world. Chapter 7 offers the first full-fledged documentation of the results of one of my problem-solving courses. A series of paper-and-pencil tests were developed to capture various aspects of problem-solving processes. These measures were used to characterize students' mathematical behavior before and after their participation in my course, and in Chapter 7 they provide clear documentation of the kinds of changes in mathematical behavior that can be induced by problem-solving instruction. Chapter 8 also looks at students' performance before and after the course, but this time with an eye toward more detailed cognitive structures (what I categorize as resources). It also offers a direct comparison of novice and expert behavior. The research indicates that experts and novices perceive different things in mathematical problem statements and that they are led by those perceptions to approach the problems differently. It shows that with practice the novices' perceptions become more expert-like.

The discussion is substantially broadened in Chapters 9 and 10. The data in those chapters are videotapes of problem-solving sessions and their transcripts (or protocols). The research question, simply put, is how to make sense of them. Chapter 9 explores the issue of verbal methods in general, looking at the reliability of analyses of verbal problem-solving sessions. Its primary focus is on the analysis of problem-solving performance at the macroscopic level, with an emphasis on control or executive behavior. A rigorous framework for such analyses is presented and discussed. Chapter 10 addresses the issue of belief systems and their influences on behavior. It begins with a before-and-after comparison that demonstrates how changes in students' beliefs about the nature of mathematics can result in changes in their performance. It continues with a discussion of the origins of students' beliefs about mathematics. Evidence for that discussion is drawn from videotapes of typical classroom instruction. The discussion of those beliefs and of their implications brings to a close this attempt to sketch, in broad terms, the dimensions of mathematical behavior and the factors that shape it. As the final section indicates, we have only taken the first few steps on the long road toward an understanding of mathematical thinking—but I think those steps are in the right direction.

Figure 0.3 provides a guide to the overall structure of the book.

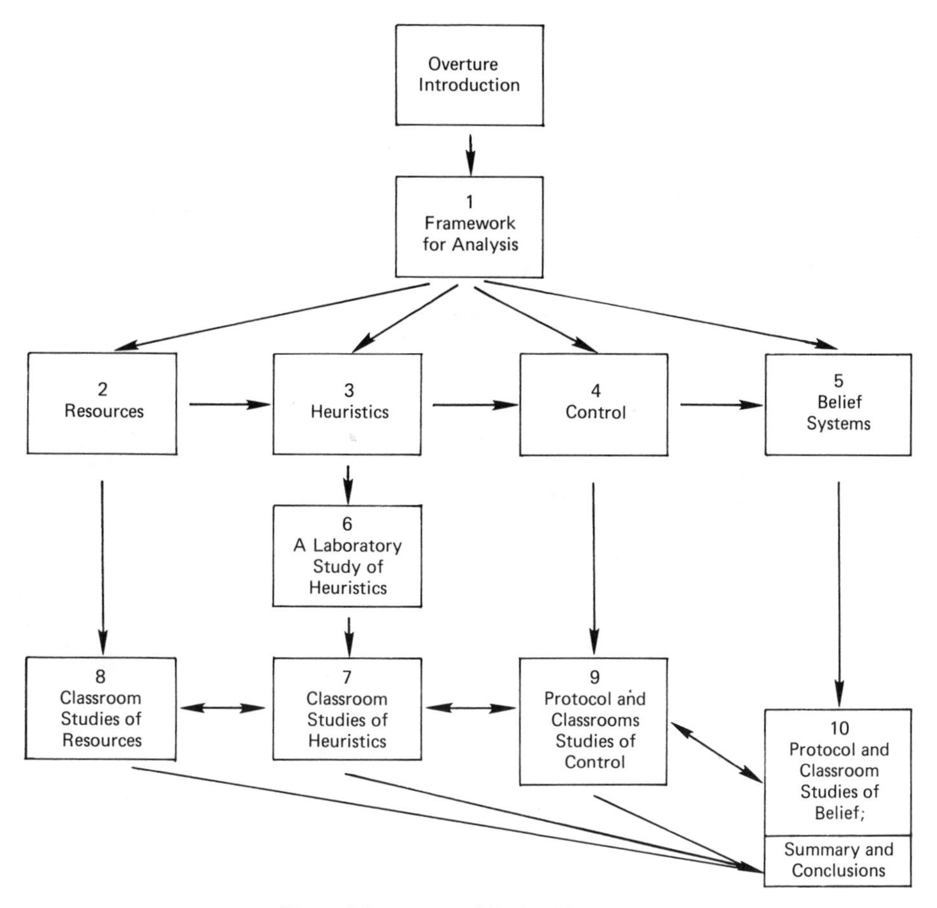

Figure 0.3 A map of the book's contents.

Part One

Aspects of Mathematical Thinking: A Theoretical Overview

1

A Framework for the Analysis of Mathematical Behavior

Overview

This chapter outlines a framework for examining what people know, and what people do, as they work on problems with substantial mathematical content. As an example, consider the following situation. Suppose that an individual or a small group of people works out loud on a mathematical problem of moderate difficulty. The problem solver does not have easy access to a procedure for solving the problem — a state of affairs that would make the task an exercise rather than a problem — but does have an adequate background with which to make progress on it; say, it might be feasible to reach a solution in a half hour or so. Moreover, the person wants to solve the problem and works actively at it. We observe the solution process as it takes place. We may interview the problem solver, or administer any tests we consider to be appropriate. The goal is to explain, as accurately as possible, what takes place during the solution attempt. What mathematical knowledge is accessible to the problem solver? How is it chosen? How is it used? Why does the solution evolve the way it does? In what ways do the aproaches taken to solve the problem reflect the individual's understanding of this area of mathematics, and what is the relationship between that understanding and the individual's problem-solving performance? And finally, what accounts for the success or failure of the problem-solving attempt?

My purpose here is to sketch out the dimensions of an explanatory framework for dealing with such questions. In a general way, the framework serves as a précis for the balance of the book. The issues raised here are treated at length in the balance of Part One (Chapters 2–5) and they serve as the background for the empirical studies described in Part Two (Chapters 6–10). But the framework stands on its own. I argue that the four categories of knowledge and behavior introduced here must be dealt with, if one wishes to "explain" human problem-solving behavior.

Some comments about my current perspective on the relationship between problem solving and understanding will help to set the stage for what follows. Early in my career I took "the ability to solve problems" as an operational definition of understanding. You understand how to think mathematically when you are resourceful, flexible, and efficient in your ability to deal with new problems in mathematics. My early problem-solving courses (the first offered to upper-division mathematics majors at Berkeley in 1976, the second to lower-division students at Hamilton College in 1978–79) reflected that perspective. Of course, any mathematical problem-solving performance is built on a foundation of basic mathematical knowledge, which I call the *resources* available to the individual. I began with the assumption that my students had nearly adequate resources for problem solving, and then I provided enough practice on introductory problems so that their skills were indeed serviceable. These basic skills were the foundation upon which the course was built. The course was designed to provide the tools for resourcefulness and efficiency. To be resourceful, students needed (I thought) to be familiar with a broad range of general problem-solving techniques known as *heuristics.* To be efficient, they needed coaching in how to manage the resources at their disposal. The course focused on providing these skills. Heuristic techniques were carefully delineated and carefully taught, and "executive" or "control" issues were dealt with through instruction in an explicit executive strategy. My test for the success of such courses reflected the operational definition of understanding as the ability to solve problems. Thus a problem-solving course was a success if, after instruction, the students showed markedly improved performance on a collection of problems that were not directly related to the problems that they had studied in the course. An even more stringent test is that the examination problems should be unlike the problems they studied in the course. My courses passed both kinds of tests with flying colors (see Chapter 7).

My notion of success for such courses has evolved in recent years. More accurately, my notion of what it means to understand mathematics (and, therefore, of what one should teach) has evolved. This change came about as a result of my research, in which I made detailed examinations of videotapes of students' problem-solving performance. By most measures, the students I videotaped were the successes of our educational system. Virtually all of them had completed at least one semester of college mathematics (calculus or beyond) with high grades. Many were volunteers for my problem-solving course. The fact that they had enrolled indicated that they were partial to mathematics (it was an optional course, fulfilling no requirements) and relatively confident about their abilities (it had a reputation for being difficult). Yet, examining the problem-solving performance of these and other students revealed some unpleasant realities. The videotapes revealed that the students' resources were far weaker than their performance (on standard tests, etc.) would indicate; typical instruction and testing provide little opportunity for students to demonstrate the breadth and depth of their misconceptions. They confirmed that students have little or no awareness of, or ability to use, mathematical heuristics. They indicated that the general issue of how one selects and deploys the resources at one's disposal—the issue of *control*—was far broader, and far more critical, than I had thought. In most testing situations students are asked to work problems similar to those they have been trained to solve. As a result, the context keeps them in the right arena, even when they are unable to solve the problems. The problems I asked students to solve were certainly within their capability and were often technically easier than problems they solved in other classes. They were not, however, put forth in a context that oriented the students toward the "appropriate" solution methods. Time and time again the students working such nonstandard problems would go off on wild goose chases that, uncurtailed, guaranteed their failure. The issue for students is often not how efficiently they will use the relevant resources potentially at their disposal. It is whether they will allow themselves access to those resources at all.

A final category of knowledge, *belief systems,* is more subtle. The close examination of these students' problem-solving performance revealed that many of them had serious misunderstandings about mathematics. Even the more successful students often held perspectives that were deeply antimathematical in fundamental ways and that had clearly negative effects on their problem-solving behavior. In some cases, students survived (often with good grades!) by implementing well-learned mechanical procedures, in domains about which they understood virtually nothing. In other cases, much of the mathematical knowledge that the students had at their disposal, and that they should have been able to use, went unused in problem solving. This was not because they had forgotten it (a matter of resources) or because they ran out of time to use it (a matter of control), but because *they did not perceive their mathematical knowledge as being useful to them, and consequently did not call upon it.* The clearest cases come from geometry, described below and extensively in Chapters 5 and 10. These examples and others demonstrate that students' problem-solving performance is not simply the product of what the students know; it is also a function of their perceptions of that knowledge, derived from their experiences with mathematics. That is, their beliefs about mathematics—consciously held or not—establish the psychological context within which they do mathematics.

These findings indicate that a view of problem solving as an operational definition of understanding is too narrow. So, too, is any positivist view that considers teaching problem solving to be equivalent to providing a set of prescriptions for students' productive behavior. Whether one wishes to explain problem-solving performance, or to teach it, the issues are more complex. One must deal with (1) whatever mathematical information problem solvers understand or misunderstand, and might bring to bear on a problem; (2) techniques they have (or lack) for making progress when things look bleak; (3) the way they use, or fail to use, the information at their disposal; and (4) their mathematical world view, which determines the ways that the knowledge in the first three categories is used.

This discussion is summarized, formally, in Table 1.1. Each of the four categories in Table 1.1 will be expanded into a full chapter (Chapters 2–5, respectively). To get a better sense of the contents of each category, and also to see the boundaries and relationships among the categories, we turn to some specifics. The next section presents some problems typical of those considered in this book and discusses some behavior typical of students and others who have worked on the problems.

Typical Problems, Typical Behavior: The Four Categories Illustrated

The following three problems are typical of those I have used in my research and are discussed extensively in the next four chapters. There is, of course, no such thing as a typical student. However, most of the students discussed here shared the following properties. They were college freshmen or sophomores who had spent a year studying Euclidean geometry in high school, who had studied at least one semester of differential calculus when they were taped (those working Problem 1.3 had in general, completed multivariate calculus), and who had received either A's or B's in their college mathematics courses. They had either volunteered for my problem-solving course or volunteered to be videotaped, both positive indications regarding their mathematical abilities and interests. Their mathematics courses had provided them, in the recent past, with the formal tools that would have enabled them to solve Problems 1.1–1.3.

Table 1.1
Knowledge and Behavior Necessary for an Adequate Characterization of Mathematical Problem-Solving Performance

Resources: Mathematical knowledge possessed by the individual that can be brought to bear on the problem at hand
- Intuitions and informal knowledge regarding the domain
- Facts
- Algorithmic procedures
- "Routine" nonalgorithmic procedures
- Understandings (propositional knowledge) about the agreed-upon rules for working in the domain

Heuristics: Strategies and techniques for making progress on unfamiliar or nonstandard problems; rules of thumb for effective problem solving, including
- Drawing figures; introducing suitable notation
- Exploiting related problems
- Reformulating problems; working backwards
- Testing and verification procedures

Control: Global decisions regarding the selection and implementation of resources and strategies
- Planning
- Monitoring and assessment
- Decision-making
- Conscious metacognitive acts

Belief Systems: One's "mathematical world view," the set of (not necessarily conscious) determinants of an individual's behavior
- About self
- About the environment
- About the topic
- About mathematics

Problem 1.1 You are given two intersecting straight lines and a point P marked on one of them, as in Figure 1.1 below. Show how to construct, using straightedge and compass, a circle that is tangent to both lines and that has the point P as its point of tangency to one of the lines.

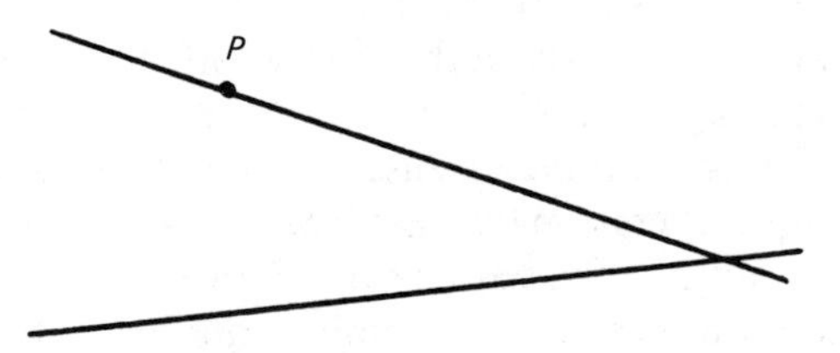

Figure 1.1

Problem 1.2 You are given a fixed triangle T with base B, as in Figure 1.2. Show that it is always possible to construct, with straightedge and compass, a straight line that is parallel to B and that divides T into two parts of equal area. Can you similarly divide T into five parts of equal area?

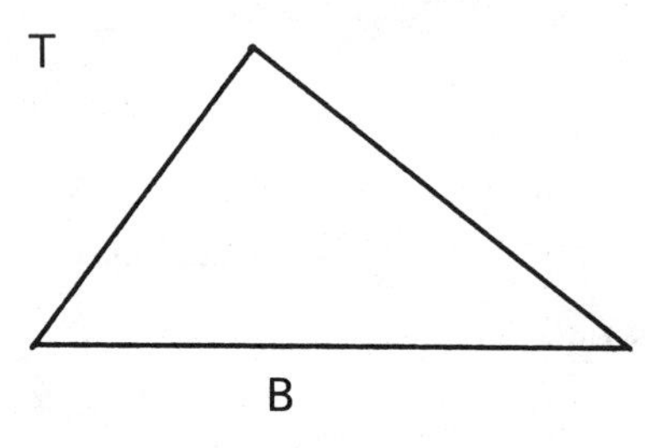

Figure 1.2

Problem 1.3 Three points are chosen on the circumference of a circle of radius R, and the triangle containing them is drawn. What choice of points results in the triangle with the largest possible area? Justify your answer as well as you can.

Prior to describing in detail what is covered by the framework, I should issue some caveats about what is not. The framework as outlined in Table 1.1 is far from comprehensive, because a complete explanation of problem-solving performance would require many other levels of analysis. At the microscopic (more fine-grained) end of the spectrum, one needs to be concerned with the processes that Monsell (1981) characterized as the "nuts and bolts of cognition" (neural processes and memory mechanisms, for example). Such processes are beyond the scope of this book. Here I assume simply that my students have memories and that those memories work reasonably well; I worry about what my students remember but not about the biological mechanisms that allow them to do so. At the macroscopic end of the spectrum, there is the broad range of social cooperative behaviors within which most "real" problem solving takes place, for example, the complex set of interactions required for the publication of this book. Those, too, extend beyond what is covered here.

A warning should be issued about the dangers of interpreting terminology in any cross-disciplinary work such as this. Many of the terms used here have different meanings in different disciplines or have connotations that suggest meanings that differ from the ones I intend. For example, the phrase *mathematical knowledge* in the description of resources might suggest that issues of classical mathematical epistemology are about to be dealt with. They are not, at least not directly. The mathematical knowledge discussed here is the knowledge possessed by the individual, which is a different matter entirely (see the following section on resources). I have tried to avoid such "loaded" terms wherever possible, and to specify the sense in which I use those terms whose use is unavoidable. Nonetheless, *caveat lector.* The four categories are briefly described below.

Resources

To understand why an attempt to solve a problem evolves the way that it does, we need to know first what "tools" the problem solver starts with. Ideally this first category, resources, provides that kind of information. It is intended as an inventory of all the facts, procedures, and skills — in short, the mathematical knowledge — that the individual is capable of bringing to bear on a particular problem. The idea is to characterize what might be called the problem solver's "initial search space." What avenues are open, at least potentially, for exploration? It is essential to understand that in discussing human problem-solving performance we are concerned with an individual, genetic epistemology rather than an abstract mathematical epistemology.*

In order to understand what someone does while working any of the problems given above, we need to have an inventory of what the individual knows, believes, or suspects to be true. We need to know how that information is organized, stored, and accessed. Here I first consider the contents of resources, and then briefly discuss issues related to accessing that knowledge.

One broad class of resources consists of the set of relevant facts known by the individual, with each fact indexed by the degree to which it is "known." As an obvious example, consider a person trying to solve Problem 1.1. Does that person (1) know for certain that, (2) think it is likely that, (3) think it might be possible that, or (4) have no idea that the segment CP in Figure 1.3 is perpendicular to the line segment VP? An attempt to solve Problem 1.1 may evolve in radically different ways depending on the answer to that question. In this case the relevant information is clearly essential for a solution. Other knowledge may be equally essential but in far less obvious ways. Such is the case with broad understandings (sometimes called propositional knowledge) regarding the nature of geometric argumentation.

* For a clear contrast of genetic and abstract mathematical epistemologies, see Beth and Piaget (1966), especially Chapters VII ff. Two comments: (1) "Genetic" is meant in the sense of the "genesis" of knowledge as Piaget uses the term, and should not be confused with hereditary (biologically genetic) traits. (2) Personally, I am more sympathetic to the epistemological stances taken by Kitcher (1983) and Lakatos (1977) than I am to the "logicism" advanced by Beth or its classical competitors (Platonism, formalism, constructivism), but that is neither here nor there. The issue here is not What is true? but What does the individual problem solver hold to be true? That is the foundation upon which the individual builds when solving problems.

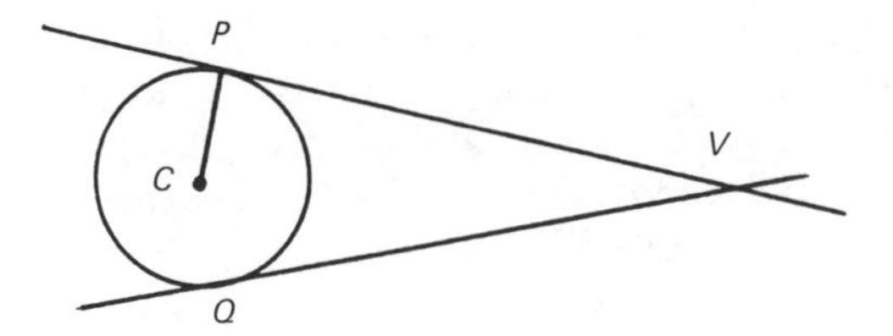

Figure 1.3

From the fact that CP and VP in Figure 1.3 are perpendicular, the mathematician derives a partial solution to Problem 1.1. This partial solution is derived from a global understanding regarding geometrical deduction, in which Figure 1.3 plays a dual role. On the one hand, Figure 1.3 is just a particular diagram, representing itself and no more—a circle of given radius tangent to two particular lines. Some statements regarding the figure (e.g., statements that depend on the size of the angle between the two given lines or on the radius of the given circle) pertain to that figure in particular and do not generalize. On the other hand, Figure 1.3 serves as a generic example representing *all* diagrams in which a circle is tangent to two intersecting lines. The generic properties possessed by that circle must be shared by all similarly structured diagrams, in particular by the desired solution to Problem 1.1. Since CP and VP are perpendicular in Figure 1.3, it must be the case that the center of the circle one wishes to construct in Problem 1.1 lies on the perpendicular drawn through P.

Moreover, the mathematician knows that additional generic information derived from Figure 1.3 yields corresponding information regarding Problem 1.1. Thus, for example, proving that the line segment CV (when drawn in Figure 1.3) will bisect angle PVQ, provides a solution to the given problem. Note, however, that one must understand the generic nature of the diagram in Figure 1.3 in order to solve the problem in this way. Lacking such an understanding, one sees only that the solution to Problem 1.1 will "look like" Figure 1.3; one does not have a means of exploiting the similarity. Thus understandings about the implications of formal procedures in a domain make up part of one's resources regarding that domain. Similar understandings of a much less formal nature have to do with one's intuitions regarding the nature of objects in a domain. Within geometry, for example, the van Hieles' work (see, e.g., Freudenthal, 1973; Hoffer, 1983; van Hiele, 1976) clearly documents the importance of the empirical, intuitive foundations that underlie the ability to deal with their formalizations as abstract mathematical objects. The van Hieles' theory of thought levels in geometry specifically posits that students need to develop intuitive understandings of the natures of geometric figures before they can deal meaningfully with the formal aspects of geometry.

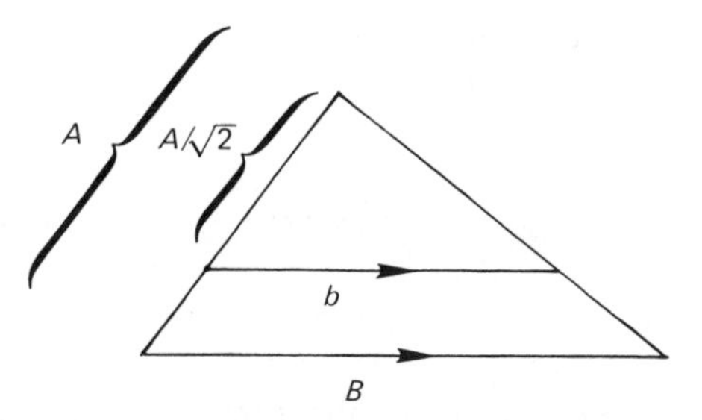

Figure 1.4

A second class of resources consists of the algorithmic procedures known by the individual. Can the problem solver construct the perpendicular to the top line in Figure 1.1 that passes through P, for example, or bisect the acute angle at V? Having determined the location of the line L that meets the conditions of Problem 1.2, can the problem solver use straightedge and compass to divide a given line segment by $\sqrt{2}$, and construct a line parallel to the base B through the designated point (see Figure 1.4)? Algorithmic procedures include all standard constructions, algebraic, manipulations, differentiation, and so forth.

Another class of resources qualitatively different from the algorithmic procedures like those just mentioned consists of what might be called routine procedures. Problem 1.3, for example, can be worked as a routine multivariate calculus problem. To solve such problems one obtains a formula for the desired quantity (here the area of a representative triangle) as a function of two or more variables and then applies the relevant techniques of the calculus. The problem solver has a fair amount of discretion about the choice of variables and about the means to obtain the formula (i.e., the solution is not algorithmic), but the outline of the procedure is prescribed and well known.

A final class of resources consists of what might be called relevant competencies. As noted above, Problem 1.1 can be transformed to a problem of geometric derivation: Can one derive a second property of the center C of the circle in Figure 1.3? Thus relevant competencies for Problem 1.1 include the abilities to produce mathematical arguments verifying that the line segments PV and QV in Figure 1.3 are of equal length and that the (not yet drawn in) line segment CV bisects the angle PVQ. Whether or not the individual takes this path to a solution, one needs to know whether it could have been taken. (If it could have been and was not, then there are interesting issues to explore; see the discussions of belief in this chapter and Chapter 5). The class of relevant competencies is quite broad. Included in the inventory of such competencies for Problem 1.1 are some procedural skills like knowing to draw in the line segments CQ and CV in Figure 1.3.

Included also is being able to implement the domain-specific heuristic strategy that calls for them, the geometric version of "taking the problem as solved." This strategy will be discussed in the section on heuristics that follows.

There is one final but vitally important aspect to the facts, procedures, and competencies in an individual's inventory of resources: *They need not be correct.* Our purpose is to explain people's behavior as they work mathematical problems. To that end we must know what they take to be true, even if it is not. The individual who firmly believes, for example, that the area of similar plane figures are directly proportional to their dimensions, may produce a solution to Problem 1.2 that only makes sense when one understands that misconception. Similarly, a student might base a geometric argument on the assumption that one can trisect an angle, or fail to solve a maximization problem, because he or she misunderstands the chain rule.

An inventory of resources provides a list of the tools and techniques that an individual could use in a particular situation, if that knowledge was called for. The issues of interest, of course, deal with which knowledge the individual does use in a given situation—and why. In general, such questions of resource allocation (e.g., selecting directions for exploration, planning a solution, deciding to pursue or to abandon a particular line of argument) are issues of control, a topic discussed extensively in this chapter, in Chapter 4, and Chapter 9. In a broad sense, all issues of knowledge access are issues of control. I prefer, however, to reserve the term *control* for discussions of active decision-making. In routine problem solving there are many situations where one just knows the right thing to do or where the right information just comes to mind. It will be useful to consider such nearly automatic access of appropriate knowledge a matter of resources. To put things simply, research indicates that for experienced problem solvers, stereotypical problem situations evoke stereotypical responses. The response may be a prepackaged solution to the entire problem, if the problem is recognized as being of a standard type; it may be a particular piece or body of relevant information in response to something specific in the problem statement. As an example of the former, one can ask any mathematician how to solve the following problem: "Find the length of the largest ladder that can be carried horizontally through two hallways that are 4 and 5 feet wide, respectively, and that meet at right angles." By the time the mathematician has finished reading the problem, it will have been characterized as a max-min problem, and the standard procedure for solving such problems will have been invoked. The details of the solution have yet to be worked out, but they are just details. Barring unforeseen complications, there is nothing problematic about this problem. It is an exercise. As an example of the latter, consider the following attempt by a mathematician to solve Problem 1.2. His solution began as follows.

Hmmm. I don't exactly know where to start.

Well, I know that the . . . there's a line in there somewhere. Let me see how I'm going to do it. It's just a fixed triangle. Got to be some information missing here. *T* with base *B*. Got to do a parallel line. Hmmm.

It said the line divides *T* into two parts of equal area. Hmm. Well, I guess I have to get a handle on area measurement here. So what I want to do . . . is to construct a line . . . such that I know the relationship of the base . . . of the little triangle to the big one.

Now let's see. Let's assume that I draw a parallel line that looks about right, and it will have base little *b*. [He draws Figure 1.5] Now, those triangles are *similar* [his emphasis].

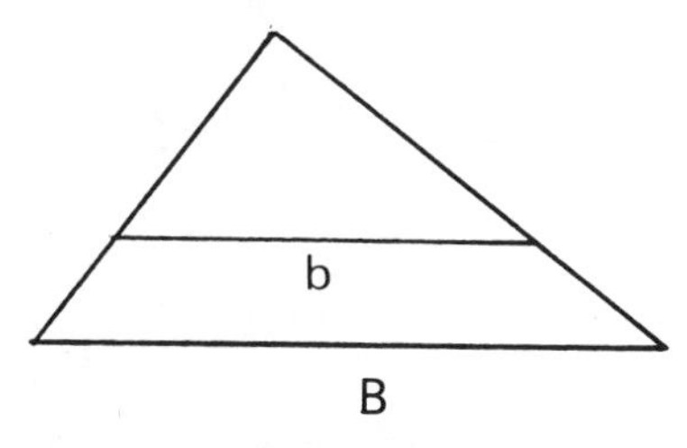

Figure 1.5

Yeah, all right than I have an altitude *A* for the big triangle and an altitude *a* for the little triangle, so I have $a/A = b/B$. . . .

It is clear that this mathematician did not know how to solve the problem as he began working on it, although he did have some sense of the kind of information he would need to solve it. It is equally clear that his recall of the relevant information was triggered by looking at Figure 1.5; having drawn it, he saw immediately that triangles were similar and almost instantaneously had at his disposal the information about proportionality that enabled him to rapidly solve the first part of the problem. In the case of proficient problem solvers, such nearly automatic responses can channel behavior into productive directions.

Things may not proceed as smoothly for individuals lacking such access to relevant knowledge. In pilot testing with a slightly more difficult version of Problem 1.2, a graduate student in physics was asked to construct two lines parallel to the base of the triangle *T* such that the resulting three areas would all be equal (Figure 1.6). Unfortunately this student, who was a good prob-

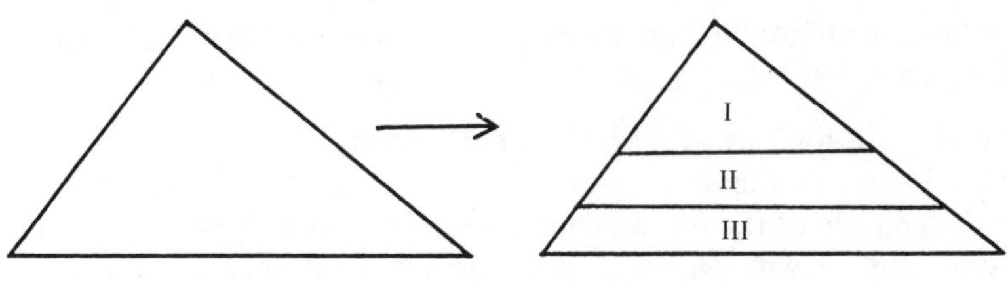

Figure 1.6

lem solver, did not perceive Figure 1.6 in a way similar to the way the mathematician perceived Figure 1.5. What he saw in the right-hand side of Figure 1.6 was a triangle and two trapezoids whose areas were to be equated (see Figure 1.7).

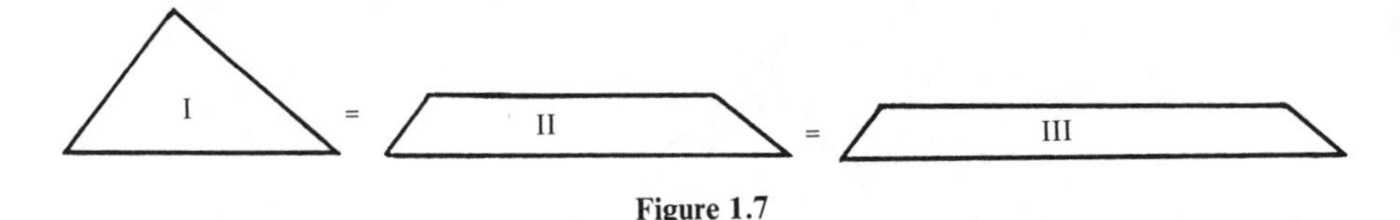

Figure 1.7

Computing and equating the areas of those three figures turned out to be quite complex, and the computations cost him so much time that he failed to solve the problem. Afterward it was pointed out that the triangles in Figure 1.6 were similar, and he then completed the problem in short order. Thus "making the right connections" is an essential component of resources. Some of the mechanisms invoked to explain how such connections are made are discussed in Chapter 2.

Heuristics

> *Heuristic, or heuretic, or "ars inveniendi" was the name of a certain branch of study, not very clearly circumscribed, belonging to logic, or to philosophy, or to psychology, often outlined, seldom presented in detail, and as good as forgotten today. The aim of heuristic is to study the methods and rules of discovery and invention. . . . Heuristic, as an adjective, means "serving to discover."*
>
> G. Pólya, How to Solve It, 1945, pp. 112–113

Ars inveniendi in the spirit of Descartes' *Method* had been dormant for about a century (the last major attempts were made by Bernard Bolzano, 1781–1848) when Pólya published *How to Solve It* in 1945, and the study of heuristic was indeed as good as forgotten. *How to Solve It* was "an attempt to revive heuristic in a modern and modest form," offering what might be considered a guide to useful problem-solving techniques. The outline of that guide is given in Table 1.2.

How to Solve It was followed by the two volumes of *Mathematics and Plausible Reasoning* (1954) and later by the two volumes of *Mathematical Discovery* (1962 and 1965), in which Pólya elaborated on the theme and on the details of heuristic strategies. Once nearly forgotten, heuristics have now become nearly synonymous with mathematical problem solving.*

Heuristic strategies are rules of thumb for successful problem solving, general suggestions that help an individual to understand a problem better or to make progress toward its solution. Such strategies include exploiting analogies, introducing auxiliary elements in a problem or working auxiliary problems, arguing by contradiction, working forward from the data, decomposing and recombining, exploiting related problems, drawing figures, generalizing and using the "inventor's paradox," specializing, using reductio ad absurdum and indirect proof, varying the problem, and working backward. There is some consensus among mathematicians that these strategies are useful. For example, heuristic thinking contributes to the solutions of each of Problems 1.1 – 1.3.

HEURISTIC APPROACHES TO PROBLEMS 1.1, 1.2, AND 1.3

One of the more frequently used heuristic strategies, "taking the problem as solved," can be informally described as follows: "In a problem 'to find' or 'to construct,' it may be useful to assume that you have the solution to the given problem. With the solution (hypothetically) in hand, determine the properties it must have. Once you know what those properties are, you can find the object you seek." That advice becomes such more specific for geometric construction problems: "Add the figure you wish to construct to the given diagram and determine the properties it must have. Those properties will provide the loci for the construction." That suggestion lies behind the solutions to Problems 1.1 and 1.2. In the case of Problem 1.1, the (rather empty) Figure 1.1 is replaced by the much more suggestive Figure 1.3. Even Figure 1.3 looks somewhat incomplete to the practiced eye; after one draws

* Curiously, that synonymy takes place only within mathematical problem solving. In other problem-solving domains such as AI, the term *heuristics* has been revived, but is generally used to refer to procedures such as means – ends analysis. In fact, heuristics of the type Pólya describes are not held in high esteem in AI. There are interesting reasons for this, but a discussion of them would take us far afield.

Table 1.2
A Guide to Problem-Solving Techniques[a]

	UNDERSTANDING THE PROBLEM
First. You have to *understand* the problem.	*What is the unknown? What are the data? What is the condition?* Is it possible to satisfy the condition? Is the condition sufficient to determine the unknown? Or is it insufficient? Or redundant? Or contradictory? Draw a figure. Introduce suitable notation. Separate the various parts of the condition. Can you write them down?
	DEVISING A PLAN
Second. Find the connection between the data and the unknown. You may be obliged to consider auxiliary problems if an immediate connection cannot be found. You should obtain eventually a *plan* of the solution.	Have you seen it before? Or have you seen the same problem in a slightly different form? *Do you know a related problem?* Do you know a theorem that could be useful? *Look at the unknown!* And try to think of a familiar problem having the same or a similar unknown. *Here is a problem related to yours and solved before. Could you use it?* Could you use its result? Could you use its method? Should you introduce some auxiliary element in order to make its use possible? Could you restate the problem? Could you restate it still differently? Go back to definitions. If you cannot solve the proposed problem try to solve first some related problem. Could you imagine a more accessible related problem? A more general problem? A more special problem? An analogous problem? Could you solve a part of the problem? Keep only a part of the condition, drop the other part; how far is the unknown then determined, how can it vary? Could you derive something useful from the data? Could you think of other data appropriate to determine the unknown? Could you change the unknown or the data, or both if necessary, so that the new unknown and the new data are nearer to each other? Did you use all the data? Did you use the whole condition? Have you taken into account all essential notions involved in the problem?
	CARRYING OUT THE PLAN
Third. *Carry out* your plan.	Carrying out your plan of the solution, *check each step.* Can you see clearly that the step is correct? Can you prove that it is correct?
	LOOKING BACK
Fourth. *Examine* the solution obtained.	Can you *check the result?* Can you check the argument? Can you derive the result differently? Can you see it at a glance? Can you use the result, or the method, for some other problem?

[a] From Pólya (1945).

in the appropriate line segments (a very specific example of "introducing auxiliary elements," another heuristic strategy), one has Figure 1.8. Figure

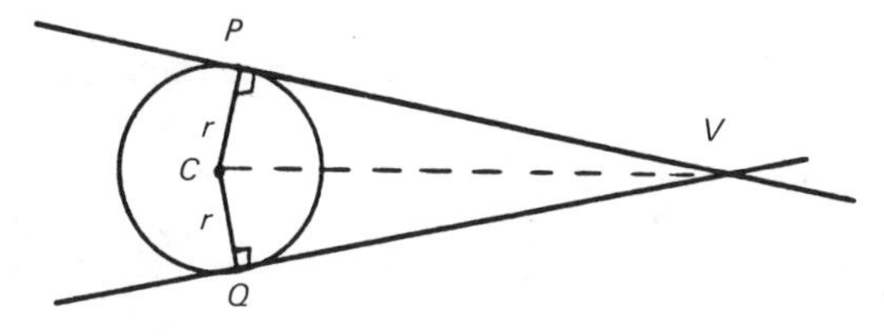

Figure 1.8

1.8 suggests two obvious conjectures, Problems 1.1A and 1.1B.

Problem 1.1A The circle in Figure 1.8 is tangent to the two given lines at the points P and Q, respectively. Prove that the two line segments PV and QV have the same length.

Problem 1.1B The circle in Figure 1.8 is tangent to the two given lines at the points P and Q, respectively. Prove that the line segment CV bisects angle PVQ.

Either of these results, combined with the fact that CP and CQ are perpendicular to PV and QV, respectively, yields a construction that is *guaranteed* to be correct. (The reason for the emphasis becomes clear in the discussion of belief systems.)

Problem 1.2 is substantially more difficult than Problem 1.1, and many students simply stare at the problem in frustration. Recall the mathematician's solution to Problem 1.2, given in the discussion of resources:

> Hmmm. I don't exactly know where to start. . . .
> Now let's see. Let's assume that I draw a parallel line that looks about right, and it will have base little b. [He draws Figure 1.5.] Now, those triangles are similar.

The mathematician was able to solve the problem because he accessed the right information when he faced Figure 1.5. But he was able to get that far in the first place because he took the problem as solved.

Heuristic strategies can be used in a number of ways to facilitate a solution to Problem 1.3. One can, for example, *assume without loss of generality* that the base of the desired triangle is horizontal. This assumption, the equivalent of mentally rotating the triangle, makes it easier to see what is important in the problem. But it also does more, in that it simplifies the remaining computations: There are two variable points in the problem, instead of three. One can also *scale the problem,* working with the unit circle instead of the given circle of radius R. This too simplifies the computations that must be performed. The resulting problem still involves a fairly messy partial differentiation but one that is far less complex than it would otherwise have been.

A different heuristic approach provides quicker and neater solutions to Problem 1.3. One can *explore the conditions of the problem, holding all but one quantity fixed* and letting that quantity vary. Suppose, for example, that only the third vertex of the triangle is allowed to vary. The resulting *easier related problem*

> **Problem 1.3A** Of all triangles in a given circle with a fixed chord as base, which has the largest area?

is trivial to solve. When all the triangles have the same base, the triangle with the largest area is the triangle with the largest altitude—the isosceles (see Figure 1.9).

It follows that the largest triangle inscribed in the circle must be isosceles, since the area of any scalene triangle could be increased by keeping one side of that triangle fixed, and building an isosceles triangle on that base. Finishing the original problem by determining the largest isosceles triangle inscribed in the circle is now a straightforward one-variable calculus problem. In fact, however, the *argument by contradiction* suggested above solves the problem quickly. If one has any nonequilateral triangle T with sides A, B, and C (say, A and B are of unequal length), the inscribed isosceles triangle with base C has an area larger than T. Thus any nonequilateral triangle can be enlarged; the largest triangle, if it exists (and a number of arguments show it does), must be equilateral.

This brief introduction to heuristic techniques provides a rough indication of their nature and some suggestion of their utility, but it leaves most of the serious questions about heuristics unaddressed. For example, How does one know when to use which strategies? What does one need to know in

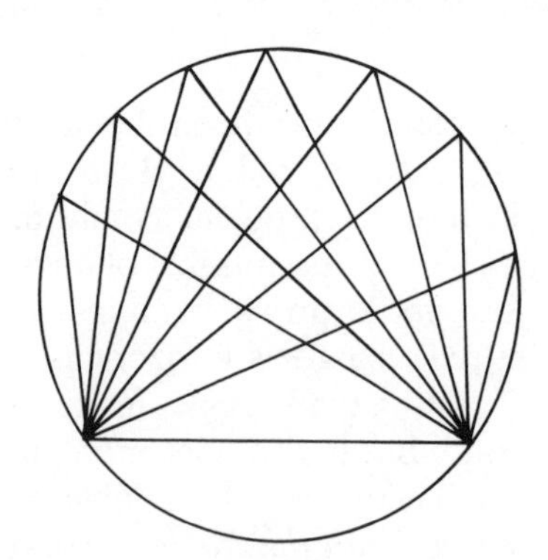

Figure 1.9

order to use them? How are heuristics related to resources or to domain-specific knowledge in general? How do mathematicians learn to use heuristic strategies since they are generally not taught directly? Can students be taught to use such strategies? These and other such issues are the focus of Chapter 3.

Control

This category of behavior deals with the way that individuals use the information potentially at their disposal. It focuses on major decisions about what to do in a problem, decisions that in and of themselves may "make or break" an attempt to solve the problem. Behaviors of interest include making plans, selecting goals and subgoals, monitoring and assessing solutions as they evolve, and revising or abandoning plans when the assessments indicate that such actions should be taken. These are executive decisions as the term is used in AI, managerial decisions as the term is used in business, or strategic (vs. tactical) decisions as the term is used in the military. The term *metacognition* has been used in the psychological literature for discussions of related phenomena.

The defining characteristic of actions at the control level is that they have global consequences for the evolution of a solution. They are decisions about what paths to take (which also, therefore, determine what paths are not taken). They are decisions about abandoning directions, which open up new possibilities but do so at the risk of curtailing efforts that might have led to success. They are, in one way or another, decisions that entail the selection and use of significant portions of the resources (including time, one of the most precious of resources) at the problem solver's disposal. Consider Problem 1.1, for example. Fifteen minutes is a generous amount of time to work that problem; many teachers might allow only 10 minutes for it on a test. Suppose that a student working Problem 1.1 conjectures that the line segment from P to its "mirror image" on the bottom line, P', is the diameter of the circle. This conjecture, illustrated in Figure 1.10, occurs quite frequently.

Suppose in addition that the student decides to test the conjecture by performing the construction. This calls for marking the point P', bisecting the line segment from P to P' to find its midpoint C, and drawing the circle with C as center and the distance from C to P as radius. Performing the construction to a reasonable degree of accuracy with a cheap compass can take more than 5 minutes, so it may take the problem solver more than half of the allotted time to discover that the conjecture is incorrect—after which too little time may remain to find a correct solution! Thus that one decison, if not reversed, can guarantee that the effort to solve the problem will fail.

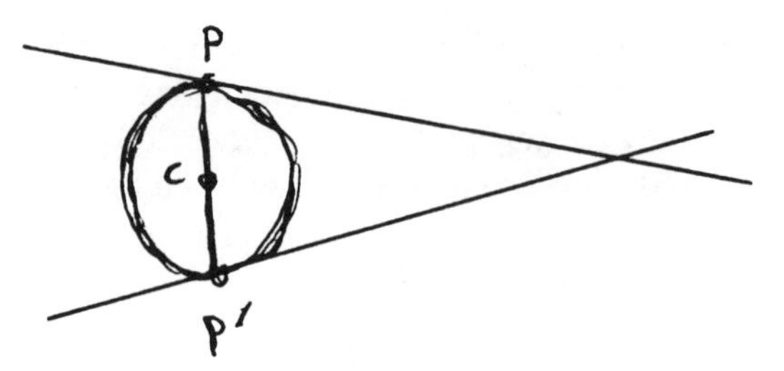

Figure 1.10

The potential impact of that one decision on the solution as a whole is what makes it a control decision. To see the effects that such decision-making can have on problem-solving performance, we briefly examine two pairs of students' attempts to solve Problem 1.3. The full problem sessions are discussed in Chapter 9.

STUDENTS KW AND AM WORK PROBLEM 1.3

The students KW and AM had completed one and three semesters of calculus, respectively. Student AM had just finished a course that included the multivariate techniques that can be used to solve Problem 1.3 and had received full credit for solving an equally complex problem on the final examination in that course. Student KW read the problem out loud. The following dialogue took place 35 seconds after the problem session began.

AM: I think the largest triangle should probably be equilateral. OK, and the area couldn't be larger than 2.

KW: So we have to divide the circumference of the three equal arcs to get this length here. [He points to a side of the equilateral triangle and begins to calculate its length, incorrectly using the arclength formula.] So 60, no, 120 arc degrees. So let's see, say that it equals *R* over *S*. . . . This radius doesn't help. [He draws Figure 1.11.]

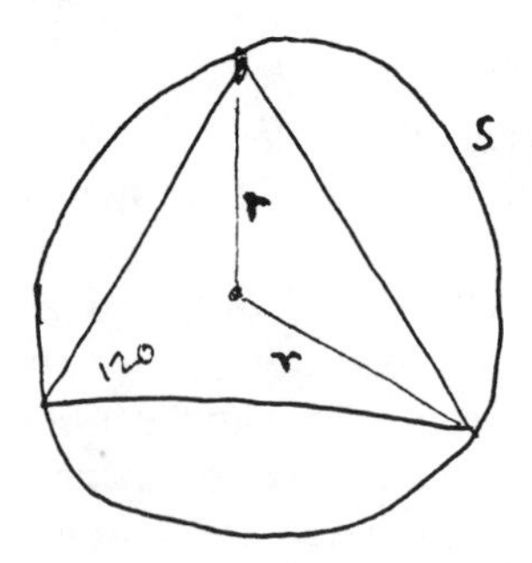

Figure 1.11

Here AM has made the obvious and almost universal conjecture. Student KW is in complete agreement and simply begins calculating the area of the equilateral. There is no discussion about the decision to undertake this calculation, a decision that AM clearly supports. As KW slows down, AM rereads the problem statement and worries about the word "justify."

AM: Do we have to . . . justify your answer as best you can. . . . Justify why this triangle [the equilateral] . . . Justify why you . . . OK, right.

KW: OK, let's . . . take a right triangle and see what we get. [He draws Figure 1.12.]

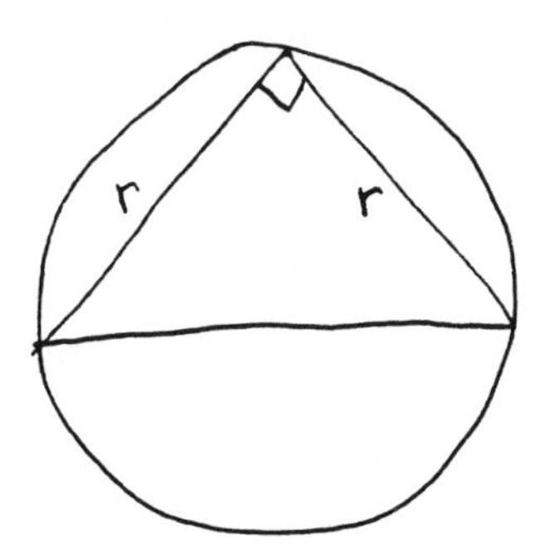

Figure 1.12

They take 15 seconds to calculate the area of the isosceles right triangle in Figure 1.12. This is a weak attempt at justifying their choice of the equilateral as the "solution" to Problem 1.3. If the area of the equilateral exceeds that of the isosceles right triangle (another archetype, and another possible candidate for the largest triangle), it is more likely to be the largest. With this calculation out of the way, AM immediately returns to calculating the area of the equilateral. Student KW joins in actively; they are both committed to the computation. There is no discussion here, or at any time during the solution, of *why* they are calculating the area. Unfortunately for KW and AM, they have chosen an inconvenient way to calculate the area of the equilateral and have inadvertently mislabeled the angles of the triangle. Their computations become rather contorted, and 7 minutes elapse. At that point their energy flags a bit, and the following dialogue takes place.

AM: There used to be a problem . . . about the square being the largest part of the area. . . .

KW: The largest area of . . . something in a circle, maybe a rectangle, something like that. . . .

AM: Ah, well . . .

KW: So, this is R [the radius of the circle], and this is 120 degrees, and . . .

This statement signals a return to the calculation of the equilateral's area, which they pursue relentlessly for the 20 minutes allotted to them. There are two more brief pauses like the one just quoted. The possibility of working the problem by calculus is casually mentioned and then dropped without serious consideration. So is the possibility of approaching the problem by some sort of variational argument. Each time, *without any discussion of the merits of the suggestion that has just arisen, or of the value of the computation they have been engaged in,* KW and AM return to computing the equilateral's area. They are still engaged in this computation when the videocassette recording their session clicks off. At that point I intervene and ask them what good it would do them to know the area of the equilateral. They are unable to provide an explanation.

STUDENTS DK AND BM WORK PROBLEM 1.3

Both DK and BM had completed three semesters of calculus when they worked Problem 1.3. This session is particularly interesting, because the previous year DK, as a freshman, had seen me demonstrate the heuristic solution to the problem.

DK reads the problem out loud, and BM immediately conjectures that the equilateral is the largest. He suggests the calculus, but DK clearly remembers that there is a clever, noncomputational solution to the problem —one that is based on symmetry in some way.

BM: Do we need calculus for this? So we can minimize, or rather maximize it.

DK: Why don't we find . . . some way to break this problem down into . . . like what would a triangle be for half the circle? . . . Why don't we find the largest triangle with the base one of the diameters, OK?

His idea at this point is to find the largest triangle in the semicircle and then to reflect it in the diameter to obtain the largest triangle in the circle. He has not made this intention clear, but BM rejects the suggestion.

BM: That would just be a family of right triangles that go like this [he draws Figure 1.13].

The largest is clearly the isosceles, but surely that is smaller than the equilateral. Student DK then makes explicit his symmetry argument. Of course, reflecting the isosceles right triangle in the diameter results in a square (see Figure 1.14).

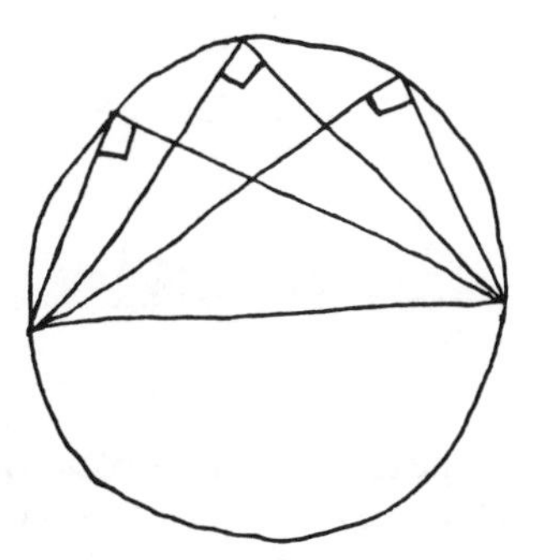

Figure 1.13

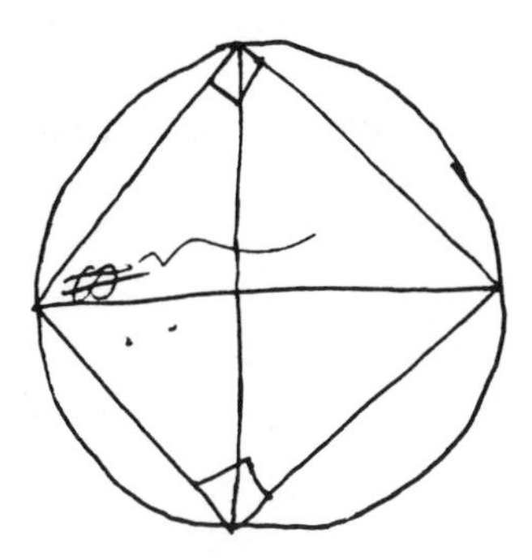

Figure 1.14

DK: Well, we can't build a diamond, so we can't build a diamond that would go like that [points to Figure 1.14], obviously you want to make it perfectly symmetrical. . . .

BM: Yeah, it is symmetrical.

DK: If it is going to be symmetrical, though, then you know this line has to be flat. . . . It is going to have to form a right angle. So all we really have to do is form a right angle. So all we really have to do is find the largest area of a right triangle [whose right angle is on the diameter, Figure 1.15] inscribed in a semicircle.

This is now a one-variable calculus problem, which they set about solving. Unfortunately a minus sign gets lost in the computations, and after 12 minutes of work the answer they get is impossible. BM tries to patch things up, without success. Then,

DK: Well, let's leave the numbers for a while and see if we can do this geometrically.

BM: Yeah, you're probably right.

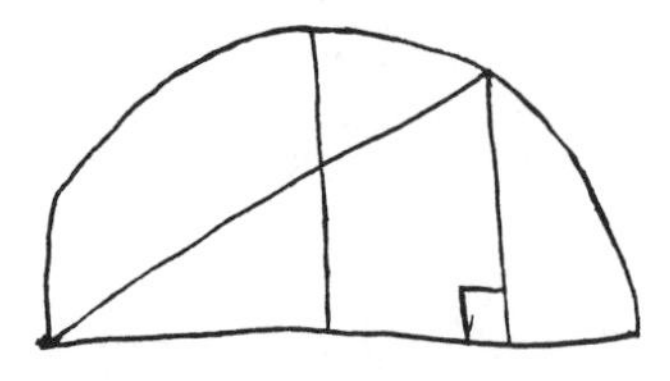

Figure 1.15

This discussion terminates the analytic attempt to solve the problem, and the solution meanders to a half from there. When they have run out of steam, I ask them for a recapitulation. BM focuses on his attempt to use the calculus.

AHS: When you do that . . . you wind up finding the area of the largest right triangle that can be inscribed in a semicircle.
DK: We determined that.
AHS: My question is: How does that relate to the original problem?
BM: Well, . . .

They had no ready answer.

DISCUSSION

Any discussion of these two problem-solving sessions must, of course, begin with an account of the resources the students had at their disposal. As indicated above, both pairs of students knew more than enough to solve Problem 1.3. The issue is not what the students knew, it is how they used that knowledge, or failed to use it. I argue here that the students' failures to solve Problem 1.1 were caused by malfunctions at the control level; poor decision-making guaranteed that both pairs of students would fail, regardless of what they "knew."

The attempt by KW and AM to solve problem 1.3 provides, at one level, a clear example of control failure. The students' absolute commitment to calculating the area of the equilateral triangle, for the full 20 minutes allotted to them, guaranteed that they would fail to solve the problem. This slavish pursuit of a particular goal—which they were unable to justify after the session was over—is an archetypal wild goose chase if ever there was one. Having said this, I should note that there is nothing wrong in itself with a decision to calculate the area of the equilateral triangle while working Problem 1.3. It is quite reasonable to try to substantiate one's intuition regarding the nature of a solution by making a few calculations before embarking on full-fledged solution attempts (and the equilateral triangle is an obvious intuitive choice). Given 20 minutes to work Problem 1.3, one might well choose to spend 5 minutes or so calculating the areas of some likely candidate for a solution. Fortified by the empirical evidence, one might then look for other methods. But 5 minutes are about all one should spend, and the decision to undertake the computations must be made with some idea of the role that the computations are to play in the solution as a whole.

Nothing resembling this happened in KW and AM's attempt. When AM made the obvious conjecture, KW's response was to begin calculating immediately—a calculation that AM endorsed. There was not a single word of discussion between the two students about the path they had chosen; there was no planning except for the tacit agreement to compute the area of the equilateral triangle, and no consideration of why it might be a reasonable thing to do. Their initial decision was not called into question once during the entire problem solution. That decision certainly set them on the road to failure.

What guaranteed their failure, however, is more subtle. Admittedly, an unjustified attempt to calculate the area of the equilateral triangle is a mistake. But after all, everyone makes mistakes. It is unrealistic to expect problem solvers to always make the right decisions; good problem solvers often make false starts when working on difficult problems. What makes them good problem solvers, however, is that they can recover from those false starts. A much more serious mistake in KW and AM's attempt is that they did not curtail their wild goose chase as it became increasingly clear that their approach was producing little of value. After 5, or 10, and certainly 15 minutes, they should have asked if they were on the right track. Even if nothing they had done up to the point of evaluation was salvageable, they might have been able to use their remaining time to some good effect. A major component of effective control consists of the *periodic monitoring and assessment* of solutions as they evolve and the curtailing of attempts that are unfavorably assessed. The absence of this control mechanism was one major cause of KW and AM's failure. Another cause of their failure was their lack of opportunism. When control works well, ideas of potential interest are examined to see if they might be useful and are pursued if they look promising. During this solution attempt, a number of different approaches came up for possible use: a related max-min problem, a variational argument, and so forth. They were all left unexplored. Such sins of oversight are costly. One or more of these approaches, properly exploited, might have led KW and AM to a solution.

The attempt by DK and BM provides similar evidence of the effects of bad control, some of it more dramatic. BM's initial suggestion about using calculus might have led to a correct solution, and DK is certainly responsible for shifting attention away from it. This too was a sin of oversight: The idea of using calculus should not have ben passed over without serious evaluation. The rejection by BM of DK's suggestion that they examine triangles "with base one of the diameters," is similarly costly. Observe that BM's argument and the figure he draws (Figure 1.12) are almost identical with the heuristic argument and diagram (Figure 1.9) that were used above to solve Problem 1.3. Thus BM's argument could be exploited. It is not, however. With the observation that the isosceles triangle, reflected in the diameter, yields a "diamond," they discard all that they have done to this point. This decision, made without any attempt to see if something they have done might be salvaged for future use, allows them to throw away a potentially useful tool.

The next major decision DK and BM make is to maximize the area of triangles like those in Figure 1.15, and then reflect the largest such triangle in the diameter of the circle to solve the original problem. In adopting this suggestion, they change the problem: The reflected triangle will be isosceles, and they will have found the largest isosceles triangle inscribed in the circle. There is no guarantee that this triangle will be the largest of all inscribed triangles (although it happens to be). In consequence, there is a fair chance that all of their effort will be wasted. With DK's caveat "if it is symmetrical" and BM's assertion "Yeah, it is symmetrical," they embark on this approach. It consumes 12 minutes, roughly 60% of their allotted time. Yet that large investment of time and energy—which can be salvaged—is casually discarded (DK: "Well, let's leave the numbers for a while and see if we can do this geometrically." BM: "Yeah, you're probably right."), and the solution degenerates from there.

These two problem sessions offer an exemplary collection of unwarranted assumptions and lost opportunities, a collection of control decisions that ensured that the two pairs of students would fail to solve Problem 1.3. Students KW and AM relentlessly pursued wild mathematical geese, ignoring possibilities that might have been of value. Students DK and BM explored inadequately justified solution paths, and the effort they made could have been completely wasted. When DK and BM did abandon the attempts they considered unsuccessful they threw them away without reflection—and in doing so they threw away all of the elements of a complete solution. These sessions provide clear evidence that control is important. That there are various gradations of control, and that control can have positive effects on a solution as well as negative, will be discussed in Chapter 4. Methodological issues regarding control, and the details of an analytical framework for the rigorous characterization of the behaviors informally described here, are discussed in Chapter 9.

Belief Systems

In the best of all possible worlds the three categories of knowledge and behavior described above would suffice to characterize mathematical problem-solving performance. Resources describe the mathematical facts and procedures potentially accessible to the problem solver. Heuristics provide the means for stretching those resources as far as possible. Control decisions determine the efficiency with which facts, techniques, and strategies are exploited. It should be possible to characterize "expert" performance by specifying the nature of well-codified resources, heuristics, and control. It should be possible to characterize naive student behavior by contrasting such behavior directly with the idealized performance of experts—that is, by describing the degree to which students' knowledge bases contain the relevant resources, the degree to which they are able to stretch what they know, and the degree of efficiency with which they succeed at these endeavors.

The literature now makes it abundantly clear that this is not the best of all possible worlds (see, e.g., Caramazza, McCloskey, & Green, 1981; DiSessa, 1983; Janvier, 1978; Lochhead, 1983; McCloskey, 1983a,b; McDermott, 1983; Neisser, 1976; Perkins, 1982; Perkins, Allen, & Hafner, 1983; Rogoff & Lave, 1984; see also Voltaire, 1759/1960). Research on "naive physics," for example (see the papers by Caramazza *et al.,* 1981; DiSessa, 1983; and McCloskey, 1983a,b), demonstrates that students who have taken a year of college physics, and who can apply the laws of mechanics correctly on textbook problems, will often analyze qualitative physical situations using Aristotelian models of physical phenomena that are flatly contradicted by their formal knowledge. Perkins' research on informal reasoning indicates that people with a "makes-sense epistemology" may fail to make careful and convincing arguments when they are asked to do so, not because they are incapable of making such arguments but because "making a good case" suffices from their point of view. Research on everyday reasoning indicates that relevant, formal knowledge and procedures may simply be ignored or supplanted in a "real-world" context; conversely, prior experience in the real world shapes learning in a formal context. Finally, the extensive literature on decision theory (Einhorn & Hogarth, 1981; Kahneman, Slovic, & Tversky, 1982) indicates that people's decision-making in various domains can be highly regular, strongly biased, and anything but "rational" (see Chapter 5 for a more extensive discussion of these issues). The point here is simply that "purely cognitive" behavior—the kind of intellectual performance characterized by discussion of resources, heuristics, and control alone—is rare. The performance of most intellectual tasks takes place within the context established by one's perspective regarding the nature of those tasks. Belief systems shape cognition, even when one is not consciously aware of holding those beliefs.

Similar statements hold in the case of mathematical problem solving. This section begins with a dramatic but typical example, in which two students' beliefs about the nature of mathematical thinking (their "naive geometry") prevent them from solving a problem that they are capable of solving with ease.

STUDENTS SH AND BW WORK PROBLEM 1.1

At the time they worked Problem 1.1, SH and BW were college freshmen who had both completed one semester of calculus. Each studied a year of geometry in high school.

Before the following session, I give them the problem statement, a cheap compass, and an old wooden ruler. Student BW reads the problem out loud.

BW: You are given two intersecting straight lines and a point *P* marked on one of them, as in the figure [Figure 1.16] below. Show how to construct, using straightedge and compass, a circle that is tangent to both lines and that has the point *P* as its point of tangency to one of the lines.

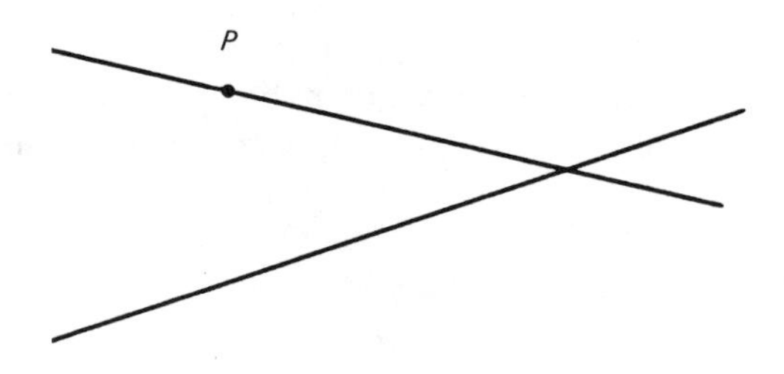

Figure 1.16

About half a minute passes in silence. SH wonders if there is anything special about the particular figure I have given them, or if the construction can be done in general.

SH: The first question is, Can you construct a circle tangent to two lines? We'll worry about that later.

BW conjectures that the line segment from *P* to its counterpart *Q* on the bottom line [Figure 1.17] is the diameter of the desired circle. She traces the hypothesized solution with her fingers.

BW: We can construct a line through *P* that makes an isosceles triangle. . . . We'll bisect that [the segment *PQ*], draw a circle in there. . . .

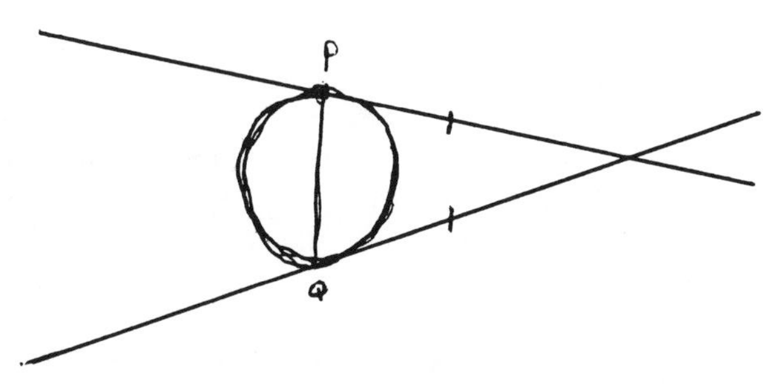

Figure 1.17

Without a word of discussion about the rationale for the construction, SH picks up the straightedge and compass and begins to perform it. The two students work slowly and meticulously with the somewhat inaccurate tools I have given them. The result is Figure 1.18, which does not look quite right. SH runs his fingers over the intersection points between the circle and the given lines.

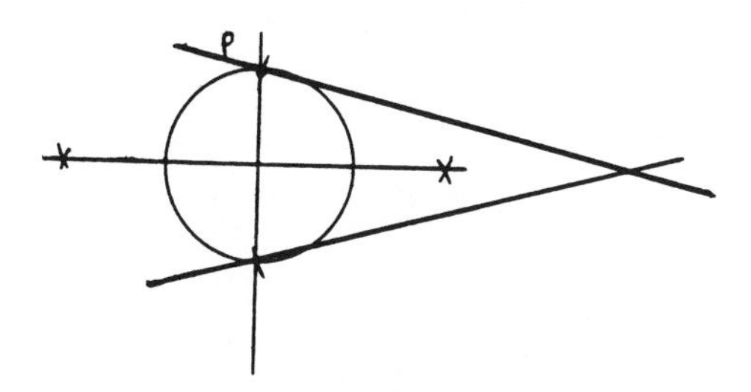

Figure 1.18

SH: It seems to cross these lines.
BW: Yeah, I think that might be the way it is because . . . it is possible that . . .
SH: It's probably just that we drew the lines . . . if our work were perfect . . .

Seeing their concern about the quality of their construction, I rummage through my desk and find a good ruler. I offer it to them, and they do the construction over with great care. They examine the new construction.

BW: It does go through . . . it does go over. . . . We have to get a line perpendicular to this one [the top one] so that it allows the . . .
SH: It can't go through, can it? It has to be tangent.
BW: Yeah, I don't know.

The comment by BW ("It does go through") indicates that the "PQ as diameter" hypothesis must be rejected. The two students return to making rough sketches, looking for solutions to the problem. BW draws Figure 1.19, substantiating the rejection of the first hypothesis. In the meantime SH makes the three sketches in Figure 1.20. BW returns to the original figure and traces an arc where the point of tangency should be, convincing

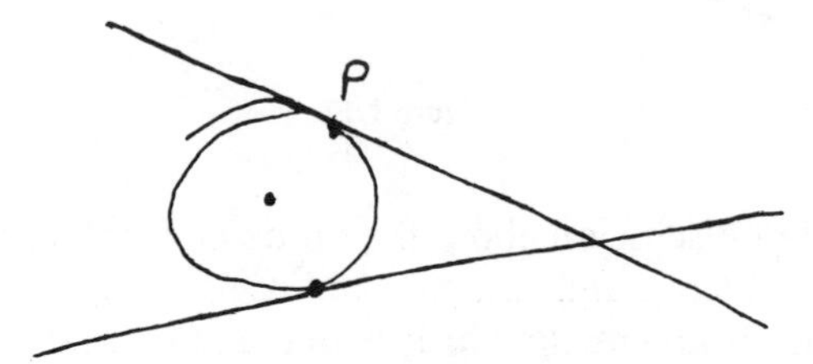

Figure 1.19

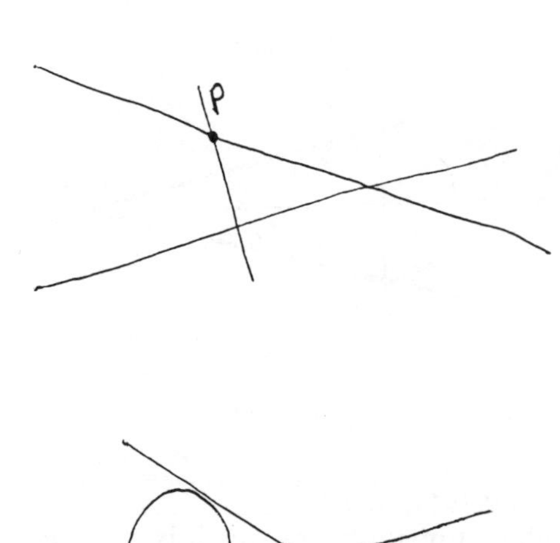

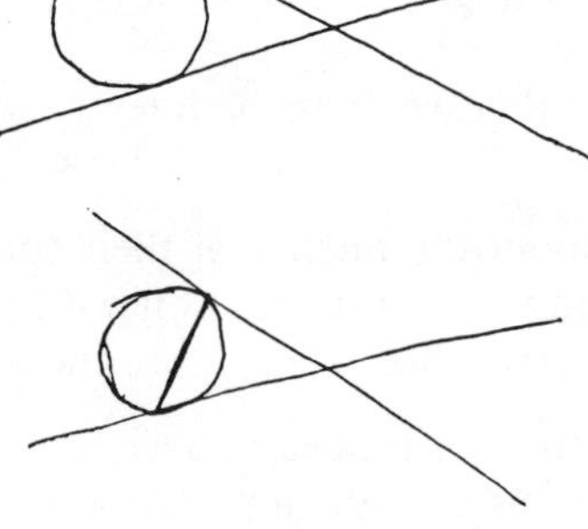

Figure 1.20

herself of the importance of the perpendicular to P, something they had failed to take into account in their first attempt. She is tracing perpendiculars on Figure 1.19 with her fingers when SH makes a suggestion.

SH: Might it involve angle bisectors at all?
BW: We can try.
SH: In that picture [pointing to the middle line in Figure 1.18] it does look like possibly an angle bisector is involved.

Student SH traces some angle bisectors and perpendiculars with his finger, while BW draws Figure 1.21.

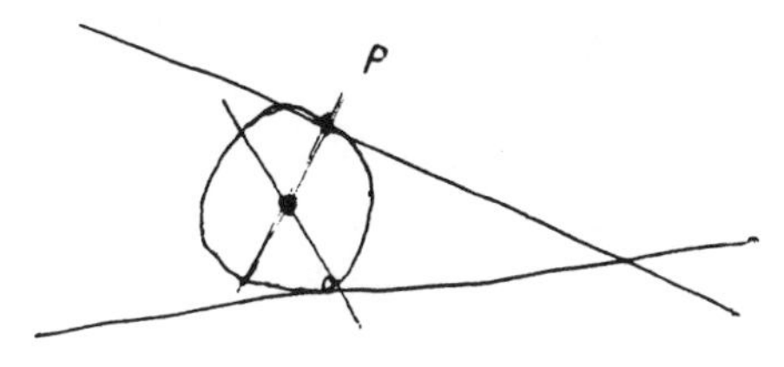

Figure 1.21

BW: OK now. Maybe we should have the line perpendicular right here [points to P] and then go here [points to the point on the bottom line the same distance from the vertex], and that would be a perpendicular there. . . . OK. Let's try it.

Now begins the construction that Figure 1.21 suggests. A few minutes pass in silence, so, seeing that she is working on the correct solution, I intervene.

AHS: Would you like to describe for the record what you are doing?
BW: We were trying to construct a perpendicular line here [indicates P] and here [indicates the point below] to get the center of the circle.
AHS: [indicating the point below] What did you do to get that point?
BW: We took a measurement and . . .
AHS: Why?
BW: To make it the same length.
AHS: And why did you want it the same length?
BW: Because it seemed that that was the way it should be.
AHS: OK. Let's let you find out whether it works and then talk about whether it should.

At this point BW returns to the construction, which SH watches actively. The result is Figure 1.22, which, unfortunately, is inaccurate.

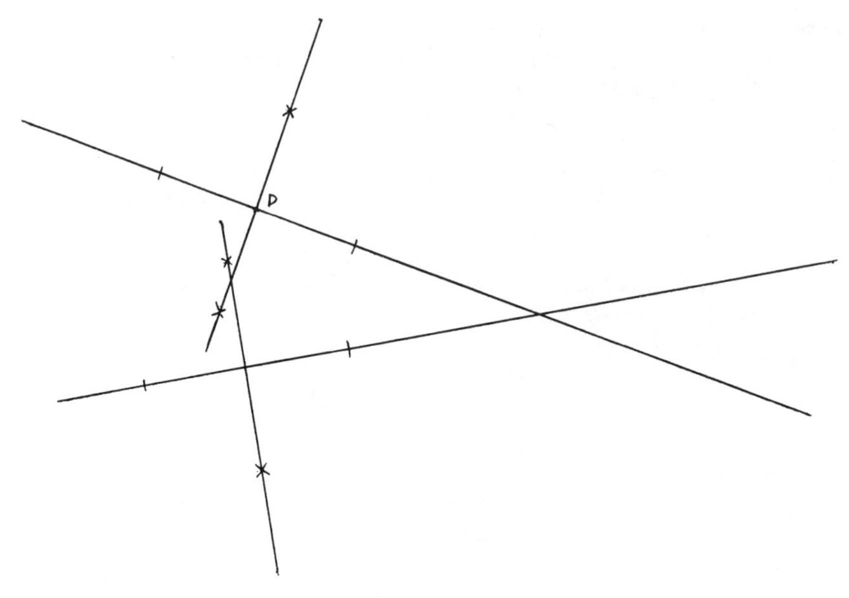

Figure 1.22

BW: [as she draws in the second perpendicular] I hope that's my construction line. [It looks bad.] Well, . . .
AHS: Is it working?
BW: No.
AHS: What's going wrong?
BW: Well [pointing to one of the two perpendicular segments], this is really short, that's why the circle . . .
AHS: Well, that's the problem when you work in the real world, your hands and your machinery don't always do what you want them to. Can you, perhaps, reason why it [the construction] ought or ought not to work, and then say "Well, if we just had a better compass . . ."
BW: Well, I already got this line way off. . . .
SH: Our mistake might be putting this point [the point opposite *P*] the same distance as the distance from these two [*P* and the vertex]. Maybe that's where we went wrong.

Out of pity at this point I tell SH and BW that their solution is indeed correct and bring the session to a halt.

DISCUSSION

One's first reaction on watching the videotape of this session is that SH and BW flounder as badly as they do because they do not know any formal mathematics. Save for the odd fact that they are perfectly comfortable with the procedures for bisecting angles and for constructing perpendiculars (indicating some sophistication), they appear to be naive empiricists. They make rough sketches to get a sense of what is important in the problem and, guided by those sketches, conjecture what a solution should be. They test their conjectures by performing and evaluating them: A hypothesized solution is accepted as correct if and only if that construction, when performed, yields a picture that meets stringent empirical standards. Roughly two-thirds of SH and BW's 15-minute long attempt are spent with straightedge and compass in hand. Even when I ask if they could reason why their construction ought or ought not to work, their response is in terms of the mechanics of their construction (BW: "Well, I already got this line way off").

As I noted above, SH and BW had each studied a year of geometry in high school. More importantly, they had not forgotten what they had learned. In class the next day, they were asked to solve Problems 1.4 and 1.5. *Each student solved both problems within 5 minutes.*

Problem 1.4 The circle in Figure 1.23 is tangent to the two given lines at the points P and Q. Prove that the length of the line segment PV is equal to the length of the line segment QV.

Problem 1.5 The circle in Figure 1.23 is tangent to the two given lines at the points P and Q. If C is the center of the circle, prove that the line segment CV bisects angle PVQ.

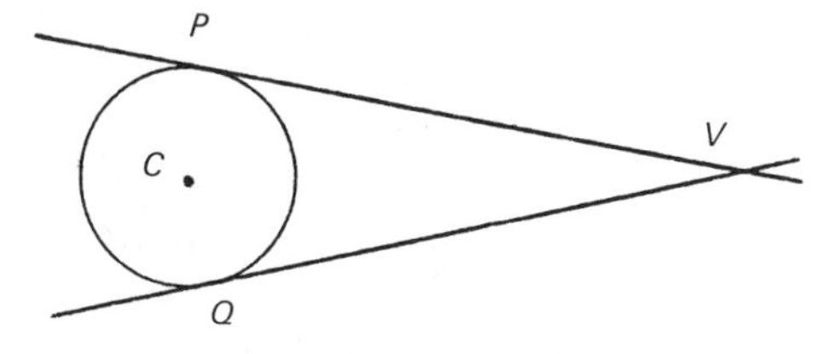

Figure 1.23

This session with SH and BW was one of the first that I recorded, and my initial reaction was that their performance was an aberration. Perhaps they were unnerved by the presence of my videotape equipment, or by the fact that I would be analyzing their performance; their minds may have "gone blank" under the stress of the videotaping session, with the result being this unusual behavior.* Much to my surprise, I discovered that their behavior was the rule rather than the exception. In that year and the following year, I recorded roughly two dozen pairs of students working Problem 1.1. All were college students. They came from school systems distributed widely across the United States (and, in a few cases, from other countries). All but two students had completed, or were enrolled in, a course in calculus. All but one had studied a year of geometry, mostly devoted to proofs, in secondary school.

In all of those videotapes only one pair of students derived an answer to Problem 1.1, although a few students did try briefly to solve it by reducing it to a known problem whose solution could be recalled from memory. Most students proceeded in a manner similar to SH and BW. Guided by rough sketches, they guessed at the nature of a solution. Every hypothesis that passed the "rough sketch test" was tested by construction. If it turned out to be wrong (i.e., the construction provided evidence for rejecting it), the students would try again with another construction. Most often a solution attempt would take between 15 and 20 minutes, with more than half of that time spent with straightedge and compass in hand. Ultimately the students would either find a construction that met their empirical standards or would report failure. Of all the cases where a solution was confirmed by construction, only once did the students provide any mathematical justification for its correctness. This was an after-the-fact observation:

> We got it, that is, it looks pretty close. [He points to the circle they have just drawn in.] . . . I think, if you center it right, they touch. OK. These are similar triangles. [He points to the two triangles in Figure 1.24.] Yeah, . . . they are equal triangles too.

The precise nature of these students' empiricism, and the effects of their beliefs on their problem-solving performance, will be discussed at length in Chapter 5. One possible explanation of such behavior was suggested in the discussions of resources above. It may be that the students lack an understanding of the generic nature of mathematical arguments and do not see any connection between Problem 1.1 and Problems 1.1A and 1.1B. In that case their difficulty would have purely cognitive origins. As research on students' "van Hiele levels" in geometry indicates (see, e.g., Hoffer, 1983), this may well be the case for a large number of students. In my experience, however, a fair number of college students (SH and BW among them) are

* There is reason to be concerned about the disruptive effects of a videotaping session on people's problem-solving performance. In some cases the effects can be quite severe; see Chapter 9 for an extensive discussion of this issue. In this particular situation, however, SH and BW, though quite shy, claimed not to be terribly bothered by the equipment.

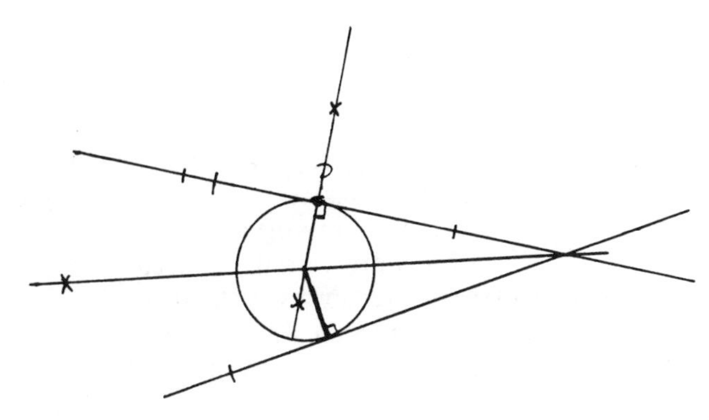

Figure 1.24

indeed aware of such connections. In this case, a second explanation will be invoked. It will be argued that such students are *naive empiricists* who do not use the mathematical tools potentially at their disposal, because it does not occur to them that such tools might be useful. That is, for these students, mathematical argumentation is not seen as being relevant to the processes of discovery needed for solving construction problems. Thus it is simply not called upon.

There are some delicate issues regarding the use of the term *belief* in this context, especially when a stronger version of the second explanation is couched in language such as "the students believe proof to be irrelevant to the discovery process." In my opinion such statements merit serious consideration. Stated below, in this stark form, are three "typical" student beliefs and the consequences, in terms of students' behavior, of having them.

Belief 1 Formal mathematics has little or nothing to do with real thinking or problem solving.

Consequence In a problem that calls for discovery, formal mathematics will not be invoked.

Belief 2 Mathematics problems are always solved in less than 10 minutes, if they are solved at all (consider the typical experience of most students).

Consequence If students cannot solve a problem in 10 minutes, they give up.

Belief 3 Only geniuses are capable of discovering or creating mathematics.

First consequence If you (a typical student) forget something, too bad. After all, you are not a genius and you will not be able to derive it on your own.

Second consequence Students accept procedures at face value, and do not try to understand why they work. After all, such procedures are derived knowledge passed on "from above."

What it means to have such beliefs (e.g., Need one be conscious of the forces that drive one's behavior?), the degree to which students actually have such beliefs, and how such beliefs develop is discussed at length in Chapters 5 and 10. The term *belief* itself is controversial and needs to be modified to some degree. The point of departure, however, should not be controversial: One's mathematical world view shapes the way one does mathematics. It is in the broad sense of a mathematical world view that the term *belief systems* will be used.

Summary

This chapter presented the outline of a framework for the analysis of mathematical behavior. Four categories of knowledge and behavior were introduced. It was argued that an explanation of people's behavior in mathematical situations, including an explanation of why they succeed or fail in their attempts to solve mathematical problems, needs to take all four of these categories into account. The four categories, which overlap and interact with each other to some degree, are as follows.

Resources are the body of knowledge that an individual is capable of bringing to bear in a particular mathematical situation. They are the factual, procedural, and propositional knowledge possessed by the individual. The key phrase here is "capable of bringing to bear"; one needs to know what an individual might have been able to do, in order to understand what the individual did do.

Heuristics are rules of thumb for effective problem solving. They are fairly broad strategies for making progress on unfamiliar or difficult problems. Examples are exploiting analogies and working backward. While the mathematics education community has generally accepted the idea that such strategies are useful and worth teaching, until recently there has been little understanding of their complexity and little evidence that students can reliably learn to use such strategies.

Control deals with the question of resource management and allocation during problem-solving attempts. In this context it is reserved for major decisions regarding planning, monitoring, and assessing solutions on-line, and the like. With good control, problem solvers can make the most of their resources and solve rather difficult problems with some efficiency. Lacking it, they can squander their resources and fail to solve problems easily within their grasp.

Belief systems are one's mathematical world view, the perspective with which one approaches mathematics and mathematical tasks. One's beliefs about mathematics can determine how one chooses to approach a problem, which techniques will be used or avoided, how long and how hard one will work on it, and so on. Beliefs establish the context within which resources, heuristics, and control operate.

2

Resources

Resources are the foundations upon which problem-solving performance is built. A description of these foundations—an inventory of what individual problem solvers know and the ways in which they access that knowledge—is essential if we are to understand what takes place in a problem-solving session.

The questions What is the nature of the knowledge that individuals have at their disposal? and How is such knowledge organized and accessed for use? are central issues in cognitive psychology and AI, to name just two fields with proprietary interests in this area. They are the primary focus of information-processing (IP) psychology and of interest to fields as diverse as linguistics, philosophy, and neurobiology. The relevant literature is immense. A full book could barely do it justice, much less a few introductory pages. My purpose here is to point to some major trends in the research as general background for the rest of the book. Three issues are discussed: (1) routine access to relevant knowledge, (2) the broad spectrum of resources, and (3) flawed resources and consistent error patterns. The reader may wish to skim this chapter rapidly at this point, returning to it later when the need for greater detail is perceived.

Routine Access to Relevant Knowledge

Much of IP psychology has been devoted to modeling competent performance in familiar domains, that is, exploring the means by which experts seem to do the right things almost automatically when working on familiar tasks. Three typical examples will be discussed, after a brief introduction to the literature.*

Simon (1979) provided a brief historical overview of IP work in problem solving, with a broad review of resource-related issues. According to Simon, the most significant mechanisms for solving well-structured, puzzle-like problems (e.g., "cryptarithmetic" tasks, the "tower of Hanoi," or exercises in symbolic logic) were fairly well understood by 1972, when Newell and Simon's classic *Human Problem Solving* was published. The major issue in such problems was to search efficiently through a well-defined but very large *problem space* (the set of options open to the problem solver) for a solution. As the tools and techniques for working in such domains became well established, IP research turned increasingly to *semantically rich* domains, that is, subject areas in which successful performance depends both on the possession of domain-specific knowledge and problem-solving skills. Models of competent performance have been developed in a range of semantically rich domains including chess, chemical engineering, thermodynamics, electronics, cost accounting, business policy, medical diagnosis, and the identification of molecules using mass spectrograph data. For the most part, the behavior that has been modeled is the highly competent, routinized (though by no means trivial!), and consistent problem-solving behavior of experts.

Greeno and Simon's (1984 draft) "Problem Solving and Reasoning" updates Simon's 1979 paper and broadens its scope. Rumelhart and Norman (1983) reviewed the literatures of psychology and AI with regard to work on the representation of knowledge. The paper provides a general description of representation in memory. It compares and contrasts propositionally based representational systems (the predicate calculus, semantic networks, schemata, and frames), analogical representations, and procedural representations, with regard to their utility and accuracy in "capturing" cognitive processes. The paper provides a great deal of technical background for the "memory issues" that are dealth with here, although such detail is not necessary to understand what follows. A summary paper by Rissland (1985) and a reaction paper of mine (Schoenfeld, 1985) present an easy introduction to work in AI that has direct implications for understanding or teaching mathematical thinking skills. Extending the discussion smewhat beyond resources, Nickerson (1981) dealt with various perspectives on what it means

* The more fine-grained analyses of cognition will not be reviewed here. A good overview of such research may be found in Monsell (1981). As Monsell noted, "Every cognitive skill draws upon part of the brain's extensive repertoire of representational subsystems, storage mechanisms, and processes" (p. 378). His review covers four such areas: the perception of objects and words; the distinction between short- and long-term memory mechanisms; the retrieval of remembered episodes and facts; and attention, performance, and consciousness.

to "understand." Nickerson's discussion of misconceptions and failures to understand serves as a backdrop for our discussion of belief systems. The paper includes a distillation of useful suggestions for instruction derived from recent research. Resnick and Ford (1981) provided the psychologist's view of mathematical "basics" for instruction. And finally, the 1983 research report entitled "Research on Cognition and Behavior Relevant to Education in Mathematics, Science and Technology," produced by the Federation of Behavioral, Psychological, and Cognitive Sciences, gives telegraphic but extremely broad coverage.

I discuss three types of proficient performance, all of which have been explored in substantial detail:

1. the ability of grand masters in chess to play as many as 40 games simultaneously, selecting within seconds the proper moves for each game;
2. the ability of mathematicians or physicists to look at problems in their disciplines that cause their college freshman students great difficulty, and to "see" the appropriate techniques for solving them "on the spot";
3. the ability that most people have to make sense, almost automatically and with great accuracy, of many situations that they encounter frequently—for example, reading a short story and correctly interpreting the information provided in it.

COMPETENT PERFORMANCE IN CHESS

A number of unusual abilities might serve to explain grand masters' extraordinary capacity for simultaneous chess play. Among the capacities that have been suggested is spatial ability. Perhaps people with a gift for chess are able to see, in their mind's eye, current and proposed positions; perhaps this enables them to play out the games mentally in a way that those lacking the ability can not. Some ingenious research by de Groot (1965, 1966) rules out that explanation. In one experiment, novices and experts were shown typical midgame positions and then asked to reproduce on nearby chess boards the positions they had been shown. Hypothetically, those people with better visual ability would be able to reproduce the positions with much greater accuracy. Indeed, the experts performed at a far better level than the novices. In another experiment, the participants were shown nonstandard configurations of chess pieces. That is, the arrangements of pieces on the board did not represent legal chess positions. On such tasks the experts did no better than the novices. In fact, when the positions they had been shown closely resembled legal configurations, the experts did worse than the novices; their knowledge interfered with their perceptions, and they reproduced the positions that they expected to see rather than the ones that they had been shown. Thus visual ability is not, in itself, the source of the experts' proficiency. Other research rules out other hypotheses, for example, that the chess masters have better organized, or larger, memories for mentally playing out games dozens of moves ahead while novices could only play out a few moves. What the research reveals is an elaboration of the results suggested by de Groot's work. The expert chess player has stored, as "chunks" in memory, an extensive "vocabulary" of chess positions. Simon described chunks as follows.

> A chunk is any perceptual configuration (visual, auditory, or what not) that is familiar and recognizable. For those of us who know the English language, spoken and printed English words are chunks. . . . For a person educated in Japanese schools, any one of several thousand Chinese ideograms is a single chunk (and not just a complex collection of lines and squiggles), and even many pairs of such ideograms constitute single chunks. For an experienced chess player, a "fianchettoed castled Black King's position" is a chunk, describing the respective locations of six or seven of the Black pieces. (Simon, 1980, p. 83.)

The research on perception in chess indicates that chess masters must (and do) have approximately 50,000 chunks stored in memory just for routine play—that is, to respond to configurations of pieces that occur with some frequency. This collection of 50,000 entities stored in memory is roughly the size of a literate person's vocabulary, and it functions much the same way. The chess master who perceives a given chunk on the board, say, the "fianchettoed castled Black King's position" Simon mentioned above, will not only recognize that aggregate as a unit, but will also have associated with that chunk at least one appropriate response to it, a response that comes to mind without the need for conscious analysis of the situation that provoked it. The situation might be compared to what happens when a literate person comes upon an injunction, such as

in the middle of a body of text. (Note that one can override the associated response, but that the response itself is nearly automatic.) To use Simon's terminology, one can think of certain kinds of knowledge as being organized in "condition–action pairs" (since familiar conditions evoke nearly automatic actions) or "productions." One can think of certain kinds of competent performance as resulting from the possession of large collections of these knowledge productions. ("Production systems" are one of the formalisms used in AI programs.) Just as the routine play of the chess master may be based on the possession of tens of thousands of these productions—the chess player's "vocabulary"—so may the routine competence of the mathematician be based on a comparable mathematical vocabulary of perceptual chunks learned through the years (recall the mathematician's immediate perception of similar triangles when confronted with Figure 1.5).

PROBLEM PERCEPTION

The situations characterized just above were of the kind where an individual produces a specific response in reaction to his or her perception of a specific situation as a chunk of information. Here we move up a level. The specific details of the situation that the individual encounters may not be familiar, but the situation itself is. The response produced by the individual is not specific to the situation at hand, but rather to the type that it represents. That is, the individual recognizes the particular situation as being of a specific type (say type X) and immediately thinks of the techniques relevant for dealing with type X situations. Details are filled in as it becomes necessary to do so. Dealing with a type X situation in this way is called having a *type X schema.*

One of the classic studies of schema-based mathematical performance was carried out by Hinsley, Hayes, and Simon (1977) in an examination of word problems in high school algebra. That paper set out to verify the following four assertions.

> (1) *People can categorize problems into types. . . .*
> (2) *People can categorize problems without completely formulating them for solution.* If the category is to be used to cue a schema for formulating the problem, the schema must be retrieved before formulation is complete.
> (3) *People have a body of information about each problem type which is potentially useful in formulating problems of that type for solutions . . .* directing attention to important problem elements, making relevance judgments, retrieving information concerning relevant equations, etc.
> (4) *People use category identifications to formulate problems in the course of actually solving them.* (Hinsley *et al.*, 1977, p. 92)

The paper offers some fairly strong evidence to substantiate the importance of problem schemata. In one of the studies they reported, the authors began by reading brief sections of a problem statement to subjects. They then asked the subjects to categorize the problem, to say what information they expected to receive, and to say what question they expected to be asked.

> [A]fter hearing the three words, "A river steamer . . . " from a river current problem, one subject said, "It's going to be one of those river things with upstream, downstream, and still water. You are going to compare times upstream and downstream—or if the time is constant, it will be distance." . . . After hearing five words of a triangle problem, one subject said, "This may be something about 'How far is he from his goal' using the Pythagorean theorem." (Hinsley *et al.*, 1977, p. 97)

There has been a great deal of substantiating evidence since that article was published. The "expert" part of the story is that competent problem-solving performance in routine domains is often schema-driven, allowing for rapid access to relevant information and to the appropriate solution techniques. The "novice" part of the story, as one might expect, is less clear. Students lack many of the problem schemata that would enable them to function efficiently. Moreover, misperceptions of problem characteristics may well send students off in the wrong directions; the ability to characterize may do harm, as well as good. For an extended discussion of this issue and some experimental results, see Chapter 8.

FUNCTIONING EFFICIENTLY IN COMPLEX BUT FAMILIAR SITUATIONS

This discussion extends the results of the previous sections one level further. Condition–action pairs and schema-based solutions are two classes of a broad range of phenomena that are used to model competent mental performance—not just in problem solving, but of all types. Much of our functioning in everyday life, from making breakfast in the morning through performing routine tasks at work and then making it safely back home again, can be attributed to the almost automatic (and often unconscious) accessing of what might be called "generic responses to generic situations." The assertions quoted above from Hinsley, Hayes, and Simon can be extended as follows:

1. *People categorize their experience into types.*
2. *People tend to classify their new experiences in ways consistent with their prior categorizations, often before the new experiences are analyzed in detail.* If the "distinguishing features" of the new experience match those of an already defined category, that category is retrieved before the formulation of the new experience is complete, and helps to shape it.
3. *People have bodies of information about categories of experience that are potentially useful in dealing with new experiences that fall within those categories.* That is, people frame their expectations of circumstances in the light of prior experience; tools and techniques that have been useful in the past are "brought to mind" for the present situation.
4. *People use their "categorical knowledge" to interpret and deal with new situations.* In fact, their view of these situations is shaped—sometimes inaccurately—by their prior categorizations.

There is ample evidence for statements 1–4, at various levels of processing. At the level of visual perception, they describe the reason for some

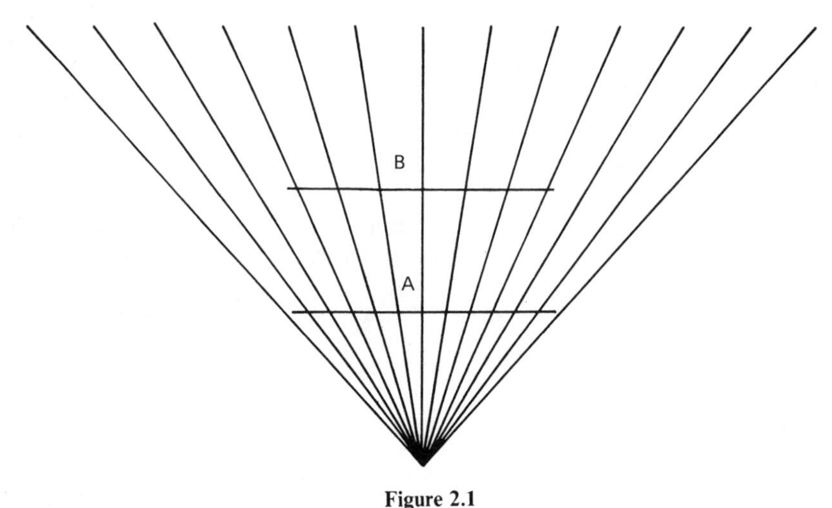

Figure 2.1

optical illusions—why Line A seems longer than Line B in Figure 2.1, for example. Similarly, expectations of what they should see led de Groot's chess masters to misperceive the almost-correct chess positions that they had been shown. But most of the time, such expectations are highly functional. In addition to being represented in the literature as schemata, these generic representations of generic situations have been instantiated in AI as *scripts* and *frames*.

One of the classic examples of a script for understanding new situations comes from Schank and Abelson's (1977) attempt to build AI programs that understand stories. Consider the following story, and question.

Story John is a reasonable and fairly generous person. After he completes a good business deal, he decides to treat himself to dinner. He goes to a restaurant he has heard about and has looked forward to trying. The waiter brings his meal, which John thinks quite well prepared. The check comes to $50. John puts the check on his credit card and leaves a 50¢ tip.

Question What was the quality of the service John received?

Assumedly your train of reasoning in answering the question went something like the following. Having completed a good business deal, John should be in a good mood (he has, after all, decided to treat himself). Since he is a generous person in good spirits, the tip he leaves should be substantial. But he only leaves a tip of 1%. So one of these three possibilities is most likely to be the case: (1) he hates his meal, (2) he is short of money and can barely pay for the meal, or (3) he is very unhappy with the service. Since John thought the meal well prepared (and a "reasonable and fairly generous" person would not blame the waiter for the kitchen's mistakes in any case), and he could have put the tip on his credit card if he wished, the first two explanations are unlikely. The remaining possibility is that the waiter offended John in some way. Thus the service was probably of low quality.

The point about this line of argument is that most of it is based on information that is not contained in the story itself. We infer from the context that John is in a good mood and that he is likely to tip well. Then, using our general knowledge, we estimate what a good tip should be. The story violates our expectations, so we then bring to mind and examine the customary reasons for leaving small tips. The story provides direct information to rule out one possibility (the meal was good) and indirect information (we must know that tips can be paid with credit cards) to rule out another. Only one plausible hypothesis remains, and thus we draw our conclusion.

Schank and Abelson argued that readers make sense of stories like this in the way I have just described, because they have scripts that supply many of the details that are not explicit in the text itself. For example, the knowledgeable reader has a restaurant script that includes the assumptions made in the argument above. There is not just one restaurant script, of course; the reader has one set of expectations for fast food and quite another for haute cuisine. But there are cues in the story (more generally, in our experience) that determine which script should be invoked.

Similar in intent but somewhat different in instantiation is the notion of a frame. According to Minsky,

> A *frame* is a data-structure for representing a stereotyped situation like being in a certain kind of living room or going to a child's birthday party. Attached to each frame are several kinds of information. Some of this information is about how to use the frame. Some is about what one can expect to happen next. . . . Much of the phenomenological power of the theory [of frames] hinges on the inclusion of expectations and other kinds of presumptions. *A frame's terminals are normally already filled with "default" assignments.* Thus, a frame may contain a great many details whose supposition is not specifically warranted by the situation. . . . (quoted in Winston, 1977, p. 180)

Here are a few examples of such default assignments. When we enter a room we expect that certain rectangular shapes on the walls will be windows, and we perceive such shapes accordingly (thus the "surprise" in some of Magritte's paintings, or in various optical illusions). Stories about dining out similarly carry default assignments quite similar to those discussed above. (These default assignments hold, however, only so long as they are not replaced or contravened by specific information. For example, we might read the short story given above somewhat differently if the opening sentences informed us that John's idea of a gourmet meal is a hamburger with all the trimmings.)

DISCUSSION

The preceding discussion skims over a huge amount of territory. Even so, it covers only a small fraction of the resources that individuals bring to problem situations. In particular, the work discussed above deals primarily with competent performance in familiar situations—of routine access to well-codified knowledge. The premise underlying such work is that stereotypical circumstances evoke stereotypical responses. Scripts, schemata, frames, and the like, are ways of characterizing behavior that is consistent, reliable, and nearly automatic. Such behavior occurs in a broad variety of circumstances, for example, when physicists solve exercises in freshman physics, when a chess master plays against a weaker opponent, or when I solve max-min problems at the blackboard in a calculus class.

Important as such behavior is, it is not *problem solving* as the term was characterized in the introduction to this book. The physicist, the chess master, and the mathematician are all working *exercises* when performing tasks like those described in the previous paragraph. By definition, problem situations are those in which the individual does not have ready access to a (more or less) prepackaged means of solution. This comment is not meant to demean such performance; routine access to such subskills when working challenging problems constitutes a major part of the support structure for problem solving. But there is much more. In the case of experts, routinely accessed formal knowledge may represent the tip of a cognitive iceberg—a large body of knowledge, including intuitions based on empirical experience, that may either remain invisible or may seem unnecessary when working routine formal tasks. The situation is far more stochastic in the case of students. Even if subskills can be executed once accessed, access to them may be unreliable. Equally important, much of what is accessed may not be reliable (in fact, it may be dead wrong). The broad spectrum of resources is addressed in the next section, and the issue of flawed resources is explored in the section that follows it.

The Broad Spectrum of Resources

This section points, briefly, to the wide range of resources that can contribute to an individual's problem-solving performance in a particular mathematical domain. Among the kinds of knowledge relevant to such performance, and sketched below, are

1. informal and intuitive knowledge about the domain,
2. facts, definitions, and the like,
3. algorithmic procedures,

4. routine procedures,
5. relevant competencies, and
6. knowledge about the rules of discourse in the domain.

Three points mentioned in Chapter 1 bear repeating. First, inventories of resources are descriptions of the knowledge possessed by individuals. The focus here is ontological rather than epistemological, at least so far as the sense of classic epistemology (explorations of the nature of abstract mathematical knowledge) is concerned. While the flavor of the current investigations will differ, they may be considered along the lines of a genetic epistemology (à la Piaget, 1970) or personal knowledge (à la Polanyi, 1962), in that they deal with the knowledge held to be true by individuals. Second, knowing is not an all-or-nothing proposition. There are various degrees of "mastery" of knowledge, even with regard to simple facts and procedures. (Bloom's 1956 "taxonomy" elaborates this point *ab extenso.*) Third, the facts and procedures that one holds to be true need not, in fact, be true; flawed resources play an important role in mathematical performance. These last two points are discussed at some length below. To illustrate our discussion, we return to a problem presented in Chapter 1:

Problem 2.1 You are given two intersecting straight lines and a point P marked on one of them, as in Figure 2.2 below. Show how to construct, using straightedge and compass, a circle that is tangent to both lines and that has the point P as its point of tangency to one of the lines.

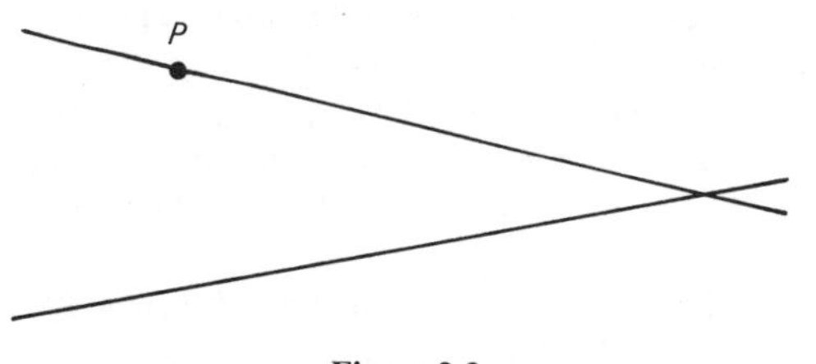

Figure 2.2

INFORMAL AND INTUITIVE KNOWLEDGE ABOUT PROBLEM 2.1

Mathematics as practiced by mathematicians and as taught in schools (at least at the secondary level and beyond) is a formal affair, one with its own highly codified language and set of meanings. In geometry courses a circle is defined as the set of all points in a plane that are a given distance from a given point in the plane, a triangle as the figure formed by three segments joining three noncollinear points, and so on. For the mathematician and for the student who is to learn to use mathematics, these are the definitions—period. Like Lewis Carroll's Humpty Dumpty, the mathematician claims that "When *I* use a word, . . . it means just what I choose it to mean—neither more nor less" (Carroll, 1960, p. 269).

Mathematics depends on the use of clear and unambiguous terminology. That terminology may have been inspired by objects and experiences in the real world, but it is now divorced from it. The student must come to accept that a circle is the set of all points in a plane that are a given distance from a given point in the plane, neither more nor less. The student must learn to manipulate formal objects in a formal way, as does the mathematician. Thus in instruction, students are often presented with the mathematical objects divorced from their informal, intuitive associations.

When one has come to accept the mathematical definitions as primitives, has worked with them, and has developed intuitions about them, the formal approach seems natural. It is easy to underestimate the importance of the empirical base, including the informal and intuitive knowledge about real-world objects, that underlies one's intuitions about the formal ones. As the work of the van Hieles and the Soviet psychologists suggests (Freudenthal, 1973; Hoffer, 1981, 1983; van Hiele, 1957; van Hiele & van Hiele-Geldof, 1958; van Hiele-Geldof, 1957; Wirszup, 1976), it is dangerous to do so. The van Hieles' hypothesis, buttressed by a fair amount of experimental evidence, is that students must pass through five qualitatively different levels of thought in learning geometry:

> *Level 1:* Recognition. The student learns some vocabulary and recognizes a shape as a whole. For example, at this level a student will recognize a picture of a rectangle but will likely not be aware of many properties of rectangles.
>
> *Level 2:* Analysis. The student analyzes the properties of figures. At this level a student may realize that the opposite sides and possibly even the diagonals of a rectangle are congruent, but will not notice how rectangles relate to squares or right triangles.
>
> *Level 3:* Ordering. The student logically orders figures and understands interrelationships between figures and the importance of accurate definitions. At this level a student will understand why every square is a rectangle but may not be able to explain, for example, why the diagonals of a rectangle are congruent.
>
> *Level 4:* Deduction. The student understands the significance of deduction and the role of postulates, theorems, and proof. At this level the student will be able to use the SAS [side/angle/side] postulate to prove statements about rectangles but not understand why it is necessary to postulate the SAS condition . . . and how the SAS postulate connects the distance and angle measures.
>
> *Level 5:* Rigor. The student understands the importance of precision of rigor in dealing with foundations and interrelationships between structures. . . .
>
> The van Hieles' research indicates that for students to function adequately at one of the advanced levels, they must have mastered large chunks of the prior levels. (Hoffer, 1981, pp. 13–14)

In particular, the pedagogical suggestion based on this type of research (made in all the studies cited above) is that intuitive understanding of geometric objects must be strengthened before one learns the formal properties of those objects; the intuitions can be used as a bridge to the formal situations.

FACTS AND DEFINITIONS: ALGORITHMIC PROCEDURES

This section is brief, given that questions of routine access to relevant knowledge—of access to the "basics" stored in long-term memory—were the focus of the first part of this chapter. The main points here are that, in analyzing students' work in a domain, access may not be routine and the knowledge may be shaky or incorrect. An inventory of resources must take these points into acount.

For example, the following Fact F is useful, although not essential, for a solution of Problem 2.1.

Fact F The two tangents drawn from a point to a circle are of equal length.

It is not difficult to imagine that an attempt to solve Problem 2.1 might evolve in quite different ways if the person working on it

1. is completely unfamiliar with Fact F and has no idea that it might be true;
2. suspects that Fact F might be true but is not certain and sees no way to test F as a conjecture or to prove that F is true;
3. suspects that Fact F might be true and is willing to accept empirical evidence as "proof";
4. suspects that Fact F might be true and looks for a geometric proof to verify it;
5. is able to recall, when asked, that Fact F is true (and perhaps to derive F if asked) but does not bring F to mind when presented with Figure 2.2;
6. thinks of Fact F immediately and with confidence when presented with Figure 2.2.

In fact, may videotapes of students working Problem 2.1 include all of these states of knowledge except the fourth. In attempting to make sense of any such tape, it is essential to know the student's actual state of knowledge at the point when the problem session begins. Thus an inventory of resources should include not only the pieces of knowledge accessible to the individual, but the kinds of access that the individual has to them. This point will be critical in our discussions of control and belief. If the student does not call on a particular body or kind of knowledge, it will be essential to know if the student could have—and if so, why he or she did not.

Table 2.1
Part of the Inventory of an Individual's Resources for Working Problem 2.1

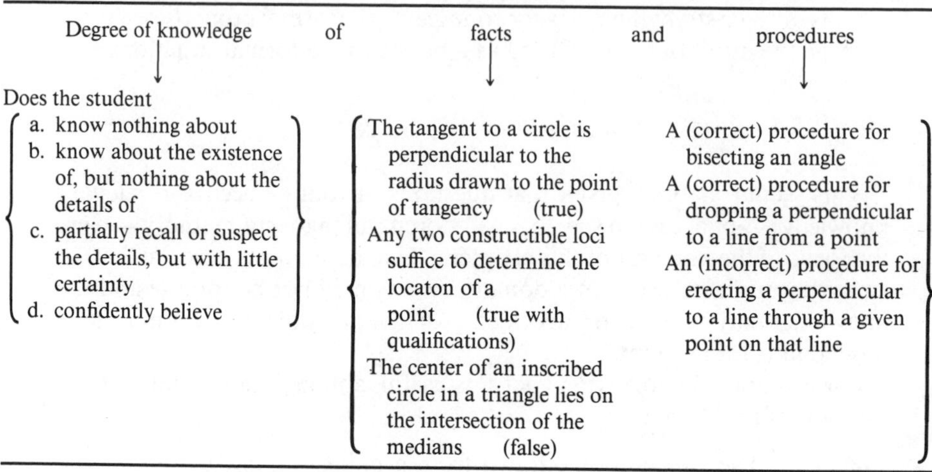

Degree of knowledge	of facts	and procedures
Does the student a. know nothing about b. know about the existence of, but nothing about the details of c. partially recall or suspect the details, but with little certainty d. confidently believe	The tangent to a circle is perpendicular to the radius drawn to the point of tangency (true) Any two constructible loci suffice to determine the locaton of a point (true with qualifications) The center of an inscribed circle in a triangle lies on the intersection of the medians (false)	A (correct) procedure for bisecting an angle A (correct) procedure for dropping a perpendicular to a line from a point An (incorrect) procedure for erecting a perpendicular to a line through a given point on that line

An inventory of resources is a catalog of what the individual brings to a problem situation, and some of what the individual brings may be wrong. In some cases, an incorrectly remembered procedure sabotages a solution. In one videotape, for example, a student confidently hands me an absurd construction for Problem 2.1 that is based on a misremembered procedure for obtaining an angle bisector. In another, a student abandons a correct approach, because its implementation depends on an incorrectly remembered procedure, and his diagram looks wrong when he tries it. In other problem sessions, false facts (e.g., the understanding that two triangles are congruent by "angle-side-side" or that the center of a circle inscribed in a triangle lies at the intersection of the medians of the triangle) suggest lines of inquiry that send students off on wild goose chases. Thus flawed resources are part of the knowledge inventory. Of course, there are various degrees of knowledge for false facts and procedures as well as for correct ones. See Table 2.1 for a capsule illustration of the discussion to this point.

ROUTINE PROCEDURES; RELEVANT COMPETENCIES

Routine procedures are just what their name suggests, well-codified but nonalgorithmic techniques for solving specific classes of problems (e.g., the general approach for solving max-min word problems in calculus). Clearly, much competency in mathematics rests on the ability to access routine procedures when they are appropriate and to carry them out competently. Routine procedures can be quite complex and far from algorithmic. Consider max-min problems, for example. Picking a useful problem representation, making a good choice of independent variable, obtaining a formula for the dependent variable, and so on, are decidely nontrivial skills. The point is, however, that all such skills are *tactical* in this context.

The distinction between strategic and tactical decisions in problem solving will be essential in our discussions of control. Broadly speaking, strategic decisions are decisions about what to do in a solution. These include decisions about planning, about what to pursue and what not to pursue, and so forth. Strategic decisions can make or break solution attempts. As we saw in Chapter 1, the decision to pursue something inappropriate can guarantee that an individual will fail to solve a problem. These kinds of problems cannot arise when routine procedures are called upon. Once a routine procedure is accessed, the problem solver is guaranteed that the general approach taken is correct (i.e., there are no strategic decisions to be made), and the only issue is whether that approach will be carried out correctly. Control is not an issue in this case, unless something goes wrong in implementation. Thus access to routine procedures is considered a matter of resources.

Overlapping with routine procedures, but more broadly defined, is the category of relevant competencies. What mathematical knowledge does the individual have that he or she might be able to bring to bear on a given problem? To return to Problem 2.1, let us consider some of the relevant competencies suggested by Problems 2.2 and 2.3.

Problem 2.2 The circle in Figure 2.3 is tangent to the two given lines at the points P and Q. Prove that the length of the line segment PV is equal to the length of the line segment QV.

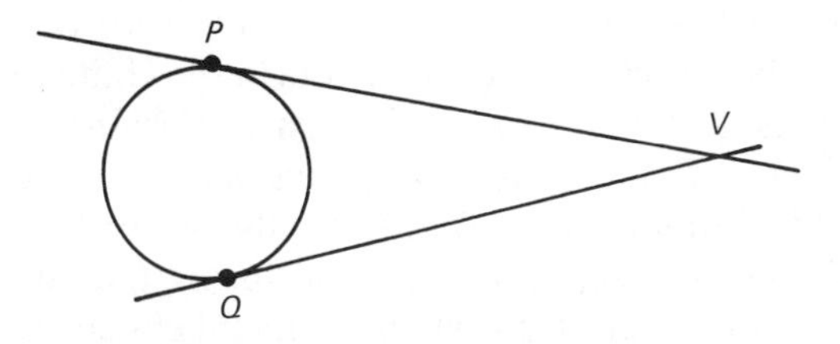

Figure 2.3

Problem 2.3 The circle in Figure 2.4 is tangent to the two given lines at the points P and Q. If C is the center of that circle, prove that the line segment CV bisects angle PVQ.

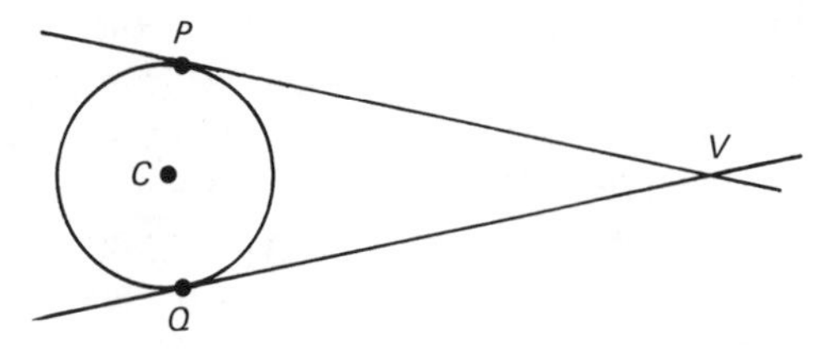

Figure 2.4

Whether (and how) an individual is capable of dealing with these two problems is clearly a significant factor in determining that person's performance in solving Problem 2.1. All of the comments made above about an individual's "degree of knowledge" of facts and procedures apply as well to abilities such as these. Have the persons asked to solve Problem 2.1 ever seen tasks like Problems 2.2 and 2.3? Do they know that the routine procedure for dealing with such problems consists of finding congruent triangles that contain, as corresponding parts, the objects that one wishes to prove equal? Are they familiar with the various ways of proving triangles congruent? Do they "see" candidates for congruent triangles when they examine Figures 2.3 and 2.4? Would the students be able (if asked) to solve Problems 2.2 and 2.3, albeit with difficulty? Or, perhaps, would they be able to solve both of them within a minute or two?

The final class of relevant competencies consists of the routine implementation of the domain-specific versions of certain heuristic strategies. Heuristic strategies were introduced in Chapter 1 and will be considered at length in Chapter 3. When discussed here, heuristics have the meaning that was intended by Pólya: The use of a general problem-solving strategy is heuristic if the problem solver is having difficulty, and there is reason to suspect that taking this particular approach might help. This use of the term needs to be discriminated from another way in which the term *heuristic* has been used in the literature. Many mathematics educators in the United States confer an invariant heuristic status on the application of certain types of problem-solving techniques. Some authors will consider *any* use of auxiliary elements in a problem solution (e.g., drawing in the line segments *CP*, *CQ*, and *CV* in Figure 2.4) as being an instance of "using the auxiliary elements heuristic." In my opinion this trivializes much heuristic usage. Students are trained to draw in certain lines as a matter of routine in geometric proof problems, and their doing so in this case reflects their subject matter training rather than the use of a general problem-solving strategy. The key word is *routine.* When a technique is closely associated with a particular domain, and it is accessed as a standard procedure, the implementation of that technique will be considered a matter of resources.

KNOWLEDGE ABOUT THE RULES OF DISCOURSE IN THE DOMAIN

Put simply, one's perception of the rules determines the way that one plays the game. For example, the claim was made in Chapter 1 that for some students, a proof may not be a generic argument. For such students, solving Problem 2.2 demonstrates the conclusion that the line segment *CV* bisects angle *PVQ* in Figure 2.3 — but perhaps not in a figure where *P* was placed at a different point on the top line, or where the angle of intersection between the two lines was substantially smaller or larger than in Figure 2.3, or possibly for any figures other than exact copies of Figure 2.3. It would not at all be surprising that a student with this perspective did not think to derive a solution to Problem 2.1. For that student, there is no connection between the two. If, however, the student does understand the generic nature of mathematical argumentation, then a different explanation for such behavior would have to be found. Similar comments can be made about virtually all mathematical domains. Consider, for example, the implications of one's understanding of the term *variable* for performance in elementary algebra, or *isomorphism* for performance in somewhat more advanced mathematics.

Flawed Resources and Consistent Error Patterns

In this section I discuss research revealing consistent patterns of errors in three different mathematical domains. The research serves to document three major points:

1. It is easy to underestimate the complexity of ostensibly simple procedures, especially after one has long since mastered them. A complete delineation of the skills required to perform an operation as simple as subtracting one three-digit number from another is much more complex than one might expect. It also points to many places where one can go wrong.
2. A large number of mistakes in simple procedures may be the result of mislearning. That is, students' flawed attempts at implementing a procedure may not be, as one might naively assume, the result of their "not having gotten it yet." Rather, the students may be implementing the same incorrect procedure over and over again.
3. Part of the support structure for resources consists of *knowledge representation.* The ways in which information is represented (structured, perceived) by an individual may determine how successful the individual can be in using that information.

We explore examples of consistent mathematical misperformance in arithmetic, algebra, and simple word problems.

ERRORS IN ARITHMETIC

A series of papers from John Seely Brown and Richard R. Burton's research group (Brown & Burton, 1978; Brown & vanLehn, 1980; Burton, 1982; vanLehn, 1981, 1983; vanLehn & Brown, 1980) documents the complex nature of apparently straightforward procedures such as elementary addition and subtraction. Equally important, these papers also call into question a broad range of naive and essentially positivist teaching and learning theories.

Burton (1982) discusses the routine algorithm for subtracting base 10 integers of four or fewer digits. It takes students quite a long time to master this procedure, and they make many mistakes in the process. The fundamental idea behind the work described here is that many of the students' mistakes are not, as is commonly assumed, random errors that occur because the student has not yet mastered the correct procedures. Rather, many of the incorrect answers that students produce result from the students' consistently applying incorrect or incomplete procedures that they have (mis)learned. Such consistent procedural errors are called bugs, in the sense of "bugs" in computer programs. The work discussed here has among its goals the diagnosis of such bugs and the explanations of their causes.

> The level of diagnosis appropriate for this paper is determining what internalized set of incorrect instructions or rules gives results equal to the student's results. *That is, we require of a diagnosis that it be able to predict, not only whether the answer will be incorrect, but the exact digits of the incorrect answer on a novel problem.* This means having a model of the students that replicates behavior on the problems observed so far and that *predicts behavior* on problems not yet done. (Burton, 1982, p. 160)

The ability to predict the incorrect answers that students will obtain, before they work a new set of problems, clearly implies that the researchers have a very good idea of what causes the students to make mistakes. Of course, one cannot expect anything approaching complete accuracy in such prediction; there are many random errors due to carelessness and the like. Even a small percentage of correct predictions would substantiate the hypothesis that the students are consistently implementing incorrect procedures. In fact, a diagnostic test enables Brown and Burton to predict, roughly half of the time, the mistakes that students will make. Moreover, the subsequent work begins to explain *why* those mistakes are made. At the level of cognitive inquiry, this kind of research suggests that it is possible to build rigorous, detailed models of covert and unarticulated cognitive processes. But this work also has strong implications for pedagogy.

Standard teaching practice is generally based on the (tacit) assumption that the student is a tabula rasa. One first shows the student how to do a procedure. Then one has the student practice the procedure until he or she masters it. If the student fails to get the correct answers, the assumption is that the student "hasn't gotten it yet"; the procedure is demonstrated once again and more practice is prescribed. This line of research indicates that such pedagogical practice is not well founded. The students' consistent error patterns indicate that they have mastered something, namely, an incorrect variant of the desired procedure. If the student is executing some procedure with consistency—quite possibly with the belief that it is the correct procedure—then demonstrating the correct procedure yet again may be of little value. The source of the student's errors may need to be diagnosed and the student "debugged," before he or she can learn the correct procedure.

ERRORS IN ALGEBRA

In a discussion of somewhat more advanced mathematics, Marilyn Matz (1982) provides similar explanations for some persistent errors in students' algebraic performance. For example, students all too frequently make the substitution

$$\sqrt{(A^2 + B^2)} \Rightarrow (A + B),$$

where $\Rightarrow$ designates "is incorrectly replaced by." According to Matz, reasonable *extrapolation* errors such as this may occur for the following reasons.

> Many common errors are shown to arise from one of two processes:
> - inappropriate use of a known rule *as is* in a new situation;
> - incorrect adaptation of a known rule to solve a new problem.
>
> These processes are termed "reasonable" because often (1) the rule that serves as the basis for extrapolation works correctly for problems that are nearly isomorphic variations of the prototype from which it was drawn, or because (2) the extrapolation techniques that specify ways to extend base rules are often useful techniques that apply correctly in other situations. The problem thus lies not in an extrapolation technique, but in the student's misguided belief in (or failure to evaluate) the appropriateness of using that technique in the particular situation at hand. (Matz, 1982, p. 26)

Consider, for example, Table 2.2. Each of the entries marked "correct"—the collection of which constitutes a large part of a student's experience in algebra—satisfies a "linear distribution law" of the form

$$f(X \cdot Y) \Rightarrow f(X) \cdot f(Y).$$

Each of the entries marked "incorrect" represents an unfortunate extrapolation of that law, generalized to

$$f(X*Y) \Rightarrow f(X)*f(Y),$$

Table 2.2
Examples of Correct and Incorrect Generalized Distribution

Correct

$$A(B+C)=AB+AC$$
$$A(B-C)=AB-AC$$

$$\frac{1}{A}(B+C)=\frac{1}{A}(B)+\frac{1}{A}(C) \quad \text{equivalently,} \quad \frac{B+C}{A}=\frac{B}{A}+\frac{C}{A}$$
$$(AB)^2=A^2B^2 \quad \text{more generally,} \quad (AB)^n=A^nB^n$$
$$\sqrt{(AB)}=\sqrt{A}\cdot\sqrt{B} \quad \text{more generally,} \quad (AB)^{1/n}=(A)^{1/n}(B)^{1/n}$$

Incorrect

$$\sqrt{(A+B)}\Rightarrow\sqrt{A}+\sqrt{B}$$
$$(A+B)^2\Rightarrow A^2+B^2$$
$$A(BC)\Rightarrow AB\cdot AC$$
$$\frac{A}{B+C}\Rightarrow\frac{A}{B}+\frac{A}{C}$$
$$2^{a+b}\Rightarrow 2^a+2^b$$
$$2^{ab}\Rightarrow 2^a2^b$$

From Matz (1982, p. 29, Table 1).

to a situation in which the extrapolation is not justified. Simarly,

> Generalization enables the formulation of a general rule from a sample problem based on the assumption that particular numbers in sample problems are incidental rather than essential. This assumption is nearly always valid, but there is a classic exception. In the solution of a problem like:
>
> $$(X-3)(X-4)=0$$
> $$(X-3)=0 \quad \text{or} \quad (X-4)=0$$
> $$X=3 \quad \text{or} \quad X=4$$
>
> although the 3 and 4 are not crucial to the procedure itself, the 0 is. Students who fail to realize the critical nature of the 0 treat it just as they do other numbers in the prototype and construct a rule . . . produc[ing] solutions of the form:
>
> $$(X-5)(X-7)=3$$
> $$(X-5)=3 \quad \text{or} \quad (X-7)=3$$
> $$X=8 \quad \text{or} \quad X=10.$$
>
> (Matz, 1982, pp. 33–34)

ERRORS IN SIMPLE WORD PROBLEMS

Problem 2.4 Using the letter S to represent the number of students (at this university) and the letter P to represent the number of professors, write an equation that summarizes the following sentence: "There are six times as many students as professors at this university."

The answer is obviously $S=6P$. Trivial, no? Well, no. The error rate for first-year *engineering* students at the University of Massachussetts is about 37%, with virtually all incorrect answers being of the form $P = 6S$. For students not majoring in science, mathematics, or engineering, the failure rate exceeds 50%. The error rate for problems with noninteger ratios ("At Mindy's restaurant there are four cheesecakes sold for every five strudels. Using C for the number of cheesecakes and S for the number of strudels. . . ") is about 65%. (See Clement, 1982; Clement, Lochhead & Monk, 1981; Rosnick & Clement, 1980.) I have been told that the error rate for academic deans at major universities approaches 100%, but the source of that data is confidential.

There are at least two plausible explanations for the frequency of the reversal error that results in writing $P = 6S$ instead of $S = 6P$. The first has to do with a "direct translation" of words into symbols, as follows:

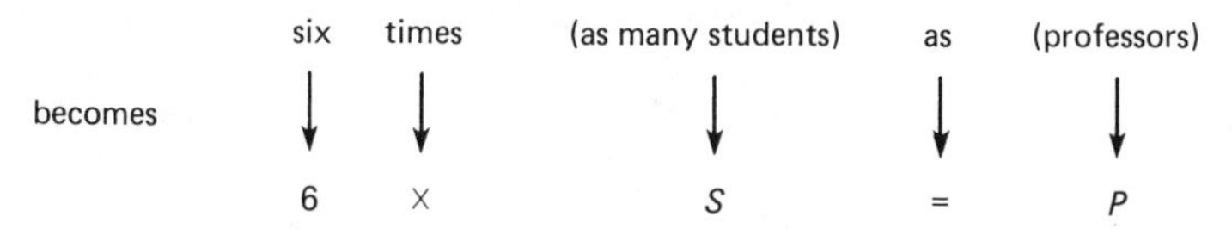

if one translates literally (syntactically) and ignores the meaning (the semantics) of the terms involved. This state of affairs is not as improbable as one would like to believe. In fact, such behavior may be—at least in part—a consequence of the way that students are taught to deal with word problems. In elementary school, for example, students are often trained to solve word problems by the "key word" method, which (in its abused form) is purely syntactic. Consider the following word problem:

John has five apples and gives three to Mary.
How many apples does John have left?

Using the key word method, the student identifies the two numbers in the problem statement (5 and 3) and the key word (in this case, "left"), which determines the choice of arithmetic operation (in this case, subtraction). Thus the procedure to implement is "5 − 3." Note that *this method provides the correct answer to this problem without the student's needing to have any understanding of the situation represented by the problem statement.* This kind of instruction occurs with some frequency. When students are trained to solve problems without understanding, they will eventually come to accept the fact that they will not understand; they will implement procedures without trying to understand. There is evidence that many students trained in the key word method will scan the problem statement to identify the two given numbers, and will then read the statement backward from the end to find the key word! Some will use subtraction for any problem that includes "left" in its statement, as in "John left the room to get more apples." Unfortunately, the students' instructional experience is not designed to provide corrective feedback. In a major textbook series, for example, the key word procedure will produce the correct answer 97% of the time. Is it any surprise that, as adults, students trained in this way will make symbolic substitutions in violation of the deep meaning of those symbols?

At more advanced mathematical levels this situation is exacerbated by the fact that the deep meaning of mathematical symbolism often differs from the meaning that is formally expressed in mathematical statements. The following examples are taken from Kaput (1979), who makes the epistemological argument in depth.

In formal terms the relationship $A = B$ is presumed to satisfy an equivalence relationship, one of whose conditions is symmetry:

$$A = B \quad \text{implies} \quad B = A,$$

and vice versa. These two expressions are formally (syntactically) equivalent. Yet in practice they may carry radically (semantically) different meanings. Consider, for example, the following two mathematical statements.

$$\frac{2}{X+3} + \frac{5}{X-2} + \frac{3}{X^2-1} = \frac{7X^3 + 14X^2 + 10X - 7}{X^4 + X^3 - 5X^2 + X - 6} \qquad \text{(Equation 1)}$$

$$\frac{7X^3 + 14X^2 + 10X - 7}{X^4 + X^3 - 5X^2 + X - 6} = \frac{2}{X+3} + \frac{5}{X-2} + \frac{3}{X^2-1} \qquad \text{(Equation 2)}$$

Despite the fact that they are formally equivalent, Equation 1 will generally be interpreted as representing the simple addition of algebraic fractions, while Equation 2 will be taken to represent the result of a complicated process, the decomposition of a complex rational function by means of partial fractions. In both cases the equal sign is read as "yields" and suggests the operation that provides the result. At a more elementary level,

$$2 \times 3 = 6$$

represents a simple multiplication,

$$6 = 3 \times 2$$

a factorization into primes, and

$$2 \times 3 = 3 \times 2$$

an embodiment of commutativity—to the cognoscenti. All three expressions are formally equivalent, and students are generally presented solely with the formal meanings of mathematical statements. Thus formal mathematical statements carry, in context, semantic meanings that contradict (or at least qualify) the formal meaning of the statements. Students are gener-

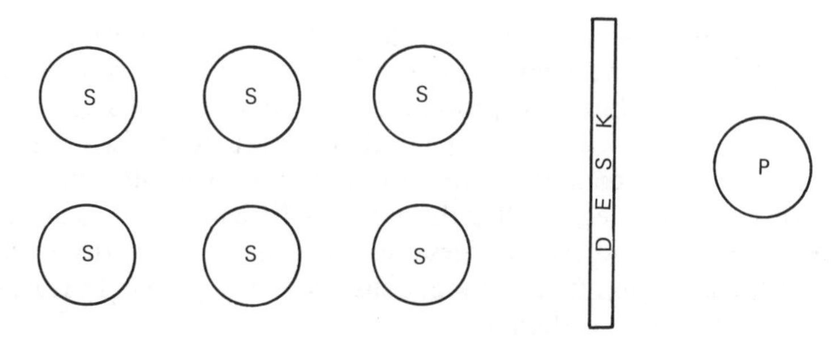

Figure 2.5

ally expected to pick up this semantic meaning on their own. Given this, and the abuses described above, it should not be surprising that students have difficulty with the semantics of mathematics.

A second cause of difficulty with the Problem 2.2 has to do with the different ways that individuals may represent the information given in the problem statement. If one pictures six students for each professor in a typical class, for example, the result is Figure 2.5. If the objects on each side of the desk are considered equal, the obvious transformation that converts the picture into symbols yields $6S = P$.

Both the syntactic and the pictorial explanations of students' behavior on Problem 2.4 illustrate that the way that one represents a problem, and interprets that representation, is a critical factor in problem solving. The issue of representation is central to any discussion of cognitive performance. (If readers doubt that the form of the representation makes a difference, I suggest that they try multiplying MCMLVXII by MMMDCCCXLIV in Roman numerals.)

Summary

An inventory of resources characterizes the knowledge possessed by individual problem solvers and the means by which such knowledge is accessed. Much work in IP psychology has been devoted to elucidating the nature of routine access to relevant knowledge, which comprises the foundation upon which competent problem-solving performance is built. The premise underlying virtually all characterizations of routine competency is that—in the case of experienced problem solvers—stereotypical circumstances evoke stereotypical responses. Formalisms that capture such routine competency include scripts, schemata, and frames.

In problem solving (which by definition includes more than routinized performance), a variety of factors shape behavior. These include the individual's informal and intuitive knowledge about the problem domain; knowledge of facts and definitions; the ability to execute algorithmic procedures; familiarity with routine procedures; the possession of a spectrum of relevant competencies; and knowledge about the rules of discourse in the problem domain. A complete inventory of resources will characterize each of the skills in these categories that the individual might be able to use; it will also describe how solid each of these resources is. (Familiarity with any relevant idea may range from the vague suspicion that it might be true, to its immediate access with certainty.)

Resources may be more than shaky, they may be incorrect. Research indicates that there are a broad range of consistent error patterns in students' behavior. This calls naive pedagogical models into question, for one may not simply assume that students "haven't gotten it yet." Rather, they may have "gotten" something that is wrong and be applying it consistently. Instead of simply demonstrating (once again) the correct procedures, good teaching may call for identifying the students' procedural bugs and then "debugging" them. Finally, representations provide the support structure for resources. The problem representations accessible to the individual (and the choice of representation for use in the problem) may determine how problem solutions proceed—and whether, in consequence, they are successful.

3

Heuristics

The present booklet is an attempt to revive heuristic in a modern and modest form.

G. Pólya, *How to Solve It,* 1945, p. 113

Introduction and Overview

Pólya's success in reviving what he called "modern heuristic" was anything but modest. A full 35 years after the declaration of heuristic intentions quoted above, the National Council of Teachers of Mathematics (NCTM) published *An Agenda for Action: Recommendations for School Mathematics of the 1980's.* Recommendation 1 of the NCTM's agenda was the following:

> PROBLEM SOLVING MUST BE THE FOCUS
> OF SCHOOL MATHEMATICS IN THE 1980'S
>
> The development of problem-solving ability should direct the efforts of mathematics educators through the next decade. Performance in problem solving will measure the effectiveness of our personal and national possession of mathematical competence. . . . (p. 2)

As part of the effort to implement its recommendations, the NCTM titled its 1980 yearbook, and devoted it to, *Problem Solving in School Mathematics.* The various chapters of the yearbook are steeped in Pólya's work, and the yearbook literally reflects his influence from cover to cover: Reproduced on its front and back endpapers is the four-stage guide to solving problems first given in *How to Solve It* (recall Table 1.2). This is only one indication of Pólya's influence. His work on problem solving is held in high regard by both mathematicians and mathematics educators. Heuristics, or the "mental operations typically useful for the solution of problems" (1945, p. 2) have been the focus of most problem-solving research in mathematics education and the foundation for most development efforts in problem solving.

This chapter explores the heuristic state of the art. This opening section provides some background. It begins with a rationale for the study of heuristic strategies, justifying some of the attention they have received. An overview of the literature indicates that the attention devoted to them has not, in general, been adequately repaid with success; attempts to teach students to use heuristic strategies have consistently produced less than was hoped for. Three hypotheses are advanced as potential explanations of that failure.

The second section establishes some parameters for the discussion that follows. The kinds of problems that are discussed, and the backgrounds of the students who are asked to solve them, are briefly described.

The third, fourth, and fifth sections elaborate, respectively, on the three hypotheses advanced in the first. In the third section an in-depth examination of two strategies indicates that a typical heuristic strategy is very broadly defined—too broadly, in fact, for the description of the strategy to serve as a useful guide to its implementation. The fourth section gives an idealized blow-by-blow solution of a problem by heuristic methods, indicating how the successful use of such strategies depends not only on access to the methods, but also on good "executive" decision-making. (This point will be elaborated in depth in Chapter 4.) The fifth section turns to the relationship between heuristics and resources, and indicates how the use of such problem-solving strategies depends on the kinds of resources accessible to the individual.

WHY TEACH HEURISTICS? A RATIONALE

A rationale for the study of heuristics, and for teaching problem solving via heuristics, is as follows:

1. As an undergraduate, as a graduate student, and then as a young professional, an apprentice mathematician solves thousands upon thousands of problems. Occasionally that person solves a problem using a technique that was successful earlier, and something clicks. (To quote Pólya's traditional mathematics professor, "a method is a device which you use twice"; Pólya, 1945, p. 208.) If that method succeeds twice, the individual may use it when faced with another similar problem. In that way a method becomes a *strategy.* Over a period of years each individual problem solver comes to

rely—quite possibly unconsciously—upon those methods that have proven useful for himself or herself. That is, the individual develops a personal and idiosyncratic collection of problem-solving strategies.

2. While the development of problem-solving strategies described in Point 1 is idiosyncratic, it is also somewhat uniform. That is, there is a substantial degree of homogeneity in ways that expert problem solvers approach new problems. (This is not terribly surprising, in that a relatively small number of techniques will allow one to be successful at mathematics. In a sense, one can say that the successful problem solvers have been trained by the discipline.) That is not to say that two experts will necessarily approach the same new problems in precisely the same ways. Rather the argument is that, if two experts grapple with an extended series of unfamiliar problems, there will be a substantial overlap in the problem-solving strategies that they try.

3. By means of introspection (Pólya's method) or by making systematic observations of experts solving large numbers of problems, it might be possible to identify and characterize the heuristic strategies that are used by expert problem solvers. This was the program that Pólya began with the short dictionary of heuristic in *How to Solve It* and elaborated in *Mathematics and Plausible Reasoning* and *Mathematical Discovery*.

4. Once the most important heuristic strategies have been discovered and elaborated, the next step is obvious. One should provide direct instruction in these strategies, thereby saving students the trouble of having to discover the strategies on their own. That is, the compilation of strategies elucidated in Point 3 should serve as a guide to the problem-solving process. If this is done, it should no longer be necessary for individual students to go through the long and arduous process of arriving at these general principles by themselves, as described in Point 1.

There is but one problem with this rationale: It hasn't worked, and not for lack of trying. The literature of mathematics education is chock-full of heuristic studies. Most of these, while encouraging, have provided little concrete evidence that heuristics have the power that the experimenters hoped they would have. For example, studies by Wilson (1967) and Smith (1973) indicate that, ostensibly, general heuristics did not, as hypothesized, transfer well to new situations. Reports by Kantowski (1977), Kilpatrick (1967), and Lucas (1974), based on examinations of problem-solving protocols, do indicate that the use of heuristics is correlated with scores on ability tests and with success on problem-solving tests. The results are far less dramatic than one might expect, however. Treatment comparison studies have consistently yielded promising but equivocal results. See Loomer (1980) and Goldberg (1974) as examples. In brief, heuristics have proven far more complex and far less tractable than had been hoped or expected. Summarizing the results of 75 empirical studies on problem-solving strategies, Begle wrote the following:

> A substantial amount of effort has gone into attempts to find out what strategies students use in attempting to solve mathematical problems. . . . No clear-cut directions for mathematics education are provided by the findings of these studies. In fact, there are enough indications that problem solving strategies are both problem- and student-specific often enough to suggest that hopes of finding one (or few) strategies which should be taught to all (or most) students are far too simplistic. . . .
>
> This brief review of what we know about mathematical problem solving is rather discouraging. (Begle, 1979, pp. 145–146)

In AI, attempts to build problem-solving programs based on heuristics like those characterized by Pólya have generally been unsuccessful. At the same time, other methods have yielded substantial results. As a result, the mathematical heuristics that are the focus of this chapter have little credibility in the cognitive science community. A sense of this perspective can be seen in recommendations made by H. A. Simon (1980):

> In teaching problem solving, major emphasis needs to be directed towards extracting, making explicit, and practicing problem-solving heuristics—both general heuristics, like means–ends analysis, and more specific heuristics, like applying the energy-conservation principle to physics. (p. 94)

The general heuristics recommended by Simon, for example, means–ends analysis and hill climbing, are those that have proven useful in AI. The more specific heuristics are subject-matter techniques or domain-specific principles. These are a far cry from the "mental operations typically useful for the solution of problems" described in *How to Solve It* (p. 2) and its successors, which are generally ignored.

To sum things up at this point, faith in mathematical heuristics as useful problem-solving strategies has not been justified either by results from the empirical literature or by programming success in AI. The main purpose of this chapter is to explore the nature of heuristic processes and to explain why attempts at implementing such strategies (either for humans or machines, but with a primary focus on the former) have not been successful. In brief, I argue that heuristics have not lived up to their promise for the reasons that follow.

Despite the fact that heuristics have received extensive attention in the mathematics education literature, heuristic strategies have not been characterized in nearly adequate detail. It is one thing to label and describe a particular strategy (e.g., "examining special cases") in such a way that students can recognize and appreciate the use of that strategy. It is quite something else to provide instructions that are sufficiently detailed so that students can use that strategy when they encounter problems for which it is appropriate. In most studies, the characterization of heuristic strategies was not sufficiently prescriptive. Not nearly enough detail was provided for the characterizations to serve as guides to the problem-solving process.

Learning to use heuristic strategies, in and of themselves, is not sufficient to ensure competent problem-solving performance. Chapter 1 illustrated the importance of control decisions and their effects on the problem-solving process. (Recall in particular DK and BM's attempt to solve Problem 1.3.) When heuristic strategies are characterized at the level of detail suggested by the previous paragraph, control becomes a major issue. With increased specificity comes an increase in the number of heuristic strategies. The number of useful, adequately delineated techniques is not numbered in tens, but in hundreds. At any time in a problem solution, any number of these techniques might appear relevant to the problem solver. The question of selecting which ones to use (and when) becomes a critical issue. If the problem solver does not have a reasonably efficient means of making such choices, or of recovering from incorrect choices when they are made, the potential benefits of heuristics can be diluted to the point where their impact is negligible.

The issue of control is the focus of Chapter 4. This chapter focuses on the use of individual heuristic strategies. It examines the degree of specificity required for an adequate characterization of such strategies, and the kinds of skills that are required in order for students to use them. As noted above, attempts to teach these strategies have generally been met with mixed success. I argue that such equivocal results have occurred because the complexity of heuristic strategies, and the amount of knowledge necessary to implement them, have been underestimated in at least three ways:

1. Typical descriptions of heuristic strategies, for example, "examining special cases," are really *labels* for categories of closely related strategies. Many heuristic labels subsume half a dozen detailed strategies or more. Each of these more precisely defined strategies needs to be fully explicated before it can be used reliably by students.

2. The implementation of heuristic strategies is far more complex than at first appears. Carrying out a strategy such as "exploiting an easier, related problem," for example, involves six or seven separate major phases, each of which is a potential cause of difficulty. Training in the use of the strategy must involve training in all of those phases, and the training must be given with at least as much care and attention as is given to standard subject matter. In general, attempts to teach such strategies have not been adequately precise or rigorous.

3. Although heuristic strategies can serve as guides to relatively unfamiliar domains, they do not replace subject matter knowledge or compensate easily for its absence. Often the successful implementation of a heuristic strategy

depends heavily on a firm foundation of domain-specific resources. It is unrealistic to expect too much of these strategies.

The next section briefly describes the sense in which *problem* is used in this book, and describes my target group of students. The three subsequent sections examine, in detail and in order, the three points raised above.

What a Problem Is and Who the Students Are

The difficulty with defining the term *problem* is that problem solving is relative. The same tasks that call for significant efforts from some students may well be routine exercises for others, and answering them may just be a matter of recall for a given mathematician. Thus being a "problem" is not a property inherent in a mathematical task. Rather, it is a particular relationship between the individual and the task that makes the task a problem for that person. The word *problem* is used here in this relative sense, as a task that is difficult for the individual who is trying to solve it. Moreover, that difficulty should be an intellectual impasse rather than a computational one. (For example, inverting a 27×27 matrix would be an arduous task for me, and I would most likely make an arithmetic error in the process. Even so, inverting a given matrix is not a *problem* for me.) To state things more formally, if one has ready access to a solution schema for a mathematical task, that task is an exercise and not a problem.

Of the definitions in the *Oxford English Dictionary,* the one I prefer is the following: "**Problem.** A doubtful or difficult question; a matter of inquiry, discussion, or thought; a question that exercises the mind."

This is the sense in which Pólya uses the term and that establishes the context for his attempt to revive heuristic strategies as tools in problem solving. That is, heuristic strategies are techniques used by good problem solvers when they need to make progress on tasks that are problems for them. As Pólya put it, "The aim of heuristic is to study the methods and rules of discovery and invention" (1945, p. 112). Roughly characterized, heuristic strategies are techniques used by problem solvers when they run into difficulty. They are rules of thumb for making sense of, and progress on, difficult problems. The examples of heuristic usage given in Chapter 1 give some sense of how such strategies can be used. It should be clear from those examples that these techniques are subtle and difficult to use—especially since they are called into play precisely when the problem solver does not have a good idea of what to do next. Learning to use heuristics calls for a (reasonably) firm foundation of mathematical resources and for a fair amount of sophistication as well. For these reasons most of our discussions will be restricted to subjects

1. who are more or less adult problem solvers, say, college students or older,[1]
2. who have reasonable mathematical backgrounds, which preferably include some familiarity with the calculus,[2]
3. who are working on problems in the sense described above, and
4. who are truly making an effort to solve the problems that are (at least potentially) within their grasp.[3]

Note 1 There is a tolerance of perhaps 2 years at the lower level. As suggested above and as is discussed at length below, heuristics are subtle, complex, and highly abstract. As discussed in Chapter 1, these are only one component of a competent problem solver's arsenal of techniques. Control strategies are even more abstract; they are, in essence, operators whose domain is the space of heuristics. The reason for the age limit suggested here is simply to help ensure that the students will have some degree of mathematical maturity.

This is not to suggest that students should first be exposed to heuristics in college. The foundations for using such strategies can and should be established during the whole of a student's mathematical career. Indeed, if such groundwork was routinely done, much of the essentially remedial work I am compelled to do at the college level would be unnecessary. Younger students can recognize, appreciate, and mimic the use of such strategies. But anyone who thinks that fourth-graders, for example, can use these mathematical strategies in the way that Pólya described them either fails to understand the complexity of the strategies or fails to understand the lessons from Piaget's work.

Note 2 None of the mathematical tasks (I am reluctant to say "problems") used as examples in this chapter—and few of those used in my problem-solving courses, for that matter—require a knowledge of calculus for solution, but all do require some degree of sophistication and some fluency with basic mathematics. The calculus prerequisite is a simple way of assuring that the students have some degree of sophistication.

Note 3 The purpose of these constraints is to allow for the exploration of heuristic behavior in relatively favorable circumstances. The intent here is not to avoid a discussion of affective issues, for affective issues are clearly a major factor determining problem-solving behavior and success. Matters will be more straightforward, however, if that discussion is postponed for a while. A direct discussion of affective issues begins in Chapter 5.

Toward More Precise and Usable Descriptions of Heuristic Strategies

This section discusses in some detail the specific nature of two general heuristic strategies: "examining special cases" and "exploiting subgoals." The analysis will reveal that such heuristic strategies are not, as is generally assumed, coherent problem-solving approaches that are used the same way across-the-boards. Rather, each general heuristic can be seen as a loose collection of somewhat related substrategies. I argue that these substrategies can be defined with enough precision so that students can learn to use them but that instruction at the level of the general strategy alone is unlikely to provide students with the tools they need to improve their problem-solving performance.

A LOOK AT A SIMPLE STRATEGY: EXAMINING SPECIAL CASES

We begin with an oft-mentioned strategy, typically useful for getting started on a problem solution:

Strategy S To better understand an unfamiliar problem, you may wish to exemplify the problem by considering various special cases. This may suggest the direction of, or perhaps the plausibility of, a solution.

The following five problems are all problems to which Strategy S can be profitably applied:

Problem 3.1 Determine a formula in closed form for the series

$$\sum_{i=1}^{n} \frac{1}{i(i+1)}.$$

Problem 3.2 Let $P(x)$ and $Q(x)$ be polynomials whose coefficients are the same but in "backward" order:

$$P(x) = a_0 + a_1 x + a_2 x^2 + \cdots a_n x^n,$$

and

$$Q(x) = a_n + a_{n-1} x + a_{n-2} x^2 + \cdots + a_0 x^n.$$

What is the relationship between the roots of $P(x)$ and those of $Q(x)$? Prove your answer.

Problem 3.3 Let the real numbers a_0 and a_1 be given. Define the sequence $\{a_n\}$ by

$$a_n = \tfrac{1}{2}(a_{n-2} + a_{n-1}) \qquad \text{for each} \quad n \geq 2.$$

Prove that $\lim_{n\to\infty} (a_n)$ exists and determine its value.

Problem 3.4 Two squares, each s on a side, are placed such that the corner on one square lies on the center of the other. Describe, in terms of s, the range of possible areas representing the intersections of the two squares.

Problem 3.5 Of all triangles with fixed perimeter P, which has the largest area? Justify.

Let us consider the way Strategy S is used in solving each of the problems.

Problem 3.1 is most often encountered in courses on infinite series. The standard solution to the problem is as follows. If one makes the clever observation that

$$\frac{1}{i(i+1)} = \frac{1}{i} - \frac{1}{i+1} \qquad \text{for all } i,$$

one sees that the series

$$\frac{1}{1 \times 2} + \frac{1}{2 \times 3} + \frac{1}{3 \times 4} + \cdots + \frac{1}{n(n+1)}$$

can be written as

$$(1 - \tfrac{1}{2}) + (\tfrac{1}{2} - \tfrac{1}{3}) + (\tfrac{1}{3} - \tfrac{1}{4}) + \cdots + [1/n - 1/(n+1)].$$

Once the series has been written this way, one can observe that the terms on either sides of the plus signs cancel. This leaves only the first and last terms after the series collapses. We say that the series "telescopes" to $1 - 1/(n+1)$.

Admittedly the preceding approach provides an elegant solution to the problem. However, possessing such cleverness is not necessary. If one employs Strategy S and computes the first few sums in the series, the first partial sums are $\tfrac{1}{2}, \tfrac{2}{3}, \tfrac{3}{4}, \tfrac{4}{5}, \tfrac{5}{6}, \ldots$. This pattern suggests what the sum will be. The conjecture can easily be verified by induction.

Strategy S is employed differently in Problem 3.2, although it appears at first that one can proceed in the same way as in Problem 3.1. Suppose one tries special cases for different values of n. Rather than deal with polynomials of arbitrary degree, one might choose to examine polynomials of degree 1, 2, 3, . . . in order. The linear case is trivial. Suppose, however, that for one $n = 2$ takes two arbitrary quadratics,

$$P_2(x) = a_0 + a_1x + a_2x^2$$

and

$$Q_2(x) = a_2 + a_1x + a_0x^2.$$

Finding the right relationship between

$$\frac{-a_1 + (a_1^2 - 4a_0a_2)^{1/2}}{2a_0} \quad \text{and} \quad \frac{-a_1 + (a_1^2 - 4a_0a_2)^{1/2}}{2a_2}$$

is far from easy, and working with general polynomials of higher degree is not any easier.

The key to using Strategy S in Problem 3.2 is to make a choice of easily factorable polynomials. Computations with polynomials such as

$$P(x) = x^2 + 5x + 6 \qquad \text{and} \qquad Q(x) = 1 + 5x + 6x^2$$

might lead to a conjecture. With a few additional convenient choices for $P(x)$ and $Q(x)$, one has good reason to suspect that the roots of the two polynomials will always be reciprocals. The task then is to prove this, and looking at the factors of the simple factorable polynomials suggests one line of proof. [Each term of the form $(ax + b)$ from P is paired with the term $(bx + a)$ from Q].

In Problem 3.3, the general computations for a_n rapidly become complex. One can avoid that complexity temporarily by considering what happens when $a_0 = 0$ and $a_1 = 1$. The sequence then becomes

$$0, 1, \tfrac{1}{2}, \tfrac{3}{4}, \tfrac{5}{8}, \tfrac{11}{16}, \tfrac{21}{32}, \ldots .$$

One may determine a formula for this sequence directly. Or by observing the differences

$$+1, -\tfrac{1}{2}, +\tfrac{1}{4}, -\tfrac{1}{8}, +\tfrac{1}{16}, -\tfrac{1}{32}, \ldots,$$

one may observe that the terms are the terms of a geometric series. That solves the special case.

As it happens, the general case given in the problem statement can be reduced to the special case computed above by means of a linear transformation; thus solving the special case solves the problem. But the solution to the special case may also provide the inspiration for a direct solution of the general case. What becomes clear in the computations above—especially if one draws a picture (another heuristic)—is that the distance from a_n to a_{n+1}

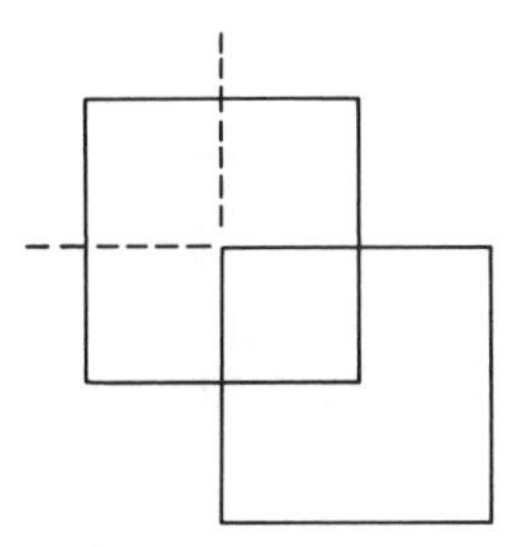

Figure 3.1

is halved with each iteration. In the general case, too, the terms are terms of a geometric series.

In Problem 3.4, one can gain some insight into the problem by considering some situations where the area of the overlap is easily calculated (see Figures 3.1 and 3.2). These suggest the answer in general, which can be derived from Figure 3.3.

In Problem 3.5, one can simplify the calculations by picking a convenient value for P. Then one might proceed by briefly calculating or estimating the areas of various triangles, including the isosceles right triangle, the equilateral triangle, and perhaps some extreme cases. The results of these computations will suggest the answer, which the problem solver can proceed to verify by analytical means. In addition to helping develop intuition, work on the special cases may suggest ways of proceeding on the problem. For example, the sketches used in comparing candidate triangles might lead the problem solver to think about a variational argument. Or perhaps the problem solver's faith in the nature of the answer—that the solution is the equilateral, which is symmetrical—might prompt an attempt to develop an argument by symmetry.

In reviewing the use of Strategy S on the solution of Problems 3.1 through 3.5, we find that the description of Strategy S given above is merely a sum-

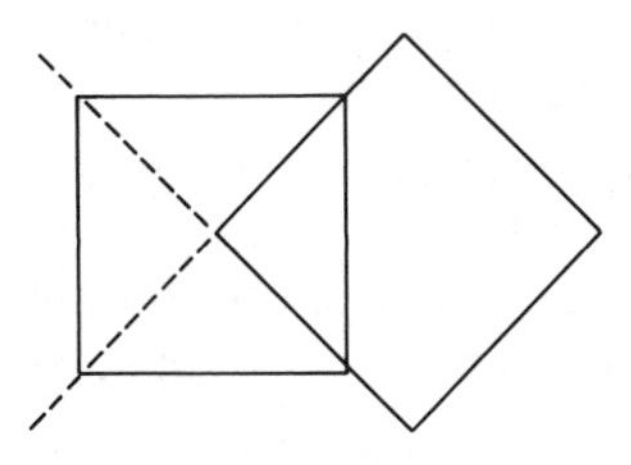

Figure 3.2

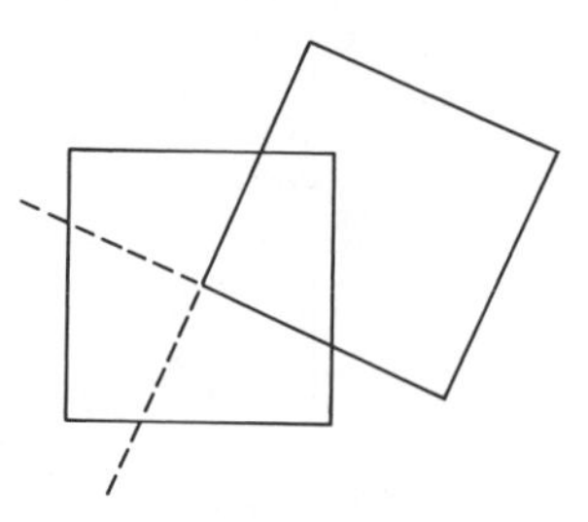

Figure 3.3

mary description of five closely related strategies, each with its own particular characteristics:

Strategy S_1 If there is an integer parameter n in a problem statement, it might be appropriate to calculate the *special cases* when $n = 1, 2, 3, 4$ (and maybe a few more). One may see a pattern that suggests an answer, which can be verified by induction. The calculations themselves may suggest the inductive mechanism.

Strategy S_2 One can gain insight into questions about the roots of complex algebraic expressions by choosing as *special cases* those expressions whose roots are easy to keep track of (e.g., easily factored polynomials with integer roots).

Strategy S_3 In iterated computations or recursions, substituting the particular values of 0 (unless it causes loss of generality) and/or 1 often allows one to see patterns. Such *special cases* allow one to observe regularities that might otherwise be obscured by a morass of symbols.

Strategy S_4 When dealing with geometric figures, one should first examine the *special cases* that have minimal complexity. Consider regular polygons, for example; or isosceles or right or equilateral rather than "general" triangles; or semi- or quarter-circles rather than arbitrary sectors, and so forth.

Strategy S_{5a} For geometric arguments, convenient values for computation can often be chosen without loss of generality (e.g., setting the radius of an arbitrary circle to be 1). Such *special cases* make subsequent computations much easier.

Strategy S_{5b} Calculating (or when easier, approximating) values over a range of cases may suggest the nature of an extremum, which, once thus "determined," may be justified in any of a variety of ways. *Special cases* of symmetric objects are often prime candidates for examination.

To appreciate the weakness of the Strategy S as a prescription for problem solving, let us consider the plight of a student faced with any of Problems 3.1 through 3.5, who has been given the suggestion Strategy S as "help." For an inexperienced problem solver, deriving the actions suggested in Strategies S_1 through S_{5b} from the 31 words of S is a very difficult task. This kind of situation is typical when students are given general heuristic strategies to apply in problem solving. Most often the statement of a heuristic strategy is quite broad and contains few clues as to how one actually goes about using it. The strategy is not, in itself, nearly precise enough to allow for unambiguous interpretation. Rather it is a label attached to a closely related family of more specific strategies. The "label" status of such strategies explains one of the apparent paradoxes about heuristics—the fact that they are used by experts but that novices find it difficult to use them. No wonder; the experts have mastered the substrategies and are seen as having "the strategy" at their disposal. The students, given only a general description of "the strategy," hardly have adequate tools at their disposal.

It should be noted that the decomposition of Strategy S into a collection of more specific heuristic strategies, as was begun above with the delineation of Strategies S_1 through S_{5b}, has not trivialized S. Each of the more specific heuristics is substantial in its own right, applies to a broad range of problems, and takes quite a bit of work to master. Decomposing such strategies in this fashion is the first step in teaching students to use them. The next step is to provide instruction in each of the more specific heuristics. For this one needs a collection of exemplary problems and a detailed set of instructions showing how the heuristic is to be implemented (see below). This is a large task even for a straightforward strategy like S. The task becomes substantially more difficult for strategies that are broader in scope.

A SECOND, MORE COMPLEX STRATEGY: ESTABLISHING SUBGOALS

As a second illustration of the point made above, let us briefly consider Strategy H, defined below.

Strategy H If you cannot solve the given problem, establish subgoals (the partial fulfillment of the desired conditions). Having attained them, build upon them to solve the original problem.

As discussed below, Strategy H can be used to solve each of the following five problems.

Problem 3.6 Place the numbers 1–9 in Figure 3.4 so that the sum of each row, column, and diagonal is the same. Such a configuration is called a magic square.

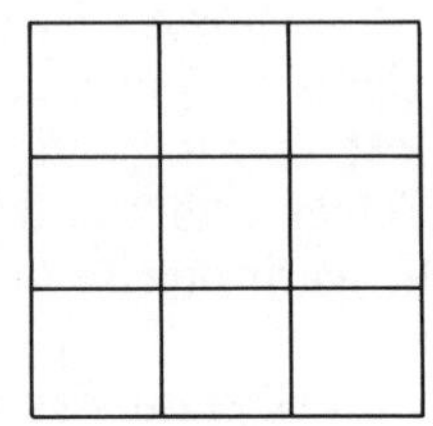

Figure 3.4

Problem 3.7 How many zeros appear at the end of 100!—the product of the integers from 1 to 100?

Problem 3.8 Find the sum of all the whole numbers from 1 to 1000, excluding those numbers that are multiples of either 4 or 11.

Problem 3.9 The positive integers are arranged in groups as follows: (1), (2, 3), (4, 5, 6), (7, 8, 9, 10), and so forth, with k integers in the kth grouping. Find the sum of the integers in the kth grouping.

Problem 3.10 Let G be a 17×31 rectangular grid, as illustrated in Figure 3.5 below. How many different rectangles can be drawn on G, if the sides of the rectangles must be grid lines? (Squares are included, as are rectangles whose sides are on the boundaries of G.)

Problems 3.6–3.10 are straightforward tasks dealing with elementary operations on whole numbers. Most of them employ, in some way or other, the formula for the sum of n consecutive integers. Such problems cover very little of the territory covered by Strategy H. Yet the use of subgoals in each of these problems is distinct.

The primary task in Problem 3.6 is to reduce the search space, given that there are 45,360 nonisomorphic ways the digits from 1 through 9 can be placed in the 3×3 box—only one of which yields a solution. Reducing the search space can be achieved by using two subgoals. The first is to determine what the sum of each row, column, and diagonal should be. The second is to

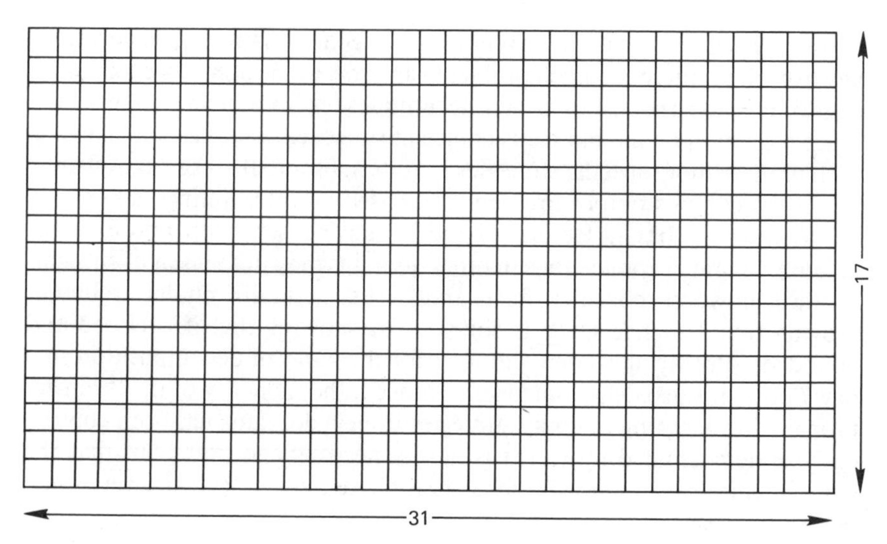

Figure 3.5

focus on which number goes into the most important square, the one in the center.

In Problem 3.7, too, one would like to reduce the amount of computation apparently required by the problem. In this case that goal is achieved by analysis: The zeros at the "end" of a number come from factors of 10. Though it may seem trivial, the major subgoal is to decompose this further. When one realizes that one only needs to count the number of 5's that appear as factors of the numbers from 1 through 100 (there are more than enough 2's), the problem is essentially solved.

Problem 3.8 can best be solved by combining two independent subgoals: (1) finding or recalling the formula for an arithmetic series and (2) reducing the problem to the point where the formula can be used repeatedly (on the integers from 1 to 1000, the multiples of 4, the multiples of 11, and the multiples of 44). Problem 3.9 also uses the formula, but the subgoal that allows one to solve it is different; if one determines the first and last terms of the kth grouping (using the arithmetic series), then one can determine the sum of that grouping (using the series again). Problem 3.10 is usually solved by a clever combinatoric argument. However, the student who is unfamiliar with such counting procedures may still solve the problem by establishing a series of subgoals: solving the simpler 1×31 problem [perhaps by induction on $(1 \times n)$ grids] and then building up a solution [by induction on $(m \times 31)$ grids, perhaps] from there.

Problems 3.6–3.10 all come from essentially the same subject domain, and each is solved by a judicious choice of subgoals. As above, however, the key to each is different. Each problem can serve as the archetype for a large class of problems solved by similar techniques (e.g., Problem 3.6 for search space reduction by the judicious choice of subgoals). And as noted above, each of these more specific strategies is quite substantial. For example, the subgoal decomposition of Problem 3.7, based on interpreting "a factor of 10" as meaning "having factors of 2 and 5," may not seem either profound or difficult. Parallel applications in other domains may not appear as straightforward, however; consider, for example, a structurally similar argument proving that a metric space is compact by showing that it is closed and bounded. Preparing students to look for this kind of decomposition in a new domain is decidely nontrivial. (Indeed, the difficulty of the "transfer problem" in learning to apply strategies in new domains has been consistently underestimated.) And it is hardly necessary to point out that the strategies discussed in the solutions of Problems 3.6–3.10 cover a very small part of the range of subgoals.

BRIEF DISCUSSION

The examples discussed in this section indicate that what is often presented as a general heuristic strategy is not in fact one coherent strategy, but rather a collection of loosely related substrategies. The strategies "examining special cases" and "exploiting subgoals" comprise as diverse a collection of techniques (to say the least!) as, say, "solving algebraic equations." Suppose one were to teach this latter topic. Clearly, one would need to decompose it. One would try to categorize different types of equations; one would try to describe various solution methods for each type and, for each solution method, describe the range of equation types for which it was likely to be appropriate. One would worry about creating sequences of practice problems so that students could learn to select the right techniques for given problems. Surely if such steps are necessary for teaching straightforward subject matter techniques, they are necessary for teaching far more nebulous techniques such as heuristics.

As the following section indicates, however, the careful delineation of heuristic substrategies is only a first step. Equally important is the careful delineation of what it takes to carry out each of these substrategies.

The Complexity of Implementing a "Straightforward" Heuristic Solution

The following Problem P is discussed on pages 23–25 of *How to Solve It*, where the solution evolves in an idealized "Socratic dialogue" with a very good student. That idealized dialogue obscures much of the difficult decision-making that is essential for obtaining a solution to that problem. In doing so, it makes the heuristic solution to the problem appear far easier than it actually is. In this section I run through an expanded, blow-by-blow—though still idealized—solution to the same problem. My discussion illustrates some of the decisions that the problem solver must make in the midst of working the problem, even when the problem solver is capable of using the appropriate heuristics without difficulty. The solution will proceed smoothly, and the appropriate heuristics will be offered at the "right" times. Even so, things will become fairly complex. The reader may wish to refer ahead to Figure 3.7 for a schematic representation of what follows.

Problem P a problem of construction. Inscribe a square in a given triangle (Figure 3.6). Two vertices of the square should be on the base of the given triangle, the other two vertices of the square on the other two sides of the triangle, one on each.

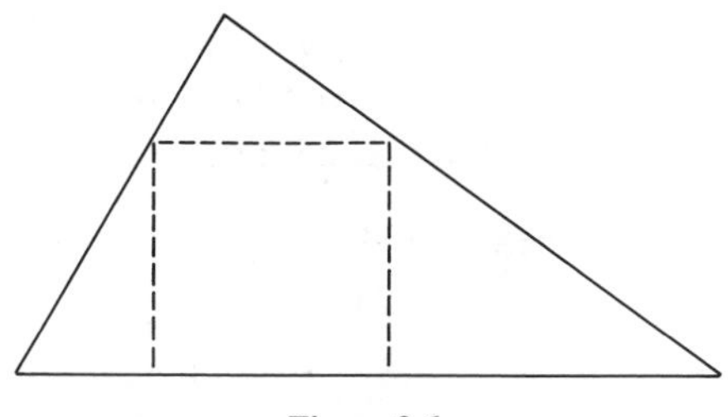

Figure 3.6

Like most good solution attempts, this one begins with an attempt to convert the problem into an exercise. The following strategy is appropriate when first working a problem.

Strategy H_1 Check to see if the problem is "standard," or reducible to a problem solvable by standard techniques.

Problem P is apparently not dispatched with so easily. A review of typical procedures, such as "the pattern of two loci," "the pattern of similar figures," and so forth (Chapter 1 of Pólya's *Mathematical Discovery*), indicates that there is no routine solution. Time for another heuristic.

Strategy H_2 If one wishes to construct an object, it is often useful to determine its properties, even by nonconstructive means. Once one knows what they are, they may then be constructible by an alternate procedure. Or,

Strategy H_2' (the domain-specific version of H_2): Determine the location of a point that you need to construct by any means available, for example, by means of algebra. Once you know where the point must be, you may be able to construct it.

Correctly interpreting H_2, the problem solver decides to work on the following modified problem:

Problem P_1 Prove there is an inscribed square in the triangle. If possible, determine the location of one of the upper vertices.

Difficulty in obtaining a direct solution to Problem P_1 may suggest the following.

Strategy H_3 If you cannot solve the original problem, try first to solve an easier, related problem. Look for known solutions to related problems.

There are, of course, various heuristics for generating the appropriate related problems for consideration, for example.

Strategy H_4 One can obtain easier problems (and a class of solutions) by relaxing one of the conditions of the problem, returning later to reimpose that condition.

Among the conditions of the problem are the fact that one wants a square that is fully inscribed in the given triangle.

Strategy H_4' (the domain-specific version for geometric constructions): One constraint that can often be relaxed in geometry problems is size. Consider constructing a figure similar to the one you need.

Once these heuristics have been implented, the problem solver faces some major control decisions. It is easy to generate a number of plausible problems related to Problem P_1. Among the related problems that one might generate are the following: (Problems $P_2 - P_6$ are listed in order of frequency, as generated by my students.)

Problem P_2 Inscribe a rectangle in the given triangle.

Problem P_3 Obtain a square with only three vertices on the triangle.

Problem P_4 Inscribe a circle in a given triangle.

Problem P_5 Draw the given triangle around a square. Better, draw a similar triangle around a square.

Problem P_6 Inscribe a square, in a special triangle such as an isosceles or equilateral triangle.

Of course, the problem solver can only work on one of these problems at any one time. Thus the wrong choice at this point can guarantee failure; with only a limited amount of time to solve Problem P, pursuing a fruitless approach may preclude the opportunity to pursue a correct one. The question of which of the subproblems P_2 through P_6 the individual should consider is an issue of conditional probabilities. There are many possible ways to proceed. Each of Problems P2–P6 represents, potentially, the first leg along a solution path to P_1. The likelihood that one will solve Problem P_1 via P_2 is

$$\text{P}(P_1 \text{ will be solved via } P_2) = \text{P}(P_2 \text{ will be solved}) \times \text{P}(P_1 \text{ can be solved given the solution of } P_2);$$

similarly for Problems P_3, P_4, P_5, and P_6. Such considerations will be sufficient to reject Problem P_4; although P_4 is a known result, that result is unlikely to be usable. (That is, the probability of being able to use Problem P_4 to solve P_1 is near 0.) But the choice among the remaining options is not trivial. And, of course, success along any of these routes only results in the solution of Problem P_1; one must then be able to use the solution of Problem P_1 to solve Problem P. Note also that different solutions to Problem P_1 may yield different amounts of usable information for solving Problem P. Thus, for example,

$$\begin{aligned}&\text{P}(\text{P will be solved via } P_2) = \\ &\quad \text{P}(\text{P can be solved given the solution of } P_1 \text{ via } P_2) \\ &\quad \times \text{P}(P_1 \text{ solved via } P_2) \\ &= \text{P}[\text{P via } (P_1 \text{ via } P_2)] \times \text{P}(P_1 \text{ via } P_2) \times \text{P}(P_2).\end{aligned}$$

These are all subjective probabilities. It is, however, essential to take them into account. Given time constraints (the fact that one can only pursue one option at a time) and given that any broken link on the way to solving P means the failure of that entire approach, a bad choice can be fatal. This set of decisions is represented schematically in Figure 3.7.

It should be noted that the control decisions for this problem are very straightforward. Once one chooses which Problem P_i to work, there is little choice regarding the order of a solution; the solution to Problem P depends on the solution to Problem P_1, which depends on the solution to Problem P_i, so the problems should most probably be worked in the order P_i, P_1, P. (That is, provided that things go well. If one runs into difficulty, there may have to be control decisions about abandoning the chosen path and taking another instead.) In general, control decisions are more complex.

As it happens, two of the paths in Figure 3.7 lead to a solution of Problem

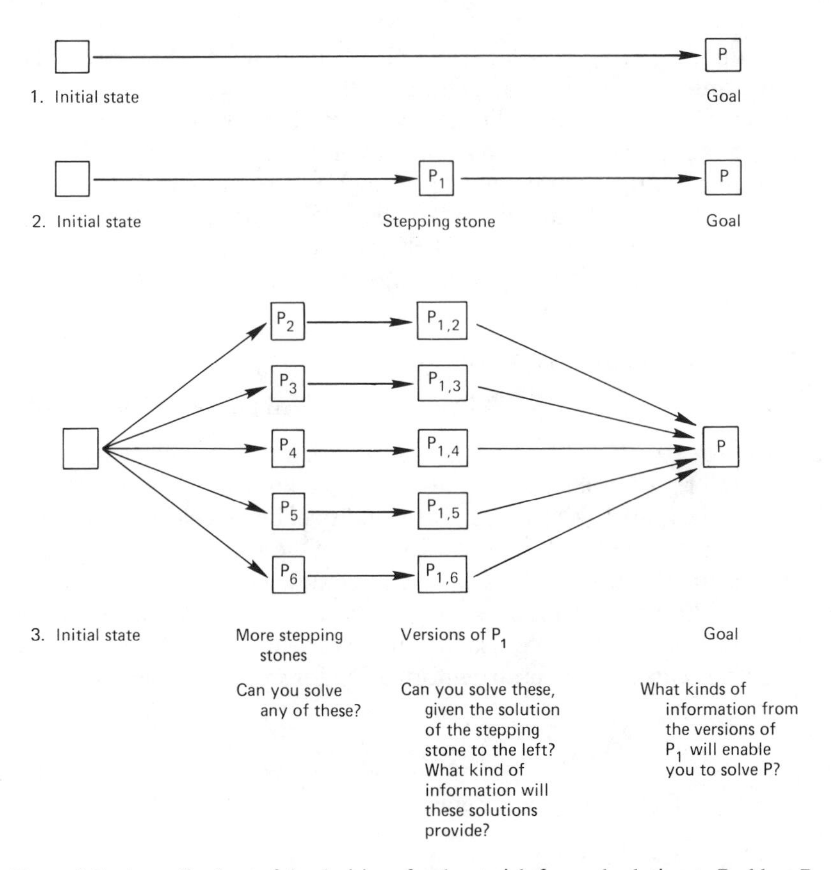

Figure 3.7 A small subset of the decisions for the straightforward solution to Problem P.

P. Problem P_6, though plausible, is no easier to solve than Problem P_2; it is a dead end. The path through Problem P_4 does not lead to a solution, and to my knowledge neither does the path that goes through Problem P_2—although the argument that proves Problem P_2 is worth noting. The set of rectangles one can inscribe in the triangle ranges from "short and fat" to "tall and thin." By the intermediate value theorem, there is a square somewhere in between these two extremes (see Figure 3.8).

Unfortunately, this classic existence proof, while solving Problem P_1, gives no hint as to the construction that might produce the square. Insofar as the solution to Problem P is concerned, this is a dead end.

Problem P_3 also gives an existence proof. The family of squares with three vertices on the triangle "grows" from the point where the fourth vertex, initially inside the triangle, passes through the opposite side. Unlike the argument that proves Problem P_2, however, this argument can be converted

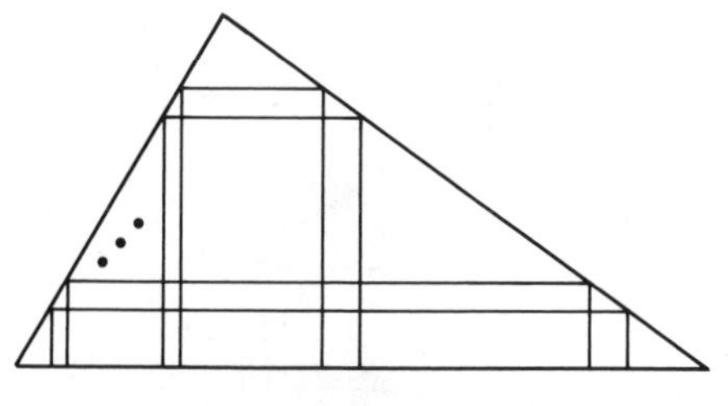

Figure 3.8

into a constructive argument. If one thinks to examine the locus of the fourth vertex (see Figure 3.9), then one can determine precisely where that locus crosses the opposite side.

Note, incidentally, that there is an element of luck in the choice of solution path. There is no a priori reason to prefer Problem P_3 to P_2, since both seem plausible and both produce existence arguments.

Finally, the domain-specific "pattern of similar figures" strategy leads to a clever solution by way of Problem P_5. Suppose one starts with a square S whose side has length h, the altitude of the given triangle. It is easy to construct a triangle around S that is similar to the given triangle. A standard proportionality construction then finishes the problem (see Figure 3.10).

DISCUSSION

A review of the heuristic solution to Problem P just offered provides a partial exegesis of the "easier, related problem" strategy, Strategy H_3, mentioned above:

> **Strategy H_3** If you canot solve the original problem, try first to solve an *easier, related problem.* Look for known solutions to related problems.

In presenting the solution to Problem P, I have provided the appropriate elaborations of the heuristics and interpreted them in the most favorable

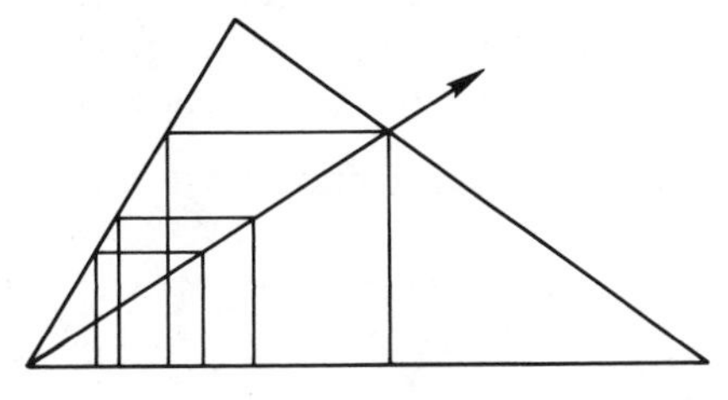

Figure 3.9

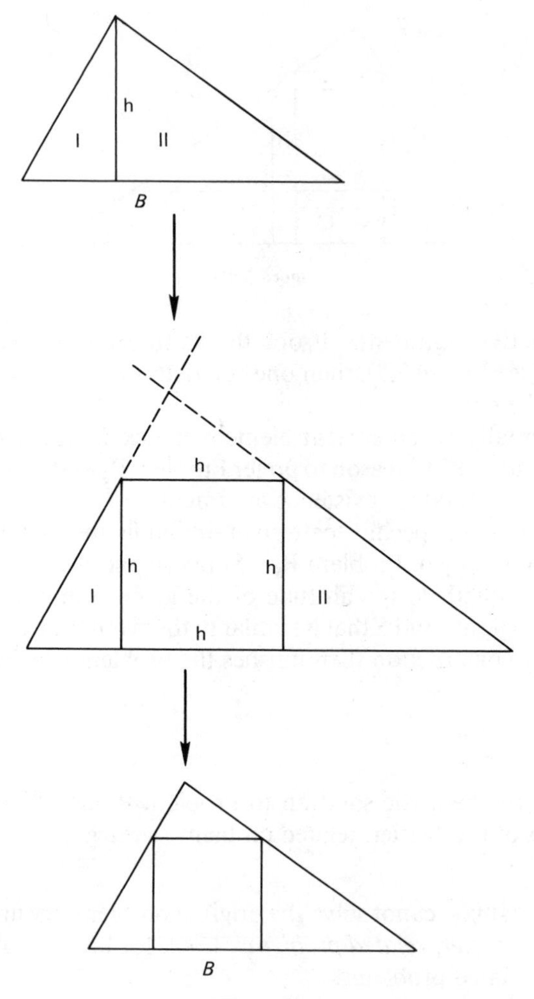

Figure 3.10

ways. Solid resources were taken for granted, and there were no false starts, blind alleys, or misinterpretations. Even so, a fair amount of sophisticated behavior was needed to solve the problem. The solution called for:

1. thinking to use the appropriate strategies in the first place, which is by no means a trivial task;
2. knowing the specific forms of the strategies that are appropriate for this problem;

3. generating a collection of appropriate easier, related problems;
4. evaluating for each problem generated in Point 3
   a. the likelihood of obtaining a solution
   b. the likelihood of being able to use that solution to find a solution to Problem P;
5. making decisions based on the probabilities in Point 4;
6. solving the easier, related problem chosen in Point 5;
7. exploiting the solution (perhaps the method, perhaps the result) to obtain a solution to Problem P.

All of the above steps—which have not been specified here in adequate detail—constitute the mechanics of implementing Strategy H_3, and a misfire in any one of them could cause an attempt to solve a problem via Strategy H_3 to fail. This complexity should be recalled when considering prior attempts to teach students to use heuristic strategies. For example, simply showing students the successful heuristic solution to Problem P would bypass all of this complexity. It would, however, leave the students completely unprepared to use the strategy—although they could still "appreciate" its use.

Heuristics and Resources Deeply Intertwined

Chapter 1 demonstrated a heuristic solution to Problem 1.2, which asked for a straightedge-and-compass construction bisecting the area of a given triangle with a line parallel to its base (Figure 3.11). The solution was based on the strategy of "taking the problem as solved." It depended on drawing in the line segment *DE* in the figure, and then determining where the segment *DE* must be in order for the two parts of the figure to have equal areas.

Finishing the solution, however, depends on a subtle bit of domain-specific knowledge. Problem solvers with the "similar triangles schema," who saw that the triangles *ADE* and *ABC* are similar, were pointed toward a more

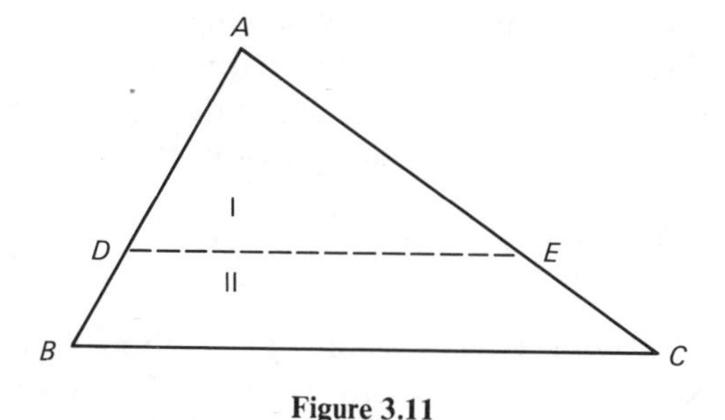

Figure 3.11

or less straightforward solution to the problem. People who lacked the schema were forced to engage in much more complex computations, and (despite their reasonable problem-solving ability) were generally not successful. This underscores the third point of this chapter: Despite the fact that their application cuts across various mathematical domains, the successful implementation of heuristic strategies in any particular domain often depends heavily on the possession of specific subject matter knowledge. While heuristics are important, their use is based on a foundation of resources. (That is, one cannot expect wonders in new domains, even if a problem solver is capable of using a broad range of heuristic strategies.) Two more examples of the interplay between resources and heuristics are given in this section. The first example comes from my students' attempt to solve the following problem:

Problem 3.11 Prove that for all sets of real numbers a, b, c, and d, $a^2 + b^2 + c^2 + d^2 = ab + bc + cd + da$ implies $a = b = c = d$.

I had assigned this problem to a class of junior mathematics majors at Berkeley. The general strategy that solves the problem goes under the label of "exploiting an easier, related problem." A more specific and more implementable version is the following.

> Consider a similar problem with fewer variables. If the problem has a large number of variables and is too confusing to deal with comfortably, construct and solve a similar problem with fewer variables. You may then be able to (1) adapt the method of solution to the more complex problem, or (2) take the result of the simpler problem and build up from there.

When my students were unsuccessful at solving Problem 3.11, I described the "fewer variables" strategy and asked for suggestions. One of the students suggested that we should set the values of c and d equal to 0 in the problem statement, to obtain the following simpler problem:

Problem 3.11A Prove that if a and b are real numbers, $a^2 + b^2 = ab$ implies $a = b$.

The class agreed and worked on Problem 3.11A without much success. This lack of success comes as no surprise, since Problem 3.11A is an inappropriate (and only serendipitously correct) modification of Problem 3.11. The correct modification that one obtains from setting $c = 0 = d$ in Problem 3.11 is

Problem 3.11B Prove that if a and b are real numbers, $a^2 + b^2 = ab$ implies $a = b = 0$.

Problem 3.11B does not resemble Problem 3.11 closely enough to be useful. To select the appropriate two-variable analogue, one must notice the cyclic nature of the terms on the right-hand side of Problem 3.11 and then construct the cyclic version of the two-variable problem statement. This yields the following:

Problem 3.11C Prove that if a and b are real numbers, $a^2 + b^2 = ab + ba$ $(=2ab)$ implies $a = b$.

Fortunately, Problem 3.11C is easily solved:

$$a^2 + b^2 = 2ab$$

implies

$$(a - b)^2 = 0,$$

so that

$$a = b.$$

With some insight, the idea behind this solution, which depends on showing $a = b$ by getting $(a - b)^2 = 0$, can be used to solve Problem 3.11. The original problem statement can be multiplied by 2, yielding

$$2a^2 + 2b^2 + 2c^2 + 2d^2 = 2ab + 2bc + 2cd + 2da.$$

This can now be arranged as

$$[a^2 + b^2] + [b^2 + c^2] + [c^2 + d^2] + [d^2 + a^2]$$
$$= 2ab + 2bc + 2cd + 2da,$$

which yields

$$(a - b)^2 + (b - c)^2 + (c - d)^2 + (d - a)^2 = 0.$$

In this expression a sum of squares is 0, so that each term in that sum must be 0. It follows that $a = b$, $b = c$, $c = d$, and $d = a$, which is what we wanted to show.

The students' difficulty in implementing the heuristic solution to Problem 3.11 is a matter of resources. The problem is to "see" enough of the structure inherent in the problem statement—here the cyclic nature of the terms $(ab + bc + cd + da)$—so that one can implement the strategy properly and construct the analogous two-variable expression $(ab + ba)$. This is nontrivial, even with fairly sophisticated and talented students.

This point is brought home again in a second example, the sequel to Problem 3.11. We had just finished solving Problem 3.11 in class and wrote up a formal argument. The last line of our argument read as follows:

Since $(a-b)^2+(b-c)^2+(c-d)^2+(d-a)^2=0$, we have $a=$ b, $b=c$, $c=d$, and $d=a$, as desired.

I left the solution on the blackboard and gave the class the following problem:

Problem 3.12 Let $\{a_1, a_2, \ldots, a_n\}$ and $\{b_1, b_2, \ldots, b_n\}$ be given sets of real numbers. Determine necessary and sufficient conditions on $\{a_i\}$ and $\{b_i\}$ such that there are real constants A and B with the property that

$$(a_1x+b_1)^2+(a_2x+b_2)^2+\cdots + (a_nx+b_n)^2=(Ax+B)^2$$

for all values of x.

I told the students that Problems 3.11 and 3.12 were related and that they should try to exploit the first in solving the second. I gave them 15 minutes and asked them to work individually. None of the students made any progress, because they saw no structural similarity between the two problems.

This is not surprising. The second equation is, after all, a morass of symbols, none of which is quite comparable to those in the first. The number of terms is different, the quantities are polynomials instead of numbers, there are subscripted variables, and the right-hand side is a quadratic polynomial with undetermined coefficients. Nonetheless, when I read Problem 3.12 (when I was selecting problems for the class to work on), I was reminded of Problem 3.11, even though a fair amount of time had passed since I had selected Problem 3.11 for discussion. In solving Problem 3.11, I had been impressed by the fact that a great deal of information is contained in an equation where a sum of squares equals 0. When I read Problem 3.12, I saw a sum of squares equal to *something.* Thus I would gain information if *something* were replaced by 0. The compact form of encoding "sum of squares equals" enable me to do this. Without this concise yet powerful means of summarizing the relationships among the symbols in those two equations, the structural similarity between the two equations is virtually impossible to see. This, too, is a matter of resources.

DISCUSSION

Although the tone of this chapter may appear to be mostly negative, my intention was not to paint a pessimistic picture about the possibility of teaching students to use individual heuristics. Rather, my intention was to paint a realistic one. The complexity and subtlety of these strategies has been consistently underestimated, with unfortunate consequences. Learning to apply such techniques in new arenas is a classic example of the "transfer problem," which is notoriously difficult. In my opinion there are such things as general heuristic strategies; competent mathematicians will indeed make progress on new problems by examining special cases, establishing subgoals, and so on. I believe they do so in unfamiliar domains, working on problems that are not simple extrapolations of ones they already know how to solve. I also believe—as is documented in Part II, specifically in Chapter 7—that students can learn to master such techniques and use them in novel situations. The point of this chapter (and the next) is that the amount of groundwork necessary for doing so is tremendous. If the conditions for transfer are established, it is much more likely that students will be able to transfer what they have learned to new domains when appropriate. These conditions include breaking down the "general heuristic strategies" into a collection of coherent substrategies, carefully delineating a fairly large sample of these, discussing the conditions under which they seem to be appropriate, and using problems from new domains for practice (with later discussion as to how and why the particular strategies used were appropriate).

Summary

This chapter discussed a series of major obstacles to the mastery of heuristic strategies. The first is that most "general heuristic strategies" are so broadly defined that their definitions are far too vague to serve as a guide to their implementation. More often than not, a capsule description of a strategy—be it of "exploiting analogy," "decomposing and recombining," "using subgoals," or any other—is a summary label that includes under it a class of more precise substrategies that may be only superficially related. An adequate delineation of such strategies calls for identifying and characterizing the major component substrategies in substantial detail and giving as much attention to these as one would any other complex subject matter.

The second is that the successful use of such strategies calls not only for "knowing" the strategies, but for good executive decision-making and an extensive repertoire of subskills. The straightforward solution of a Pólya's geometry problem by means of an easier, related problem called for (1) knowing to use the right strategy, (2) knowing the appropriate versions of it for that problem, (3) generating appropriate easier, related problems, (4) assessing the likelihood of being able to solve and then exploit each of the easier problems, (5) choosing the right one, (6) solving the chosen problem, and (7) exploiting its solution. Learning to use the strategy means learning all of these skills.

A third point is that one cannot expect too much of heuristic strategies. One's success in any domain is based on a foundation of one's resources in that domain, and even a good mastery of heuristics cannot be expected to replace shaky mastery of subject matter.

Despite the focus on obstacles in this chapter, the intention was not to paint a pessimistic picture. The issues raised here point to the amount of work that needs to be done if such general skills are to be mastered. Evidence indicating that they can be are given in Chapter 7.

4

Control

The problem solving skills . . . attributed to the executive in many theories of human and machine intelligence: predicting, checking, monitoring, reality testing, and coordination and control of deliberate attempts to solve problems . . . are the basic characteristics of thinking efficiently in a wide range of learning situations. . . . [T]he use of an appropriate piece of knowledge or routine at the right time and in the right place is the essence of intelligence.

Ann Brown, 1978, pp. 78–82

Introduction and Overview

Somewhat paradoxically the decomposition of general heuristic strategies into their constituent parts, as suggested in Chapter 3, results in a new set of difficulties for the problem solver. The short dictionary of heuristic in *How to Solve It* offers roughly a dozen broad categories of problem-solving strategies. This is a substantial but still manageable number. Selecting the right strategy in the midst of working a problem is by no means a trivial task, but if the search space is limited to a dozen or so techniques, the task is not likely to prove so unwieldy that it will severely hamper performance. Consider what happens, however, when each general heuristic strategy is elaborated into its constituent collection of detailed strategies. Now the problem solver's collection of potentially useful techniques includes many dozens and perhaps hundreds of strategies. Recall the complexity of Figure 3.7, which represents a small part of the decision tree for implementing just one heuristic

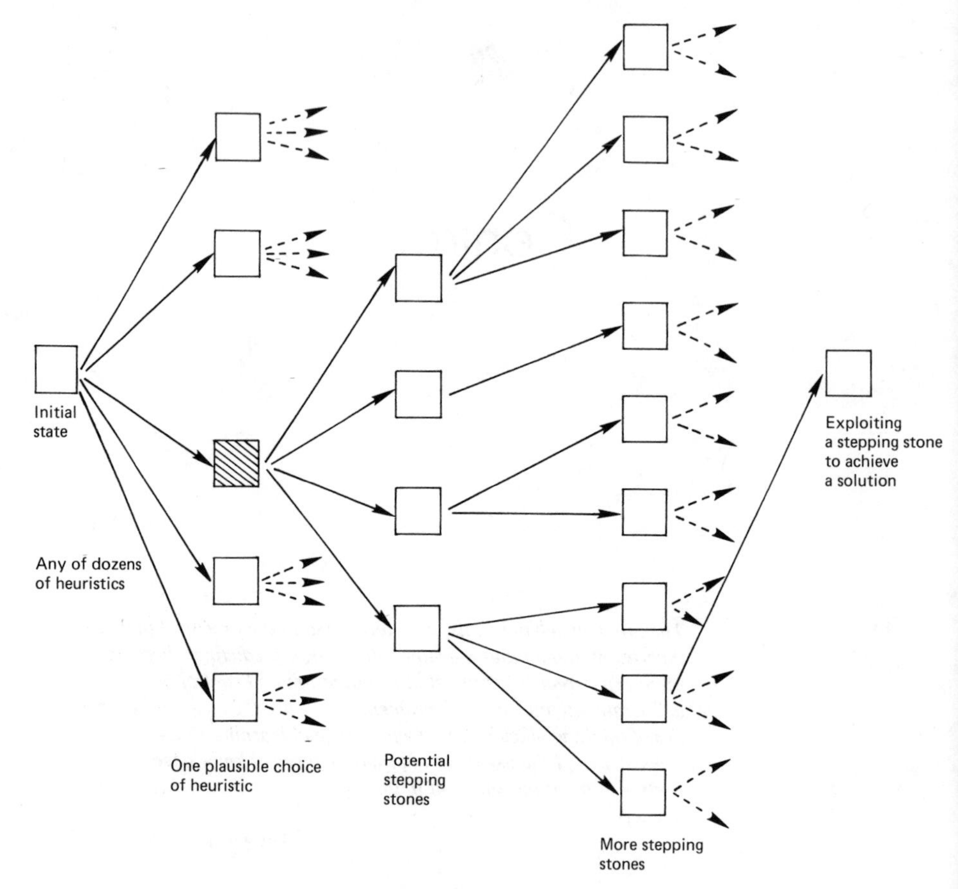

Figure 4.1

strategy. As seen in Figure 4.1, all of Figure 3.7 represents only one branch of the very large number of branches of the search tree faced by a problem solver who starts a problem from scratch. As the person begins to work on a problem, it may be the case that some of the heuristic techniques that appear to be appropriate are not. Conversely, the strategies that actually help to solve the problem may not appear to be useful. If the problem solver chooses to implement one or more inappropriate strategies and pursues them to the exclusion of other possibilities, then the problem-solving attempt will fail. But the attempt will fail as well if potentially valuable approaches are abandoned before they can bear fruit. In consequence, having a mastery of individual heuristic strategies is only one component of successful problem solving. Selecting and pursuing the right approaches, recovering from inappropriate choices, and in general monitoring and overseeing the entire problem-solving process, is equally important. One needs to be efficient as well as resourceful. In broadest terms, this is the issue of control.

This chapter is divided into two parts. The first part provides a discussion of control within mathematics. The second part is a review of selected parts of the literature that bear on the issue of control.

The first part deals with three issues of increasing complexity and is divided into three sections. The first section offers a study of control in a simple mathematical microcosm—techniques of integration as studied in first-year calculus. The argument is advanced that, even in this simple mathematical domain, students do not develop efficient control strategies and that the students' performance suffers substantially in consequence. This being the case, a prescriptive control strategy—one that helps students to select appropriate techniques—might enhance their performance. One such strategy, how well it has worked, and the implications of its effects on performance, are discussed.

The second section returns to the topic of problem solving via heuristic strategies. The focus remains prescriptive, and the line of reasoning put forth is an extension of that in the first section. In general problem solving, there is a quite large number of potentially useful heuristic strategies, most of which are far more complex than the straightforward techniques of integration. Thus if the poor management of strategic resources in integration can inhibit performance, then comparably poor managment in general problem solving could have far more severe effects. Conversely, if a prescriptive control strategy could improve performance in integration, then perhaps a prescriptive control strategy could help students to reap the benefits of heuristic techniques as well. The broad outline of an attempt at such a strategy is discussed. (Early versions of my problem-solving courses were explicitly based on that strategy. The results of the instruction are discussed in detail in Chapters 7, 8, and 9.)

In the third problem section I leave the prescriptive arena to discuss the issue of control in general. The object is to explore the kinds of impact that different kinds of control behavior can have on problem-solving performance. Examples from transcripts of problem-solving sessions are given, demonstrating how control can help or hamper problem-solving performance. These discussions provide some of the background for the analyses of control behavior given in Chapter 9.

The second part of the chapter summarizes some related work from a variety of disciplines. It begins with the notion of *control* as the term is used in AI. The discussion then moves to the issue of metacognition and briefly touches on the role of social cognition in the development of internal control strategies.

On the Importance of Control: A Look at a Microcosm

This section explores the role of strategic decision-making in a straightforward domain: the solving of indefinite integrals in elementary calculus. By all rights, indefinite integration should cause students little difficulty. The prerequisite algebraic and differentiation skills required for integration are mechanical. The tools of the trade (techniques of integration such as substitution, parts, and partial fractions) are straightforward algorithmic procedures. Indeed, students do learn to perform these techniques reliably. They can use them well when they know which technique they are supposed to be using—for example, when working exercises at the end of a particular section of text. The issue of strategy selection should not be a major factor in performance; there are a dozen or fewer techniques to choose from, and the appropriate technique for a problem is often suggested by the form of the integrand itself.

PILOT STUDIES INDICATE THAT CONTROL IS A PROBLEM

A typical calculus textbook concludes its chapter on integration with a set of 100 to 200 practice problems. It is generally assumed that students will, by working these problems (and often many others from other texts), come to develop efficient selection mechanisms. The author of one of the most widely used calculus texts, George Thomas, makes this assumption explicit in his instructions to students:

> Perhaps a good way to develop the skill we are aiming for is for each student to build his own table of integrals. He may, for example, make a notebood in which various sections are headed by standard forms. . . . Making such a notebook probably has educational value. But once it is made, it should rarely be necessary to refer to it! (1960, p. 281)

Evidence to the contrary was provided by a colleague, who, in order to bolster his students confidence, had deliberately placed the problem

$$\int \frac{x\,dx}{x^2 - 9}$$

at the beginning of an hour examination. This problem can be solved in less than a minute using the substitution

$$u = x^2 - 9,$$

an elementary and—one would assume—obvious approach. Unfortunately, 44 of his 178 students, noting that the term $(x^2 - 9)$ in the denominator could be factored, decided to work the problem by using the technique of partial fractions. This approach does yield a solution, but it takes students from 5 to 10 minutes to apply. Even worse, 17 chose to work the problem using the trigonometric substitution $x = 3 \sin \Theta$. This approach also yields a solution, but it takes even longer to implement. These students each spent 10–15 minutes on a problem that should have been trivial.

The difficulty that students had with this problem was not in finding a successful approach, or in implementing that approach; it was in selecting a reasonable approach. Their behavior violates a cardinal rule of control in problem solving: Never implement difficult or time-consuming procedures unless you have checked to see whether other, far simpler procedures will work.

Pilot studies for the prescriptive integration strategy described below included making tape recordings of students solving integration problems out loud. A week before the final examination in her second semester of calculus, the student IB was asked to work

$$\int \frac{dx}{e^x + 1}.$$

IB was doing well in the course and went on to earn an A in it. She had already reviewed integration for the final exam and made the following comments as she worked this problem:

Well, right now I'm just thinking of the various methods I know. There is substitution, which I like more than any of the others, and integration by parts, and partial fractions. Maybe if I use substitution . . . I'll play with that. . . .

After a first unsuccessful attempt at the substitution

$$u = (e^x + 1),$$

she looked for ways to use integration by parts and then tried to use partial fractions. She also used the same techniques, in the same order, on a large number of other problems.

As it happens, the text used in IB's calculus class presents, in order, the following techniques of indefinite integration: "some basic facts," *the substitution technique,* using a table of integrals, *integration by parts,* rational functions, and *partial fractions.* The italicized sections offer implementable techniques. Not at all coincidentally, these were the techniques tried, in order, by IB. She had acquired a control strategy by default. For obvious reasons, this particular strategy—trying a series of techniques in a particular order—can result in remarkably inefficient problem-solving performance.

One must be careful in saying that IB was making use of a particular control strategy, because (1) it is not clear that she was consciously aware of implementing it and (2) if asked, she probably would not have been able to elucidate it. Nonetheless her behavior was remarkably consistent, and the statement quoted above indicates that her consistency was no accident. She certainly behaved as though she had a particular control strategy.

The pilot studies indicated that IB's consistency was somewhat unusual, but that two aspects of her behavior were absolutely typical. First, students are generally inefficient in their strategy selection. They will often make their first attempt on a problem using a complicated and time-consuming strategy, without checking to see whether simpler and faster techniques might be appropriate. As a result they waste much time and effort, even when they solve problems correctly. Second, students do not (in general) learn to key their strategy selection to the characteristics of the problems they are working. The quotation from Thomas suggests that the form of an integrand may suggest which techniques are likely to prove useful in solving it. Yet good students who had already studied for their final examinations did not take advantage of this fact when trying to solve problems.

MODELING A CONTROL STRATEGY FOR INTEGRATION

As is described in the second part of this chapter, issues of planning and strategy selection are major foci of work in AI. Methodologies have been developed in AI for making systematic observations of people in the process of solving problems, and for abstracting from those observations the consistent patterns of useful behavior that lead to problem-solving success. The idea explored here (in 1975–1976) was to turn things around and have AI serve as a metaphor for human problem solving.* Perhaps systematic observations of a competent human problem solver would suggest patterns of strategy selection that other humans (rather than machines) could learn to use.

Simply put, this approach has two phases. The first is to look for consistent patterns of technique selection on the part of competent problem solvers (experts) in the domain. The second is to construct a strategy that yields

* The work on integration was intended as an existence proof of sorts, a demonstration that prescriptive control strategies can improve problem-solving performance. The long-term goal at that point was to develop a similar prescriptive control strategy for mathematical problem solving via heuristic strategies. In that domain, the comparison between human and machine problem solving at the control level is truly metaphorical. In the case of integration, the ties between human and machine implementation are much closer, and there was work in AI of direct relevance. Slagle's (1963) SAINT was an early AI program that worked integrals at the level of a good MIT freshman. For a variety of reasons SAINT's approach was inappropriate for human implementation. However, its success suggested strongly that the development of comparable control strategies for humans was feasible.

Table 4.1
A Broad Overview of the Integration Strategy

Proceed from one step to the next when the techniques of that step fail to solve the problem. Always look for easy alternatives before beginning complicated calculations. If you succeed in transforming the problem to something easier, begin again at Step 1.

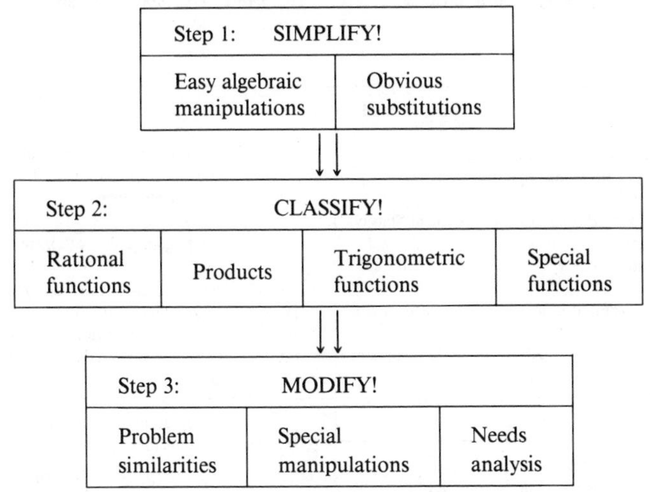

similar patterns of technique selection. It should be noted that the strategy need not be the same one as that used by the experts; it need only produce the same results.

The data for strategy abstraction consisted of my solutions to the more than 150 miscellaneous integrals at the back of a standard text. Technique selections were monitored as the problems were worked. It was by no means apparent how techniques were being selected, but there was no doubt that the selections were systematic; in more than three-fourths of the problems, a method had been chosen, and implementation begun, within 20 seconds. Looking over the solutions made it clear that there were two components to the pattern of strategy selections. The first component was domain-specific: The form of the integrand often points to one or more appropriate techniques of integration. The second component had to do with efficiency and was a procedural embodiment of the general rule discussed above: One should never do anything difficult and time-consuming until one has checked to see that simpler and faster alternatives do not suffice.

The full strategy was written up as a set of instructional materials (Schoenfeld, 1977), which are outlined in Tables 4.1 and 4.2. The domain-specific component, based on pattern recognition, was built into the separate steps of the procedure. In Step 1, configurations that lent themselves to simplifica-

Table 4.2

An Elaboration of the Integration Strategy (Supplemented by a Text and a Solutions Manual)

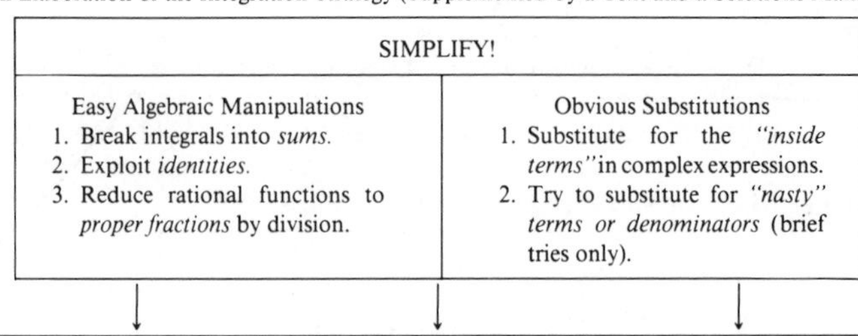

SIMPLIFY!

Easy Algebraic Manipulations	Obvious Substitutions
1. Break integrals into *sums*. 2. Exploit *identities*. 3. Reduce rational functions to *proper fractions* by division.	1. Substitute for the *"inside terms"* in complex expressions. 2. Try to substitute for *"nasty" terms or denominators* (brief tries only).

↓ ↓ ↓

CLASSIFY!

Integrating Rational Functions

1. Reduce to "proper fractions" by division.
2. Factor the denominator.
3. Decompose by *partial fractions* into a sum of "basic" rational functions.
4. If the denominator is $(ax + b)$ or $(ax + b)^n$, use the substitution $u = (ax + b)$.
5. If a quadratic denominator does not factor easily, complete the square. For the terms
   a. $(a^2 + u^2)$, integrate directly to obtain a logarithm and/or arctangent.
   b. $\pm(u^2 - a^2)$, break into a sum and use the formula on p. 17, or use partial fractions.

Integrating Trigonometric Functions

1. Exploit *twin pairs* to prepare for substitutions. Try to obtain integrals of the form

$$\int f(\sin x)(\cos x\, dx), \text{ etc.}$$

2. Use half-angle formulas or integration by parts to reduce powers of trigonometric functions in the integrand.
3. As a last resort, the substitution $u = \tan(x/2)$ transforms rational functions of $\sin x$ and $\cos x$ to rational functions of u. (see p. 32)

Integrating Products

Consider *integration by parts*. The formula is

$$\int u\, dv = uv - \int v\, du,$$

and your choice of u and dv should be governed by two things:

1. You must be able to integrate the term dv.
2. You want $\int v\, du$ to be easier than the original integral.
   This often happens when u is simplified by differentiation.

Integrating Special Functions

1. *If the integrand includes terms of the form*

$$(a^2 - u^2)^{n/2},\ (u^2 - a^2)^{n/2}, \text{ or } (a^2 - u^2)^{n/2},$$

   a. draw a right triangle,
   b. place a and u so that the third side of the triangle is the term you want,
   c. "read" the substitutions from the triangle.
2. If the integrand *is a rational function of* e^x, make the substitutions

$$e^x = u \quad \text{and} \quad dx = \frac{1}{u}\, du.$$

3. If the integrand is *a rational function of* x *and* $\sqrt[n]{ax + b}$, make the substitutions

$$\sqrt[n]{ax + b} = u;\ x = \frac{1}{a}(u^n - b), \text{ and } dx = \frac{n}{a} u^{n-1}\, du.$$

↓ ↓ ↓

MODIFY!

Problem Similarities	Special Manipulations	Needs Analysis
1. Look for easy problems similar to the one you are working on. 2. Try to reduce the difficult problem to the form of the easy similar problem. 3. Try the techniques you would use on the similar problem.	1. Rationalizing denominators of quotients. 2. Special uses of trigonometric identities. 3. "Common denominator" substitutions. 4. "Desperation" substitutions.	1. Look for a term, or a form of the integral, that would enable you to solve it. 2. Try to modify the integral to produce the term or form you need. 3. Try to introduce the term you need; compensate for it.

tion by algebra or by substitution were suggested. In Step 2, the form of the integrand was used to suggest the relevant technique of integration. If the techniques of Step 2 did not suffice to solve the problem, the form of the integrand was used in Step 3 to suggest the kind of desperation attempts that might help one make progress.

Efficiency was built into the strategy as follows. Step 1 of the strategy offered simple procedures that, in general, should be considered early in a solution. They are quick, and a problem is solved easily if they work. Step 2 offered procedures that are more complex but still standard. Step 3 offered heuristic procedures that are often quite difficult and may or may not yield success when implemented. As suggested by the instructions in Table 4.1, one should only consider techniques at a particular level of difficulty when techniques at lower levels have been considered and are found wanting. Whenever progress is made on the problem — that is, the current integrand is transformed to a simpler one — the procedure begins again, with a search for simple ways from Step 1 to dispatch the simpler integrand. In short, at each pass through the problem one considers easy techniques before difficult ones so that the likelihood of violating the rule italicized two paragraphs above is relatively small.

HOW THE STRATEGY WORKED

The instructional materials, which were designed as a self-contained package, worked rather well. Four days before the instructor of the calculus class gave an examination on techniques of integration, he agreed to distribute the materials to a randomly chosen half of his class. His test had been written before he saw the materials. No attempt was made to coordinate the materials with, or adapt them to, the text used for the class. They were simply handed out with the request that the students who received them should prepare for his examination by using the materials rather than by solving the numerous miscellaneous exercises in their text. The materials had to stand on their own, because no further help would be available. No promises regarding their effectiveness were made. It was stressed that they had never been tested and that we had no way of knowing a priori whether they would help or hinder the students. The students were promised that the final grades for the two halves of the class would be determined separately if either group outperformed the other.

Students were asked to keep track of the time they spent studying for the exam. Those who used the materials averaged 7.1 hours of study time, while those who prepared the conventional way averaged 8.8 hours each. There were nine questions on the exam, seven of which were indefinite integrals

Table 4.3
Scores on Indefinite Integrals

	Question							
	1	2	3	4	5	6	7	Average
Test group	82.7	56.4	80.0	74.6	83.6	31.8	33.5	63.2
Control group	70.0	62.0	61.3	60.0	80.0	19.4	20.5	53.3
Difference	+12.7	−5.6	+18.7	+14.6	+3.6	+12.4	+13.0	+9.9

Table 4.4
Scores on Other Questions

	Question		
	1	2	Average
Test group	48.2	74.2	61.2
Control group	59.1	64.5	61.8
Difference	−10.9	+9.7	−0.6

and two of which tested other matters. The scores for these problems are given in Tables 4.3 and 4.4.

The scores on problems that did not test integration were about the same for the two groups, suggesting that the two groups were of roughly comparable ability. On integration, however, the group that used the materials did far better than the other. As often happens, the statistical significance of the difference in performance depends on one's choice of tests. The worst results indicated that the students who used the materials were better at $p < .15$, the best results at $p < .01$. More details can be found in my *Monthly* article (Schoenfeld, 1978).

Modeling a Control Strategy for Heuristic Problem Solving

The work discussed in the previous section indicates that even in a domain as straightforward as integration, the absence of efficient control behavior can have a significant negative effect on students' problem-solving performance. Such negative effects are likely to be much stronger in the arena of general mathematical problem solving via heuristic techniques. While the student must only "manage" a dozen or so techniques of integration, the number of heuristics the student must manage is larger by an order of magnitude. The techniques of integration are algorithmic and, if appropriately chosen, are guaranteed to produce results. In contrast, heuristic strategies are far more complex to implement, are not algorithmic, and, even if seemingly appropriate, carry no guarantees of success with them. Since inefficient control dilutes the benefits of a few straightforward techniques of integration, it could easily wash out the potential benefits of heuristic techniques altogether.

Similarly, the success of a prescriptive control strategy for integration can be seen as lending credence to the idea of developing a managerial strategy for general problem solving via heuristic techniques. The procedure is essentially the same. By making detailed observations of good problem solvers in the process of working difficult and unfamiliar problems, one might see some uniformity in their strategy selection. That uniformity could then be captured in a prescriptive strategy. As in the integration materials, the strategy would (1) point individuals to particular domain-specific techniques wherever possible; and it would do so (2) within a global framework for being efficient, for trying easier techniques before more difficult ones, and so forth. I developed such a strategy in 1975 and 1976, and it served as the foundation for two courses in problem solving. The first was for upper-division mathematics majors and was offered at Berkeley in 1976. The second was for freshmen liberal arts majors and was offered at Hamilton College from 1978 through 1980.

The bare bones of the strategy are given below. As a matter of convenience (and metaphor), the strategy was given in flowchart form. This form should not be interpreted, however, as a "program" that students were supposed to implement mechanically. Rather, it was intended as a default strategy—a guide to use when the student did not know what to do next and could use guidance in selecting from among the heuristic techniques that might be appropriate. At a global level the strategy worked in a fashion similar to the integration strategy. If the student knew what to do when working a problem, then the student should do it. (After all, the strategy was developed independently of them as individuals and without any particular problem in mind; it should not interfere in processes that proceed well without it.) If a student did not know what to do, the strategy suggested which heuristics might be appropriate and roughly in what order. If the student made substantial progress and reduced the original problem to a more manageable one, then the process began again with the new problems.

The following excerpts are taken from a handout that the students received the first day of class. They provide a brief overview of the strategy.

* * *

The Strategy: An Overview

On the following pages you will find a brief description of the problem-solving strategy we shall be using this term. The strategy represents what may be called the *ideal* problem solver, or the most systematic behavior of good problem solvers. As far as I can tell, most good problem solvers use the techniques indicated here and in an order more or less like the one given here.

The strategy is given in the form of a flowchart indicating the major stages of the problem-solving process: analysis, design, exploration, implementation, and verification. We shall study each of these in some substantial detail, beginning with analysis later today. First, let me introduce you to a word I shall be using a great deal. The word is *heuristic.* Its origins are Greek, and roughly translated it means "serving to know or understand." A heuristic strategy is a technique or suggestion designed to help you better understand a problem—and if you are lucky, solve it as a result. Examples of such strategies are "draw a diagram," "consider easier, related problems," and so forth. A list of the most important problem-solving heuristics is given in Table 4.5 and a rough outline of our strategy in Figure 4.2. They should be interpreted as follows:

You begin with an analysis of what the problem really calls for. This means getting a feel for a problem: for what is given, what is asked for (the goals), why the "givens" are there and whether the goals seem plausible, what major principles or mechanisms seem relevant or appropriate to bring to bear, what mathematical context the problem fits into, and so on. Of course you read the problem carefully. Which heuristics (if any) are appropriate to use during analysis may depend on both the problem and who is solving it. But examples of appropriate usage of some heuristic strategies at this stage of problems solving are as follows:

1. Drawing a diagram, even when the problem appears amenable to a different kind of argument (say an algebraic one). Pictures often help you to see things.
2. Exemplifying the problem (examine special cases) with the result that you either solve it for special cases or see empirically determinable patterns. If you are asked to show something "for all triangles," does it hold for isosceles, or equilateral, or right triangles?
3. Looking for preliminary simplifications. In the problem "Find the largest area of any triangle that can be inscribed in a circle of radius R," you might (1) consider first the unit circle, (2) note that without loss of generality the base of the triangle can be assumed horizontal, and (3) examine a variety of sketches to guess a plausible answer before jumping into an analytic solution.

Design is, in a sense, a "master control." It is not really a separate box on the flowchart but something that pervades the entire solution process; its function is to ensure that you are engaged in activities most likely (as best as you can tell at that point) to be profitable. Most generally, it means keeping a global perspective on what you are doing and proceeding hierarchically. You should outline a solution of the problem at a rough and qualitative level and then elaborate it in detail as the solution process proceeds. For example, you should not get involved in detailed calculations or complex operations until (1) you have looked at alternatives, (2) there is clear justification for them, and (3) other stages of the problem solution have proceeded to the point where the results of the calculations are either necessary or will clearly prove useful. (How painful it is to expend time and energy solving a differential equation, only to discover that the solution is of no real help in the balance of the problem!)

Table 4.5
Frequently Used Heuristics

Analysis

1. Draw a diagram if at all possible.
2. Examine special cases:
   a. Choose special values to exemplify the problem and get a "feel" for it.
   b. Examine limiting cases to explore the range of possibilities.
   c. Set any integer parameters equal to 1,2,3, . . . , in sequence, and look for an inductive pattern.
3. Try to simplify the problem by
   a. exploiting symmetry, or
   b. "without loss of generality" arguments (including scaling).

Exploration

1. Consider essentially equivalent problems:
   a. Replace conditions by equivalent ones.
   b. Re-combine the elements of the problem in different ways.
   c. Introduce auxiliary elements.
   d. Re-formulate the problem by
      (i) change of perspective or notation,
      (ii) considering argument by contradiction or contrapositive,
      (iii) assuming you have a solution and determining its properties.
2. Consider slightly modified problems:
   a. Choose subgoals (obtain partial fulfillment of the conditions).
   b. Relax a condition and then try to re-impose it.
   c. Decompose the domain of the problem and work on it case by case.
3. Consider broadly modified problems:
   a. Construct an analogous problem with fewer variables.
   b. Hold all but one variable fixed to determine that variable's impact.
   c. Try to exploit any related problems that have similar
      (i) form,
      (ii) "givens",
      (iii) conclusions.

   Remember: when dealing with easier, related problems, you should try to exploit both the *result* and the *method of solution* on the given problem.

Verifying Your Solution

1. Does your solution pass these specific tests?
   a. Does it use all the pertinent data?
   b. Does it conform to reasonable estimates or predictions?
   c. Does it withstand tests of symmetry, dimension analysis, and scaling?
2. Does it pass these general tests?
   a. Can it be obtained differently?
   b. Can it be substantiated by special cases?
   c. Can it be reduced to known results?
   d. Can it be used to generate something you know?

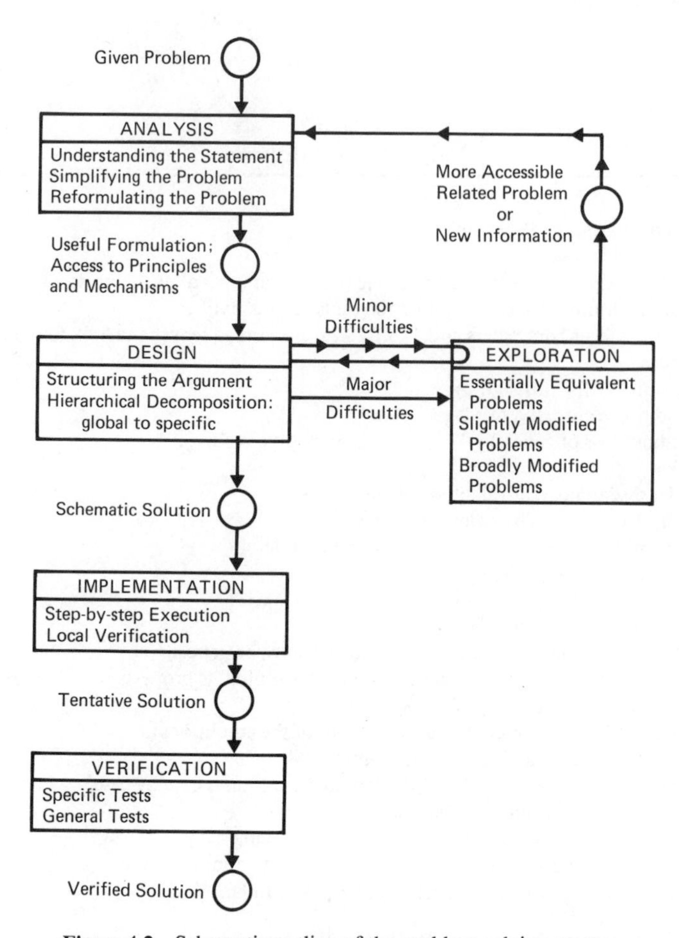

Figure 4.2 Schematic outline of the problem-solving strategy.

Exploration is the heuristic heart of the strategy, for it is in the exploratory phase that the majority of problem-solving heuristics come into play. As seen in Table 4.3 and Figure 4.2, exploration is divided into three stages. In general, the suggestions in the first stage are either easier to employ or more likely to provide direct access to a solution of the original problem than those in the second stage; likewise for the relation between Stages 2 and 3. All other factors being equal, when you get to exploration you would briefly consider those suggestions in Stage 1 for plausibility, select one that seemed appropriate and try to exploit it. If the plausible strategies in Stage 1 prove insufficient, you proceed to Stage 2; if need be, when Stage 2 has been exhausted you try the strategies in Stage 3. If you make substantial progress at any point in this process, you may either return to *design* to plan the balance of the solution, or may decide to reenter *analysis,* with the belief that the insights gained in *exploration* can help recast the problem and allow you to approach it differently.

Table 4.6
The Analysis Stage of the Process

Preliminaries: Transcribe the problem in *shorthand,* drawing a *diagram* or making a rough sketch if you can. If someone took away the original problem statement, would you have enough to work with? Have you really identified all the essential information in the problem statement?

1. Make sure you *understand the statement* of the problem!
   a. *Examine the conditions* of the problem—what you are given and what you want to get.
   b. *Look at some examples,* to get a "feel" for what the problem really asks of you.
   c. *Check for consistency.* Do you have enough information? Does the result seem plausible? Reasonable? Do you have too much information?
2. *Try to simplify the problem!* Look for ways to make the problem easier, before you get immersed in details. Depending on the problem, you might appeal to
   a. *symmetry,*
   b. *scaling* (this can reduce the number of variables you might have to consider),
   c. *"without loss of generality"* arguments, where the whole problem is reduced to a special case.
3. *Reformulate the problem in the most convenient way possible.* Now that you have a solid grasp of what the problem asks of you, you want to express the problem in a concise and convenient mathematical form that can be comfortably manipulated.
   a. *Choose the perspective* you will use. Asked to prove that two lines have the same length, you might choose to use Euclidean geometry (show they are parts of congruent triangles), trigonometry (express them in terms of known quantities like sides and angles), or perhaps a distance formula.
   b. *Choose your focus.* Decide what the most important quantities or systems are.
   c. *Rewrite the problems.* Use a notation which is consistent with the perspective you have chosen. Make sure important quantities are clearly labeled. Use *diagrams* if possible.

Implementation needs little comment, save that *it should (usually) be the last step in the actual problem solution. Verification* (checking), on the other hand, should be stressed. Students rarely check over their solutions, and that can be *very* costly. At a local level, you can catch silly mistakes. At a global level, by reviewing the solution process you can often find alternative solutions, discover connections to other subject matter, and, on occasion, become consciously aware of useful aspects of the problem solution that you can use elsewhere and that can help you become a better problem solver.

* * *

HOW THE STRATEGY WAS USED

More detailed information on the strategy was offered during the term. When we discussed what to do when first working a problem, for example, the students received a handout on analysis (Table 4.6). Of course, the students also worked dozens of problems, and we discussed the role of analysis in the solutions of those problems. Similarly, exploration was expanded as in Table 4.7.

Classroom discussions were not nearly as rigidly structured as the flowchart of the discussion or "stages" in exploration might suggest. The purpose of the flowchart was to serve as a guide when no other resources were

Table 4.7
The Exploration Stage of the Process

Preliminaries: Have you seen it before? If you have solved a similar problem, consider using the same method. If you know an analogous problem, can you infer the result of this one by analogy? Can you adapt the technique used to solve that problem? *Warning:* You are looking for quick results here. If an idea looks promising, pursue it. If not, hold off.

Phase 1: The Problem and Equivalents

1. What to look at: equivalent problems.
   a. Replace givens or goals with equivalent conditions (e.g., "closed" with "complement of open" or "contains all its limit points," "parallelogram" with "opposite sides equal and parallel," etc.).
   b. Try to reformulate the problem, using
      (1) a more convenient notation or a different perspective.
      (2) a logically equivalent form (e.g., argument by contrapositive).
   c. Reorganize the problem by
      (1) arranging things in a different way (e.g., infinite series).
      (2) introducing something new (lines in a diagram, e.g.).
2. What to try.
   a. Consider standard procedures first:
      (1) "break up" the problem by
         (a) trying to establish subgoals.
         (b) decomposing and then recombining the domain (e.g., vector analysis).
      (2) Eliminate alternatives, by a systematic reduction of the search space (anything from classic "trichotomy" arguments to systematic trial and error).
      (3) "build up" a solution by
         (a) induction (if an integer parameter is anywhere in sight).
         (b) synthesis techniques (e.g., integration).
   b. Ask general questions abut the givens and goals:
      (1) Do problems with the same or similar goals suggest appropriate subgoals, possible procedures, or auxiliary elements you might introduce?
      (2) What can you normally get from the givens? What is usually done with information like this? Can that help you?
      (3) Do you know anything related to both the givens *and* goals? Can it serve as a bridge between the two?

Phase 2: Slightly Modified Problems

1. Try to solve an *easier* related problem, either by
   a. adding a condition or piece of information to the givens, or
   b. removing, or trying to partly fulfill, a condition in the goals. You may get a family of partial solutions, from which you can determine the particular one in which you are interested.
2. Try to solve a *harder* related problem, either by
   a. weakening or removing a restriction in the givens, or
   b. generalizing, and trying to prove more (!) than the problem calls for. In (a) you may discover the role that each condition in your argument is supposed to play. ("Do I really need compactness? What if I just assume that it is closed?") In (b), we have what Pólya called the "inventor's paradox"; in trying to prove more, you may see things more clearly because they are not obscured by details (e.g., rather than showing that something converges, figure out what it converges to).

Phase 3: Desperation Attempts (Examining Any Related Problems for Inspiration)

1. Can you think of any problem with *similar* givens or goals? What techniques were used? Do any of them suggest plausible approaches here?
2. Can you obtain the result you are interested in for some subclass of the class you are interested in? If you want to show that something is true for all real numbers, can you prove it for all integers? all rationals?
3. Can you reverse some conditions? Try to prove the converse, and see what happens. If you are to show that X and Y imply Z, do X and Z imply Y, and so on?

(*Phase 4: Dealing with Failure. . . .*)

available. Discussions in class elaborated on its spirit more than on its details. When the class (as a group) worked a difficult problem, I would ask "What approaches or techniques might we try?" If a student made a suggestion that was clearly inappropriate, the class might explore it for a while until it was clear that the suggestion made little sense; at that point it would be pointed out that it helps to understand a problem before one jumps into its solution. The class was encouraged to generate a number of plausible ideas before committing itself to any particular one. When a few ideas had been generated, the class discussed which to pursue, and why. Does Approach A look easier than Approach B? How far will A get us? What about B? If we are successful with A, how shall we proceed? And the same for B. Classroom discussions focused explicitly on making informed judgments while solving problems: "We've been doing this for 5 minutes. Is it working? Should we try something else? If so, is there something in this we can salvage?" When there was good reason to do something—whether it conformed to the control strategy or not—we did so. In the absence of such good reason, the control strategy was used to suggest what might be tried next.

HOW WELL THE PRESCRIPTIVE INSTRUCTION WORKED

The early versions of the problem-solving course, at Berkeley in 1976 and at Hamilton College in 1978 and 1979, adhered closely in spirit to the brief description of the strategy given above. In terms of the narrow prescriptive goals described for the strategy—to enable students, at the end of the course, to solve a broad range of problems not necessarily related to those studied in the course—the instruction was rather successful. The results of the instruction are discussed at substantial length in Part II of this book and are only stretched briefly here.

A series of paper-and-pencil tests was designed to measure students' mastery of heuristic processes. The tests indicated that students can indeed learn to employ a variety of such problem-solving techniques. As one would expect, the students' performance on test problems closely related to those studied in the course showed significant improvement. Their performance also improved on a set of "transfer" problems that did not closely resemble those we had worked in class but that were solvable by the same techniques. The most impressive result is that the students did quite well on a set of problems that had been placed on their exam precisely because I did not know how to solve them! Details are given in Chapter 7. The students' performance was examined in other ways as well. A collateral study of problem perception indicated that, prior to the course, students focused on irrelevant surface details in problem statements; after the course, they tended to focus more on the underlying "deep structure" of the problems. Details are given in Chapter 8. Finally, there were some clear differences in students' control processes as they went about solving problems. Protocol analyses indicate that the students spent more time after the course analyzing problems than they had before and that they were more systematic in both selecting and in abandoning the techniques that they used. In general, their behavior more closely approximated the idealized control strategy given in Table 4.1. Details are given in Chapter 9.

Despite all of these positive signs, however, subsequent work indicated that teaching a prescriptive control strategy dealt with only a small part of the role of control in problem solving. Some of the larger issues are discussed in the next section.

Toward a Broader View of Control

OVERVIEW

The question that framed the prescriptive approach to heuristic problem solving described in the previous section can be phrased briefly as follows: What does it take to get students to learn to use a variety of heuristic strategies in nontrivial mathematical problem-solving situations? The answer, hypothetically, was twofold. First, heuristic strategies had to be delineated in much greater detail than they had been. Second, students needed an efficient control strategy in order to be able to use their heuristic resources. The data that are given in Part II of this book tend to strongly substantiate these hypotheses. With the help of the "executive strategy" described above, the students' problem-solving performance showed significant improvement. Control was demonstrated to be an important component of a prescriptive problem-solving strategy.

Here we turn to a much broader view of control. Control in a positive sense, and the role it plays in promoting problem-solving success, are important. But equally important, inefficient control behavior can cause individuals to squander the problem-solving resources potentially at their disposal and thus fail to solve problems that are easily within their grasp. The issue is not just the use of one's heuristic knowledge; it is how all of one's mathematical knowledge is called into play.

As motivation for the discussion in this section, let us briefly recall the attempts by two pairs of students, described in Chapter 1, to solve Problem 1.3.

Problem 1.3 Three points are chosen on the circumference of a circle of radius R, and the triangle containing them is drawn. What choice of points results in the triangle with the largest possible area? Justify your answer as well as you can.

The students KW and AM had both recently completed a multivariate calculus course that provides a standard solution to Problem 1.3. They had done well on the final exam and could have solved it easily. Instead they focused on computing the area of their conjectured solution (the equilateral triangle) and devoted all of their solution time to it. This one wild goose chase guaranteed their failure. The students BM and DK also had adequate backgrounds for dealing with the problem. But a series of bad planning decisions and the decision to discard an almost-successful approach that could have been salvaged guaranteed that they would fail to solve the problem. In both cases, bad control prevented the students from using the knowledge that they possessed.

The purpose of this section is to explore the range of ways that control can effect the evolution of a problem solution. Broadly stated, the issue of interest is the following:

How do a problem solver's decisions at the control level affect the ways that that person's knowledge—resources, heuristics, or anything else that might be brought to bear on a problem—is used? In what ways do such decisions inhibit or promote problem-solving success?

We explore the range of effects that control can have on a problem solution by looking at four attempts to solve Problem 1.2, which is reproduced below. The types of behavior described in the solution attempts are outlined in Table 4.8.

Problem 1.2 You are given a fixed triangle T with base B. Show that it is always possible to construct, with straightedge and compass, a straight line that is parallel to B and that divides T into two parts of equal area. Can you similarly divide T into five parts of equal area?

Problem 1.2 is of some interest because it consistently causes people more difficulty than it "should." There are variations, but the essential ingredients to a solution of the first part are these. Suppose that in Figure 4.8, one draws in the desired line (call it b), as in Figure 4.3. Then the triangle t that comprises the "top half" of T must be similar to T. By similarity, $a/A = b/B$. Since the area of t [$=(ab/2)(\sin X)$] is half the area of T [$=(AB/2)(\sin X)$], we have $2ab = AB$. It follows that $a = A\sqrt{2}/2$; similarly $b = B\sqrt{2}/2$ and $c = C\sqrt{2}/2$. A way of performing the construction that determines the point P is suggested by Figure 4.4. Finally, b can be con-

Table 4.8
The Effects of Different Types of Control Decisions on Problem-Solving Success: A Spectrum of Impact

Type A.	Bad decisions guarantee failure: Wild goose chases waste resources, and potentially useful directions are ignored.
Type B.	Executive behavior is neutral: Wild goose chases are curtailed before they cause disasters, but resources are not exploited as they might be.
Type C.	Control decisions are a positive force in a solution: Resources are chosen carefully and exploited or abandoned appropriately as a result of careful monitoring.
Type D.	There is (virtually) no need for control behavior: The appropriate facts and procedures for problem solution are accessed in long-term memory (LTM).

structed in one of three ways: (1) by locating P as suggested in Figure 4.4 and constructing a line through P that is parallel to base B, (2) by locating the point Q with a construction similar to that for P and then drawing PQ, or (3) by drawing the altitude H of T, and drawing the perpendicular to H through the point at distance $H\sqrt{2}/2$ down from the top vertex.

That, at least, is the mathematics of the problem. Solution attempts of types A, B, C, and D, respectively (recall Table 4.8), are given below.

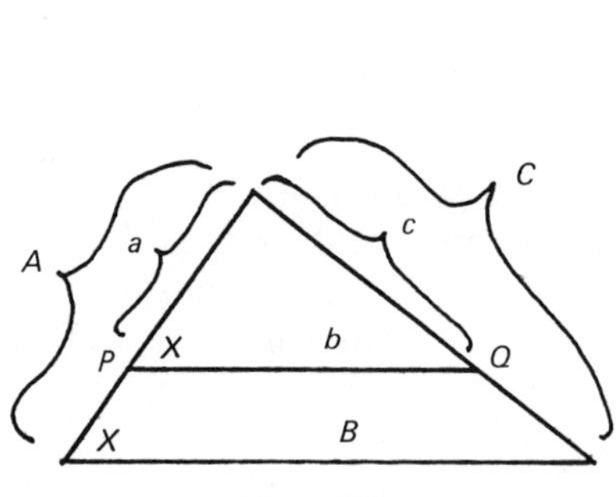

Figure 4.3

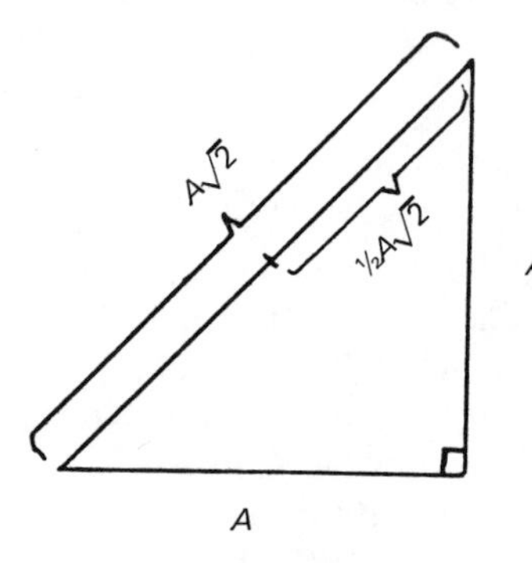

Figure 4.4

TYPE A EXAMPLE: BAD CONTROL CONTRIBUTES TO FAILURE

The students AM and JM were college freshmen who had each finished a semester of calculus. The session began as follows.

AM: [Taking the sheet and reading the problem] I remember doing this; this is grade seven.
JM: [Making this statement before he reads the problem] The problem is you're not supposed to solve it that way. It's no fun to solve it with those really technical theorems. He doesn't expect us to know those. He obviously has a different way to prove this. . . .
AM: Put the compass down on a line. And you bisect this line. Then from there draw a distance across which is as long as that so that it intersects it up there. You know that one point there, and you do it the same for the other side. And then you draw the line across, and this is equal to that. [More comprehensibly, AM conjectured that the line segment joining the midpoints of the two sides of T will divide T into two parts of equal area.]

The students reached for a straightedge and compass and set about performing the construction. This took a few minutes, and the results looked bad: The top triangle was too small. To make certain of the construction, they checked their bisections with the ruler. It confirmed the accuracy of the bisections so the conjecture was rejected. With this failure, JM looked for an elegant solution. He focused on halving the triangle, and looked for lines—any lines—that bisected its area. He conjectured that the median from one vertex of the base to the side opposite would split the triangle in half and worried about that for a while. A few minutes passed. Out of ideas, they reread the problem.

JM: Let's just underline the important parts of the problem. OK, we have to construct a straight line parallel to the base.
AM: Do you remember how to do that?

With this comment they focused on the parallel construction, which neither of them recalled. A few minutes passed, and they realized that they could perform the construction using two right angles. Then they realized that they did not know where to do the construction. This led them to discuss halving the triangle once again, and they were examining the median for the second time when time ran out.

Discussion

Since this solution attempt quite closely resembles the two attempts to solve Problem 1.3 that were discussed in Chapter 1, this discussion will be very brief. At the beginning of the attempt, JM revealed his expectation that there would be a slick solution to the problem. This is a matter of belief, but it also affects control; the search for a clever solution deflected him from more useful pursuits. In the few seconds it took JM to read the problem, AM made his (incorrect) conjecture. Without any discussion JM concurred, and they set off to perform the construction. This is a direct parallel to KW and AM's calculation of the equilateral triangle's area in Problem 1.3. Here as in that problem session, there was an impetuous leap into a solution without examination of its merit. Here AM and JM lost only 5 minutes, but much more time could have been lost if the construction were more complex. Following the parallel construction, there were long excursions in directions that, with a little reflection, could have been seen to be of no value in solving the given problem. Much time was squandered, and in general the solution meandered without direction. The general absence of control, along with a few bad control decisions, contributed directly to these students' failure.*

TYPE B EXAMPLE: CONTROL AVOIDS DISASTER, BUT IT DOES NOT REALLY HELP

The students ED and FM were undergraduate mathematics majors within a semester of graduation. Student FM was a solid student who was graduating with honors. Student ED was a mediocre student, despite having a respectable record. FM read the problem out loud.

ED: Well, I was gonna make a guess that the line would bisect the other two sides of the triangle.

FM: Divide it into equal area. OK, but it's drawn kind of weird [the base of the triangle was not quite horizontal in the sketch]. Let's look at the triangle, see if that makes sense. [He draws Figure 4.5.] I don't think so. That doesn't seem to divide it into equal areas. It isn't that simple. Show it's possible to construct with straightedge and compass.

* One might hypothesize that the meandering was a result of JM and AM's grasping at straws. With no idea of how to solve the problem, they may have been trying whatever they thought of in the hope they may have gotten lucky. In fact, subsequent discussions indicated that they knew more than enough to solve the problem; they simply failed to use that knowledge. This theme will be dealt with at length in the next chapter.

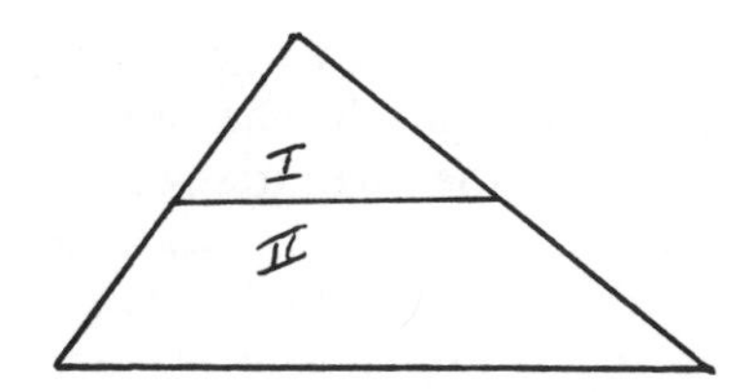

Figure 4.5 Regions I and II are not equal in area.

ED: Straightedge and compass. Then we should look for circles. The most obvious one I can think of is the circle determined by the two vertices.

FM: Right, of the triangle.

ED: That only tells us that B is a chord. It doesn't tell us the diameter or anything; that's not too exciting. It doesn't seem to help us much. So let's pick a vertex. We could draw circles with the radius being a side. Or it's possible to draw one with the diameter being a side. So let's think of those, with the radius or diameter [draws Figure 4.6].

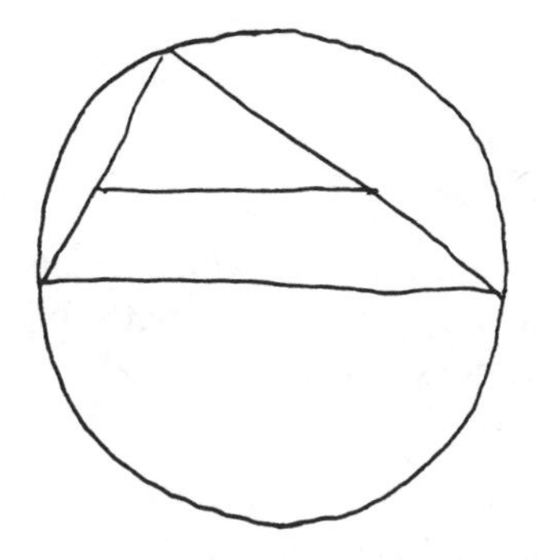

Figure 4.6

FM: I think this is a bad way to go about it. I think what we should do is try to figure out where the line is going to be and then construct it.

What ED suggests makes little sense, but she is gearing up for a major effort. If her sugestion is pursued, this would cost the two students a great deal of time and quite possibly guarantee their failure. However, FM's executive action curtails ED's impulsive speculations. As a result the two students at least have a chance to find a reasonable approach to the problem.

For some reason they continue to work on ED's sketch. They draw in the altitude h of triangle T (Figure 4.7a), noting that T's area is $\frac{1}{2}hB$ and that the area of the little triangle would thus be $\frac{1}{4}hB$.

FM: So let's draw a line in. I'm not sure where it is. We'll call it a line Base C. We've got another little height where this new one is. We'll call it j. OK; j is the height of the small triangle, and h is the base.

ED: That's a small triangle with its base drawn parallel to line B (Figure 4.7b).

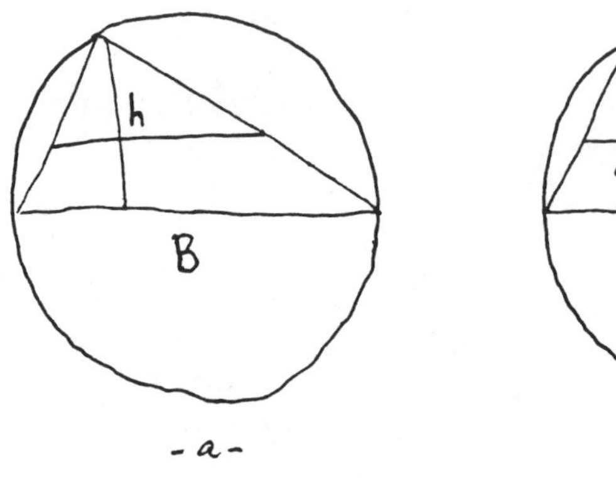

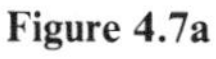

Figure 4.7a

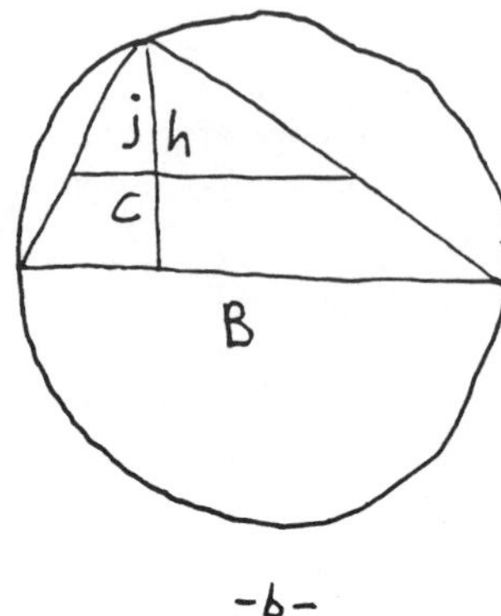

Figure 4.7b

They compare the two areas.

FM: So this new area is equal to $\frac{1}{2}jC$, which equals $\frac{1}{4}hB$. . . . That means $2jC = hB$. Now, we've got . . . That's full of variables, that's kind of strange. That doesn't look too good. We should look for other relationships such as "j is to h as . . . "

Only at this point do the two students realize that the two triangles are similar, and they use that fact to set up the relevant proportions. Solving for B in terms of C, they discovered that $B = C\sqrt{2}$. With "that's interesting; that should tell us something" from FM, they returned to the equations and discovered that $j = h\sqrt{2}$.

ED: We know where the altitude and the new line intersect, so what does that tell us about the sides, now? We should be able to figure out something about the sides, shouldn't we?

They begin to set up some equations, modifying Figure 4.7b to include new letters to represent the sides of the triangles. In the midst of this, FM realizes that since the two triangles are similar,

FM: Everything's proportional, so the sides have to be in the . . . from the vertex opposite B to the intersection is $\sqrt{2}/2$ times the original length. Whatever it is. OK. Remember $\sqrt{2}/2$ is the cosine, the sine

> of a 45-degree angle, so somehow we're gonna get something involving a 45-degree angle. . . .

There are then some rather odd meanderings from which the students do not recover. The audiocassette recording then clicks off as the half-hour point of the problem-solving session has passed.

Discussion

The full transcript of this problem-solving session makes for more painful reading than the excerpts presented here, because there was a good deal of meandering and backtracking during the solution attempt. Control was certainly not a positive force in the solution, in that the students did not actively take advantage of what they knew. Rather, they seemed (as often as not accidentally) to run into things that might be useful and then to follow them up in some way. They almost tripped over similarity, for example, after realizing that comparing the areas of the two triangles gave them only one equation in four variables.

The significant difference between this solution attempt and the type A example discussed above is that some reasonable control decisions guaranteed that FM and ED had the opportunity to make the discoveries that they did. The intervention by FM kept ED from pursing the irrelevant constructions with circles that she had suggested and allowed the students enough time for pursuing other options.

It is important to discuss the nature of FM's intervention at the point when he interrupted ED's playing with circles. If FM had his own clear ideas about a solution at that point, his intervention might not have been a true control decision; rather, he might simply have been imposing his problem solution on ED. That was not the case, however. When he put a stop to ED's wild goose chase, FM had no idea of what further calculations might lead to; he had not yet discovered that the triangles in his figure were similar. He was not, therefore, pushing the solution in a direction he knew would be useful. Rather, he was trying to avoid going off in directions that he strongly suspected would be a waste of time and effort. This is a good example of how effective control decisions can prevent disasters. As their meandering indicates, however, control did not help to push their solution forward.

TYPE C EXAMPLE: CONTROL IS A POSITIVE FORCE IN A SOLUTION

Subject GP was a mathematician who, when he worked Problem 1.2, had not done plane geometry for many years. It is clear that he was rusty, for he did not remember the relevant relationships and had to derive almost all of them "on line." By some standards his solution would be considered clumsy and inelegant. (In a department meeting it was held up for ridicule by the rather smug colleague who produced the type D transcript given in the next section.) From my perspective, however, it provides an archetypal example of problem-solving skill—a tour de force of control.

We saw the beginning of GP's solution in the discussion of resources in Chapter 1 (p. 21). Taking the problem as solved, he drew in the desired line segment and then noticed that the two triangles in his diagram were similar. The transcript continues as follows.

Yeah, all right, then I have an altitude for the big triangle and an altitude for the little triangle [he draws Figure 4.8] so I have $a/A = b/B$. So what I want

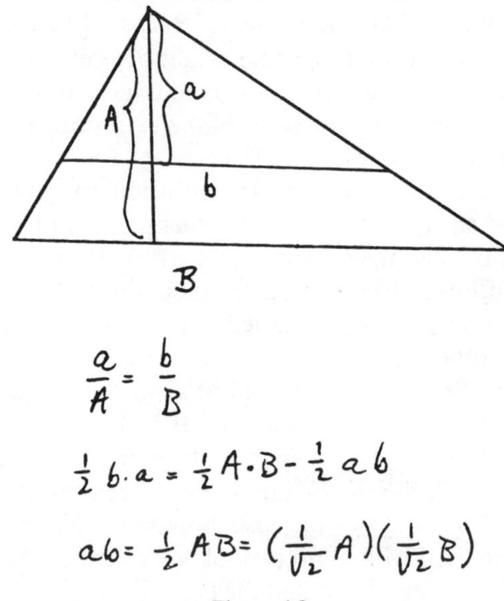

Figure 4.8

to have happen is $\frac{1}{2}ba = \frac{1}{2}AB - \frac{1}{2}ba$. Isn't that what I want? Right! In other words I want $ab = \frac{1}{2}AB$. Which is $\frac{1}{4}$ of A times . . . [confusedly mumbles]; $A/\sqrt{2}$ times $B/\sqrt{2}$. So if I can construct $\sqrt{2}$, which I can! then I should be able to draw this line . . . through a point which intersects an altitude dropped from the vertex. That's $a = A/\sqrt{2}$, or $A = a\sqrt{2}$, either way.

And I think I can do things like that because if I remember I take these 45° angle things and I go 1, 1, $\sqrt{2}$ [he draws Figure 4.9]. And if I want to have a $\sqrt{2}$ . . . then I do that . . . mmm, wait a minute. . . . I can try to figure out how to construct $1/\sqrt{2}$.

OK. So I've just got to remember how to make this construction. So I want to draw this line through this point, and I want this animal to

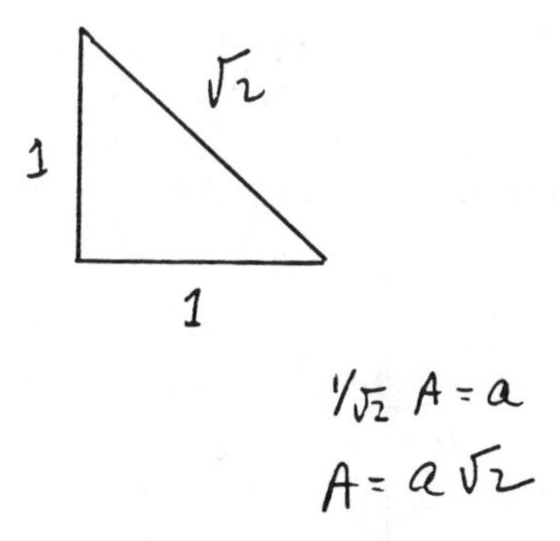

Figure 4.9

be . . . $1/\sqrt{2}$ times A. I know what A is—that's given. So all I've got to do is figure out how to multiply $1/\sqrt{2}$ times it.

Let me think of it. Ah hah! Ah hah! Ah hah! $1/\sqrt{2}$ . . . let me see here . . . umm . . . that's $\frac{1}{2} + \frac{1}{2} = 1$ . . . So of course if I have a hypotenuse of 1 . . . [he draws Figure 4.10]. Wait a minute: $1/\sqrt{2} \times \sqrt{2}/\sqrt{2} = \sqrt{2}/2$. . . . That's dumb!

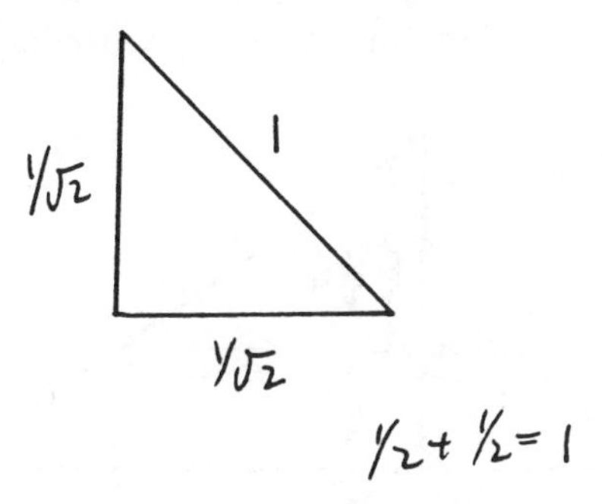

Figure 4.10

Yeah, so I construct 2 from a 45, 45, 90. OK, so that's an easier way. Right? I bisect it. That gives me $\sqrt{2}/2$ [he draws Figure 4.11]. I multiply it

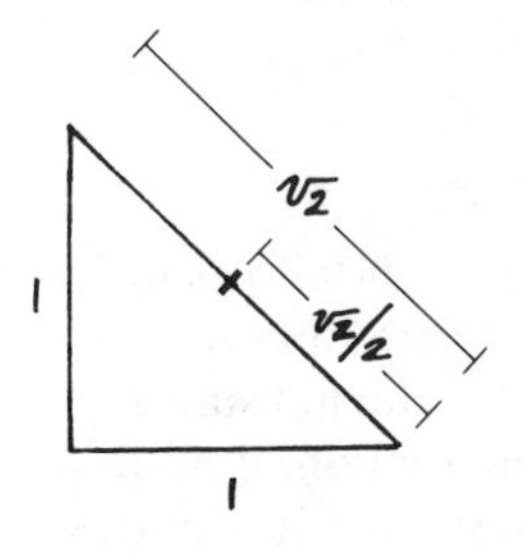

Figure 4.11

by A. Now how did I used to do that? Oh heavens! How did we used to multiply times A. That . . . the best way to do that is to construct A . . . A . . . then we get $A\sqrt{2}$ and we just bisect that and we get $A\sqrt{2}/2$. OK. [He draws Figure 4.12.] That will be . . . what! mmm . . . that

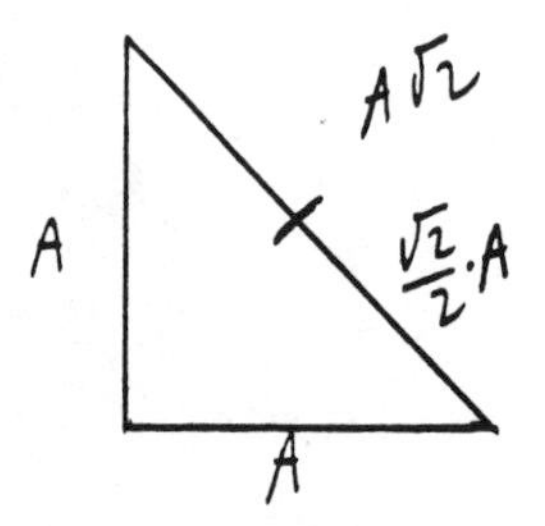

Figure 4.12

will be the length . . . now I drop a perpendicular from here to here. OK, and that will be . . . ta daa . . . little a.

So I will mark off a as being $A\sqrt{2}/2$. OK, and automatically when I draw a line through that point . . . I'd better get B times $\sqrt{2}/2$. OK. [He draws Figure 4.13.] And when I multiply those guys together I get 2 over 4 times A

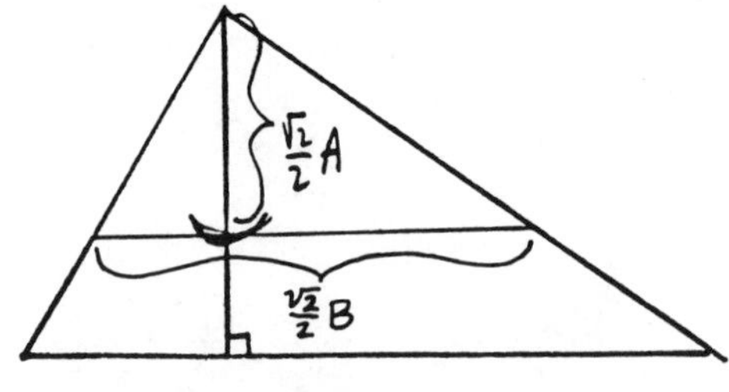

Figure 4.13

times B. So I get half the area . . . what? yeah . . . times half . . . so I get exactly half the area in the top triangle so I'd better have half the area left in the bottom one. OK.

Discussion

Observe that GP's solution follows the plan outlined by FM in the previous solution: "try to figure out where the line is going to be, and then construct it." Though GP did not state this plan out loud (he was asked not to "explain" for the microphone), he followed this tacit plan closely, in strongly goal-directed fashion. It is quite clear from the transcript that he did not know specifically what he was looking for as he worked through the problem; he expressed surprise and pleasure as he made discoveries and expressed chagrin when he realized that he had been clumsy and could have done things more easily. Subject GP's goal-directed behavior paid off; he kept himself on track and did not meander in the ways that ED and FM did in the previous attempt. Note also that he was constantly evaluating the status of his attempt. There would be no wild goose chases, because no more than a minute passed between "status checks," or reviews of how things were going. GP also checked his solution carefully. He was in firm control of this part of the solution from beginning to end.

The second part of Problem 1.2 (to divide the triangle into five equal parts) is more complex, and there GP ran into more difficulty. He was not sure whether the problem could be done and was more tentative in his approach. In working this part of the problem, GP generated enough potential wild goose chases to distract an army of problem solvers, much less just one. Yet he avoided those pitfalls. He did so by monitoring things closely as they evolved, even to the point of assessing ideas as he generated them (e.g., "[I'm] trying to remember my algebra to knock this off with a sledgehammer"). A full transcript of GP's problem session is given and analyzed in Chapter 9. However, a brief summary of his control behavior and its impact is as follows: GP evaluated new ideas as they occurred to him, either (1) acting on them, (2) postponing them temporarily, or (3) discarding them, depending on his perception of their promise. He evaluated the paths he took as the solution proceeded, revising his assessments of their promise in the light of what they "delivered." Potential wild goose chases were rapidly curtailed with the result that GP had ample time for positive developments to take place. But equally importantly, GP actively looked around for interesting and relevant ideas as he worked the problem. Once such ideas were found, they were explored and exploited if they seemed useful. His use of control did far more than guarantee that he did not get lost; it ensured that he made the most of the resources at his disposal. In this sense, control was a very positive force.

TYPE D EXAMPLE: NO NEED FOR CONTROL

What follows is the complete transcript of the mathematician JA's work on the first part of Problem 1.2. His solution of the second part went as smoothly.

The first thought is that the two triangles for the first question will be similar.

And since we'll want the area to be one half, and area is related to the product of the altitude and the base, we want the area of the smaller triangle to be one half. . . .

And corresponding parts of similar triangles are proportional. We want the ratio of proportionality between the altitudes and the bases both to be $1/\sqrt{2}$.

So I will draw a diagram, . . . and I'm drawing that parallel, and checking the algebra. [He draws Figure 4.14.] I hope you can hear the pencil

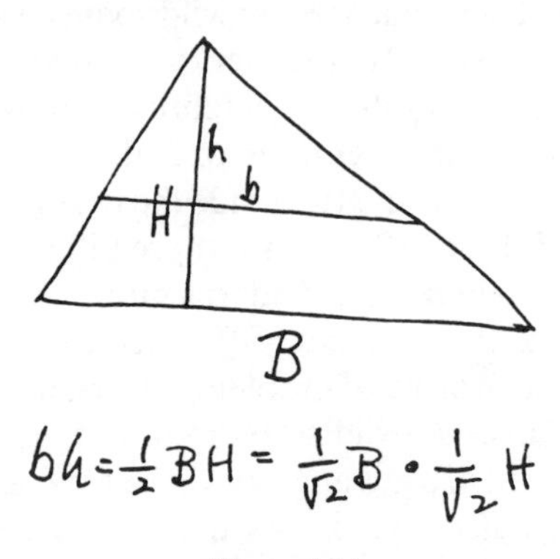

Figure 4.14

moving, because that's what's happening at this point.

And now I'm multiplying out a bunch of letters on my diagram and multiplying them together . . . leaving the "$\frac{1}{2}$" out, of course, . . . and I want that [points to *bh*] equal to half of that [points to *BH*].

So that certainly seems like a reasonable solution. So all I have to be able to do is construct $\sqrt{2}$. And I can do that with a 45° right triangle, and then given a certain length, namely the altitude to base *B*, which I can find by dropping a perpendicular, I want to construct a length which is $1/\sqrt{2}$ times that, and I can do that with the ordinary construction for multiplication of numbers. So, I can do the problem.

Asked about the details of the constructions, he explained that he does all of them in a geometry course he offers annually; they can be taken for granted.

Discussion

In this protocol we see an archetypal example of what many researchers in AI and in IP psychology would define as expertise: JA's access to the right approach was nearly automatic, and the solution proceeded smoothly from beginning to end. One can hardly disagree that this is an expert solution. My point here is that the automatic access of the correct solution paths renders control superfluous, as the term *control* is used in this book. If you know precisely what you are doing, there is no need to worry about being efficient. The reason for stressing this point is that there is a tendency in some of the literature to define expertise exclusively as the type of behavior demonstrated in this protocol, and to brand as "nonexpert" anything but such routinized competent performance. When you do so, you overlook the importance of Type C behavior, and the issue of control as discussed here is defined out of existence. Problem 1.2 was more of an exercise than a problem for JA; in his transcript we see a display of proficiency. In GP's solution we see true problem-solving skill. By effectively marshalling the resources that were at his disposal, he managed to solve a problem that many individuals who began their work on the problem with a larger collection of resources failed to solve.

Literature Related to Control

There is a large body of literature that deals with issues related to control, although there is very little in that literature that deals directly with issues of resource allocation during problem solving as those issues were described in the previous section. Most of the literature on control comes from AI and its psychological companions, IP and cognitive psychology. The term *control* is used in various ways in the AI and IP literatures. It is a central element in the structuring of AI programs and has a highly technical meaning, related to search, in that context. It is also used to describe executive decision-making and planning, generally in a prescriptive sense. Tied in with the AI and IP literature, developers of "intelligent tutoring systems" (computer-based tutorial devices) have grappled with issues of control in trying to develop competence in a variety of domains. A broader version of the issue is dealt with under the heading "metacognition" in the psychological literature, and there are some recent studies of metacognition in mathematical problem solving. Finally, there is a sparse but suggestive literature on the role of social interactions in the development of internal control strategies. These topics are covered briefly, in the order listed here.

ISSUES OF SEARCH AND CONTROL IN ARTIFICIAL INTELLIGENCE

Once they have been appropriately represented and structured, many problems in AI become problems of search through certain kinds of graphs or trees. The nodes of the graphs represent various problem states, and the paths that join them represent procedures that carry the problem solver from one problem state to another. A classic example, adapted from Nilsson (1980), serves to illustrate the notion of search through a problem space.

The 8-puzzle consists of eight movable tiles, each marked with a different digit from 1 through 8, set in a 3×3 frame. One cell of the frame is empty,

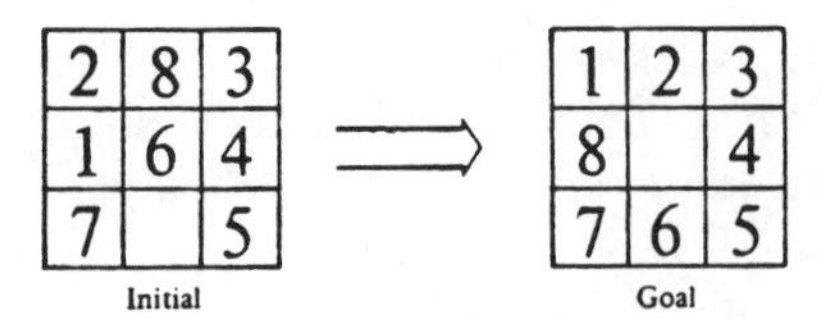

Figure 4.15

so that any of the tiles adjacent to the empty cell can be moved into it. The problem is to find a sequence of moves that will transform one configuration into another, for example, to specify the transformation suggested in Figure 4.15 (taken from Nilsson, 1980, p. 19, Fig. 1). Note that the initial configuration in Figure 4.15 can be changed in any of three ways: The tiles containing the digits 5, 6, and 7 can be moved into the blank space. Graphically, these transformations can be represented as in Figure 4.16.

Figure 4.16 represents the first level of the search tree for the 8-puzzle. Each of the positions in the second row can be transformed by moving a tile into the blank space. The positions that result can be similarly transformed, and so on. A small part of the problem space for the task posed in Figure 4.15 is given in Figure 4.17 (taken from Nilsson, 1980, p. 28, Fig. 1.4). This kind of configuration is called a search tree. Once a problem domain has been represented in this way, the task of finding a sequence of transformations that converts the initial state into a goal state is a search problem. One must find a path that leads from "start" to the goal state (or possibly one of many such states, where there are multiple solutions). The darkened line in Figure 4.17 represents a successful search, demonstrating one solution to the given problem.

The 8-puzzle illustrates some aspects of problem representation in AI. In

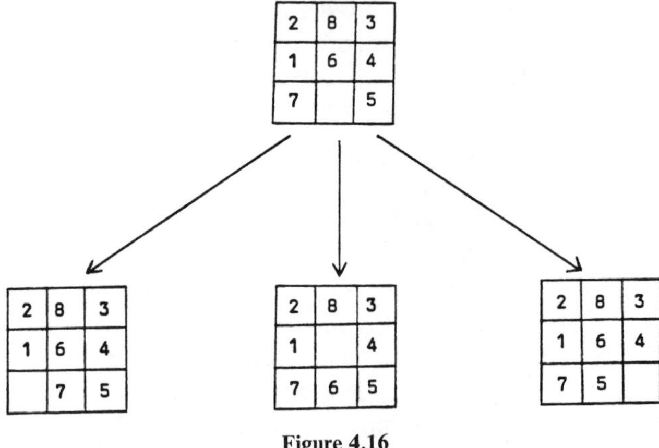

Figure 4.16

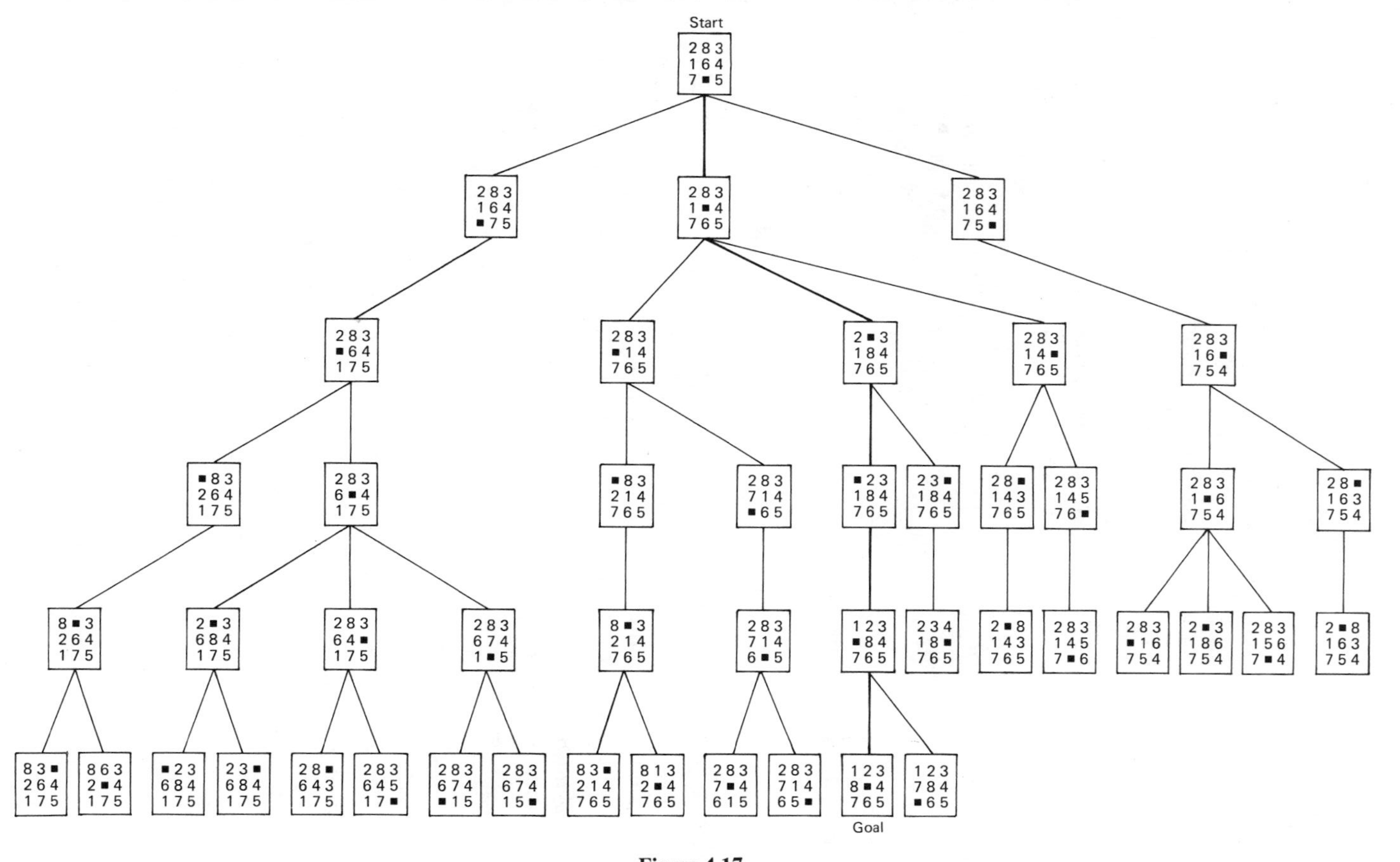

Figure 4.17

the problem one has an initial state (or to use Nilsson's term, an "initial database"), a collection of rules for transforming states (production rules for transforming the database), and a set of termination conditions that allow one to determine when the goal state has been attained. Many problems can be structured this way, using the following schematic representation:

Problem
IS: initial state;
TR: transformation rules;
TC: termination conditions.

Consider, for example, Problem A, a typical integration problem in freshman calculus (see Slagle, 1963); Problem B, a proof problem in symbolic logic (see Newell & Simon, 1972); and Problem C, a construction problem in Euclidean geometry. These might be represented in the following way:

Problem A
IS: an indefinite integral, say

$$\int \frac{x^4\,dx}{(1-x^2)^{5/2}};$$

TR: a specified set of the techniques of algebra and calculus for replacing algebraic and integral expressions with equivalent forms (the new state is the result of the manipulation);
TC: an algebraic expression that does not contain an integral sign.

Problem B
IS: the expressions

$$Q \longrightarrow R, \qquad P \vee Q, \qquad \sim(R \wedge S) \qquad S;$$

TR: the rules of symbolic logic for transforming symbolic expressions (the database is augmented by the result of any such transformation);
TC: any database that includes the expression P.

Problem C
IS: Figure 4.18 and a well-specified knowledge base of facts and procedures related to Euclidean geometry;

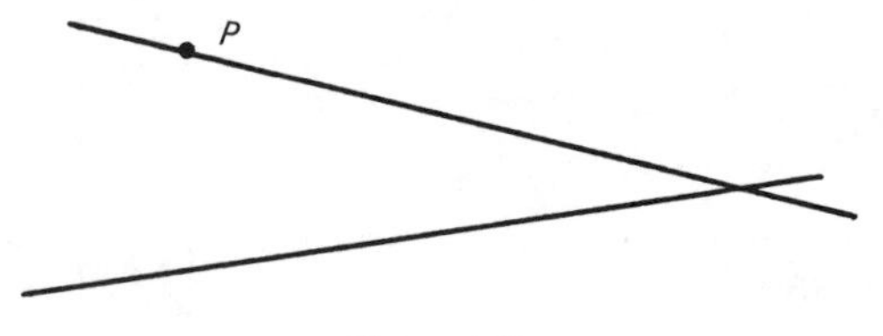

Figure 4.18

TR: rules for augmenting the knowledge base (including modifications to figures) that are consistent with the deductive tenets of Euclidean geometry;
TC: a figure that contains Figure 4.19 as a subfigure.

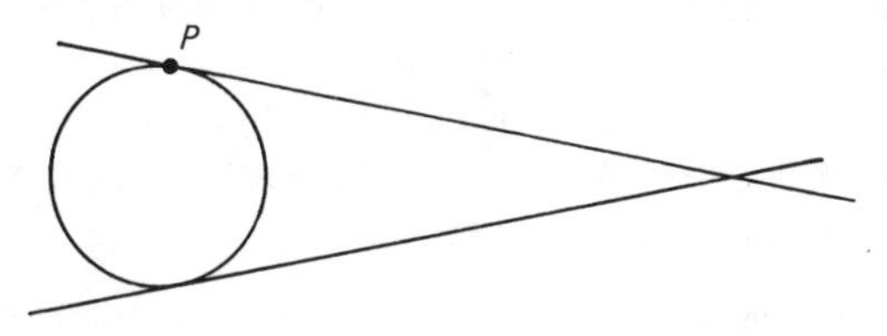

Figure 4.19

These examples indicate that with the proper formalization the problem domains for indefinite integrals, for symbolic logic, and for our very own Problem 1.1 can be converted to search spaces. However, navigating through a search space is no easy task. In the 8-puzzle there are $9! = 362,880$ different configurations of tiles in the 3×3 frame. These are just the different states in the problem space. The number of paths through the problem space is much larger. The 8-puzzle is a simplified version of the 15-puzzle, in which the numbers 1–15 appear on movable tiles in a 4×4 array. In the 15-puzzle there are $16! = 20,922,789,888,000$ possible problem states, of which precisely half can be obtained from any starting configuration. The number of paths through this search space is incredibly large. Yet this problem space is minuscule when compared to those for Problems A, B, and C above! The primary difficulty in AI search, then, is to control the search through the problem space in a way that is efficient and that allows one to make progress toward (and ultimately find) the goal. There is a large AI literature on search and control because—after the huge hurdle of finding an appropriate representation for a problem is overcome—these become the central issues in AI problem solving.*

Researchers in AI and IP psychology draw strong parallels between the two domains, and ideas or techniques useful for machine intelligence are

* Good introductions to this area may be found in Gardner (1981), Newell and Simon (1972), Nilsson (1980), and Winston (1977). The reader should be prepared for some disparity between the general description of control issues and the specifics of their implementation when dealing with machine-executed search. For example, Nilsson defined control knowledge as follows: "The knowledge about a problem that is represented by the control strategy is often called the *control knowledge.* Control knowledge includes knowledge about a variety of processes, strategies, and structures used to coordinate the entire problem solving process" (page 48). Yet the listing for "control strategy" in the index begins with the entries "backtracking, for decomposable systems, for game-playing systems, graph-search, irrevocable, of a production system, for resolution refutations, for rule based deduction systems. . . ."

generally considered likely to be useful for human intelligence (and vice versa). Thus we find Newell and Simon's 1972 book, devoted to AI models of cognition, entitled *Human Problem Solving.* The fundamental problem-solving strategy for the general problem solver (GPS), means-ends analysis, was derived and refined from observations of humans performing specific kinds of tasks (cryptarithmetic problems, symbolic logic, chess). When GPS was successful, this success was used to support the claim that means-ends analysis is an effective human problem-solving strategy. Recall the following quotation from Simon (1980):

> The lessons for pedagogy that can be extracted from my remarks are clear. . . . there is a small arsenal of general problem-solving procedures that have been identified from research in psychology and artificial intelligence. In this chapter I have particularly stressed one of them — means-ends analysis. (p. 93)

Such strategies do play an important role in certain kinds of human problem solving, although one should be very careful to examine the circumstances in which the strategies are useful. Means-ends analysis is useful for particular types of well-defined search problems in AI. Thus in comparable human problem-solving endeavors, in which the problem solvers (1) are working within the context of a clearly defined search space, (2) are clearly aware of the tools at their disposal, and (3) have a clearly defined goal structure, one can expect such strategies to play an important role. (See, e.g., the "novice" solution to kinematics problems in Simon & Simon, 1978.) Contrast this nearly ideal situation, however, with the situations confronted by the students who produced the Type A and Type B solution attempts that were discussed in the previous section. These students had to define the problem space for themselves. Some of their resources were shaky and access to them was uncertain. The students had to choose an appropriate representation for the problem, to decide for themselves what constituted "legitimate" techniques, and to create their own goal structures. In these circumstances the connections between machine performance and human performance are fairly tenuous, and one must be careful in suggesting that techniques that are adequate in one will be adequate in the other.

CONTROL AND PLANNING

The term *control* is used in a broader sense by those researchers in AI who build planning programs or who model planning processes. Even so, control in this sense is often rather constrained. A typical AI planning task is to specify an efficient order for carrying out a sequence of actions, such as assembling a piece of equipment. In such a situation some of those actions may preclude the possibility of performing others. A classic example is given in a painting problem: "Paint the ladder and paint the ceiling" (Sacerdoti, 1977). Performing the former forestalls the latter, at least for quite a while.

Sacerdoti's Nets of Action Hierarchies (NOAH) may be the best-known planning program. The NOAH program implements a successive refinement approach to temporal planning. High-level goals are specified first. These are expanded into subgoals, which are further expanded, and so on. The NOAH program avoids the kinds of temporal conflicts mentioned above by leaving sequences of actions unspecified for as long as possible and checking for potential conflicts as decisions are made. In the example given above, a decision to paint either the ladder or the ceiling would not be made until it was forced. When it was discovered that using the ladder was essential for painting the ceiling, this conflict would dictate that the ceiling be painted first. Eventually this kind of top-down, hierarchical approach fleshes out a plan as a fully specified sequence of elementary actions. Most AI work in planning is simlarly top-down and hierarchical. (See, e.g., Ernst & Newell, 1969; Fahlman, 1974; Fikes, 1977; Sussman, 1973.) In another area where planning is important, writing computer programs, the structured programming approach has been similarly hierarchical (see Dijkstra, 1976; Dahl, Dijkstra, & Hoare, 1972).

One can hardly quarrel with the notion of a highly structured, orderly, hierarchical planning process. It may be that such processes are natural in some domains, for example, machine assembly or certain kinds of temporal planning. Successful AI programs in such domains certainly demonstrate that hierarchical planning processes do work. And in messy domains such as computer programming, the idealized representation of the design process as being top-down and hierarchical can serve as a useful goal state for novice programmers. One should keep in mind, however, that idealized behavior is precisely that. In some domains, expert behavior may not appear to be highly structured at all. And idealized behavior may or may not serve as a useful template for the examination of "real" behavior, especially when the real behavior is severely flawed.

One of the few models of the planning process that does not make strong hierarchical assumptions, and that seems to capture the partially structured nature of problem-solving processes in many domains, is the Hayes-Roths' (1979) "opportunistic" planning model. The kind of behavior modeled by the Hayes-Roths appears in the following example. An individual is to plan a day's activity. That person is given a schematic map of a city (hypothetically the city that he or she lives in) and a list of tasks that should, if possible, be accomplished that day: to pick up dog medicine at the vet, buy a fanbelt for a refrigerator, meet a friend for lunch, see a movie, order a book, buy fresh vegetables, order flowers for a friend in the hospital, and so on. It is clear that all the tasks on the list cannot be accomplished in one day. In consequence, the individual must select the most important tasks and plan the day's errands so that essential things get done, and a large number of reasonably important ones get taken care of. The assumption underlying the Hayes-Roths' model is described as follows.

> [P]eople's planning activity is largely *opportunistic.* That is, at each point in the process, the planner's current decisions and observations suggest various opportunities for plan development. The planner's subsequent decisions follow up on selected opportunities. Sometimes these decision processes follow an orderly path and produce a neat top-down expansion. . . . However, some decisions and observations might suggest less orderly opportunities for plan development. For example, a decision about how to conduct initial planned activities might illuminate certain constraints on the planning of later activities and cause the planner to refocus attention on that phase of the plan. Similarly, certain low-level refinements of a previous, abstract plan might suggest an alternative abstract plan to replace the original one. (p. 276)

In brief, the notion explored by the Hayes-Roths is that competent problem-solving behavior is not simply hierarchical and "plan driven"; it is often "event driven" as well. As an individual works out the details of some errands in a particular neighborhood, he or she may notice that the vet is near to a flower shop. This fortuitous observation may reveal that there is no need to go to a flower shop in another part of the city, as had been planned; in consequence, there will be a major revision. Or, in listing essential tasks and finding where they can be carried out, the planner may notice a cluster of shops in which a large number of minor tasks can be completed. It may be best to iron out the details of this neighborhood trip so that planning can proceed without further concern about those tasks or that block of time. This too is nonhierarchical. There are obvious parallels in mathematical problem solving. While carrying out some minor computations, for example, an individual may observe symmetry in the equations being manipulated. This may suggest that symmetry plays an important role in the original problem and may call for a revision of the entire approach. Or the individual may decide to get some detailed computations out of the way early in the planning process so that things do not get cluttered later on. As such, much competent mathematical behavior is opportunistic in the sense discussed by the Hayes-Roths. The key in both situations lies in monitoring the state of a solution as it evolves and taking appropriate action in the light of new information.

ISSUES OF CONTROL IN COMPUTER-BASED TUTORIAL SYSTEMS

The perspective taken in this book is that it is useful to think of *resources* and *control* as two qualitatively different, though deeply intertwined, aspects of mathematical behavior. This distinction raises delicate issues, for discussions of resources must include questions of access and attention that are, in a broad sense, issues of control. Some researchers AI (e.g., Simon) argue that the qualitative separation is not essential; that, ideally, production system models of cognition do not employ a separate control structure, but handle issues of control automatically as a result of the organization of the productions (condition–action sequences) in the models. This issue of automaticity, or of automatic versus controlled processes, has a long history.

> A two-process approach to thinking predates information processing models. A notably lucid description of the distinction between automatic and controlled processes was made by James (1890) who stressed the freedom from attention and effort that automatization provides: "The more details of our daily life we can hand over to the effortless custody of automatism, the more our higher powers of mind will be set free for their own proper work" (p. 122). (Brown, Bransford, Ferrara, & Campione, 1983, p. 111)

Many researchers in AI have found it useful to make this distincton, not just as a programming convenience but as a characterization of underlying cognitive processes. The distinction becomes particularly important when dealing with student behavior, where (as we saw in earlier parts of this chapter) automatic access to resources can hardly be taken for granted. The researchers who have had to deal with this issue most directly are those who work on the development of computer-based learning systems for sophisticated domains. Machine-based learning environments are the descendants of early attempts at computer assisted instruction (CAI), but they have evolved radically through the years. Intelligent tutoring systems, as they are now called, have been developed for teaching subject matter such as electronic troubleshooting, interpreting nuclear magnetic resonance spectra, medical diagnosis, coaching in informal gaming environments, and program–plan debugging. The core of the relevant literature on intelligent tutoring systems is covered in Sleeman and Brown (1982).

Designers of intelligent tutoring systems are often quite explicit about the separation of resources and control, and about the fact that different tutorial strategies may be appropriate for these different kinds of knowledge. Many tutorial systems focus on the student's plans or strategic decision-making, and interact at that level. For example, Miller's discussion of programming skills reflects the perspective of the structured programming movement.

> What is it that the proficient programmer knows, that the beginner does not? The neophyte lacks knowledge of style, of strategy, of how to organize work on a large project, of how programs evolve, and of how to track down bugs.
>
> This paper reports on an investigation of this knowledge, and describes a model which formalizes it as sets of rules. In accord with the above intuitions, only the lowest level rules in the model deal with the constructs of particular programming languages. The most important rules deal with plans (which are independent of the

> detailed form of the code), debugging techniques, solution-order (search) strategies for choosing what to work on next, and methods for exploring interactions among design alternatives. (Miller, 1982, p. 119)

The most sophisticated discussion of strategic versus tactical knowledge comes from Brown, Burton, and de Kleer's report on SOPHIE, an environment to develop electronic troubleshooting expertise.

> There are many different kinds of knowledge that have to be interwoven in order to establish a sound framework for expert troubleshooting. Troubleshooting *strategies* are responsible for controlling the sequence of measurements to be made. . . . Closely allied to strategies are the troubleshooting *tactics* which concern the ease of making one class of measurements over another. . . .
>
> In addition, the good troubleshooter must have sufficient *understanding of electronic laws, of circuit components, and of the overall functional organization of the device* to enable him to draw conclusions from his measurements. (Brown, Burton, & de Kleer, 1982, p. 230)

The expert debugger in SOPHIE is designed to

> operate at a function-block level, to rely on qualitative measurements, to utilize multiple strategies and make measurements which are teleologically significant. . . .
>
> The expert's debugging strategies rely only on qualitative measurements. The sort of causal reasoning promoted by this qualitative approach develops the student's tendency to think logically. . . .
>
> We emphasize that our expert is not committed to any single strategy, but to several. . . . By offering [multiple] approaches, we expose a student to alternative ways of attacking a problem letting him witness an expert that is flexible enough to use different strategies for specific reasons. (Brown, Burton, & de Kleer, 1982, pp. 233–234)

Two other chapters in *Intelligent Tutoring Systems* point to the need for a separation of resources and control in many domains, and to the limitations of purely top-down or bottom-up models of cognition in such domains. Clancey (1982) discussed the development of GUIDON, a tutorial system for medical diagnosis. The GUIDON system is based on MYCIN, a knowledge-based expert system for infectious disease diagnosis and therapy. The performance of MYCIN in selecting antimicrobial therapy for meningitis and for bacteremia has been evaluated as being comparable to the performance of the infectious disease faculty at the Stanford University School of Medicine, so there is little doubt about the effectiveness of its knowledge representation—a production system using sets of condition–action pairs. This is very much a bottom-up, resource-driven model of cognition. Clancey's finding was that this type of knowledge organization was not sufficient for tutorial purposes. In building GUIDON he found it necessary to include two other levels of organization: a "support level" that provides the rationale for individual rules and an "abstraction level" that organizes rules into patterns.

If a bottom-up, resource-driven model of cognition like MYCIN is at one end of the spectrum, then a top-down, hierarchical, script-driven model in the first version of the program WHY is at the other. The latter was designed to tutor students about the causes of rainfall. The first version of WHY was found lacking. While the top-down approach to knowledge embodied in its script-based structure was found useful (it provided the macrostructure for tutorial interactions), it was discovered to be too narrow in many ways. According to Stevens, Collins, and Goldin (1982), the scriptal structure is sensitive to students' misconceptions at a macrolevel, catching errors that are indicated by missing or extra steps in an argument. "It is sensitive to some student errors. However, it typically misses the causes of these errors, correcting the surface error but failing to diagnose the underlying misconception that the error reflects" (p. 15). The authors concluded that representing knowledge about the particular domain they studied (knowledge about rainfall) requires multiple representational viewpoints. The hierarchical, scriptal view embodied in the first version of WHY was only one such viewpoint, and it was not sufficient in itself.

In sum, developers of intelligent tutoring systems find it useful (if not essential) to make a qualitative separation between resources and control. Moreover, no single approach to control—either top-down and hierarchical, or bottom-up and knowledge-based—is considered adequate for tutorial purposes.

METACOGNITION

The umbrella category that subsumes much of the discussion in this chapter is *metacognition*. In an early paper, Flavell (1976) characterized it as follows.

> Metacognition refers to one's knowledge concerning one's own cognitive processes or anything related to them, e.g. the learning-relevant properties of information or data. For example, I am engaging in metacognition . . . if I notice that I am having more trouble learning A than B; if it strikes me that I should double-check C before accepting it as a fact; if it occurs to me that I had better scrutinize each and every alternative in a multiple-choice type task before deciding which is the best one. . . . Metacognition refers, among other things, to the active monitoring and consequent regulation and orchestration of those processes in relation to the cognitive objects or data on which they bear, usually in the service of some concrete [problem solving] goal or objective. (p. 232)

The literature on metacognition has been somewhat more narrow than Flavell's description would suggest. According to Brown, Bransford, Ferrara, and Campione (1983), that literature splits into two parts; knowledge about cognition and regulation of cognition. "Knowledge about cognition refers to the relatively stable, statable, often fallible, and late-developing information that human thinkers have about their own cognitive processes and those of others" (Brown *et al.*, 1983, p. 107). A major focus of this work has been on what is called metamemory, or individuals' knowledge of how they store and retrieve information. Much of the work of metamemory has been "developmental" in nature, that is, paying specific attention to the way that metamemory grows and changes as children mature. For an overview of the metamemory literature, see Flavell and Wellman (1977). Various papers in the area deal with issues related to those considered here, although the significance of much of this work is methodological rather than direct. Markman (1977, p. 986) deals with "the question of how people become aware of their own comprehension failure." Lachman, Lachman, and Thronesbery (1979) discuss the importance of metamemory in using one's knowledge to cope with the environment and discuss ways to measure its accuracy and efficiency. Ringel and Springer (1980) discuss the importance of self-monitoring as a component of effective control behavior in memory tasks. In general, the research indicates that knowledge and awareness of one's own thinking strategies develop with age in children (i.e., young children have a poor sense of their own thinking strategies, and this improves as the children mature) and that performance on many tasks is positively correlated with the degree of one's metaknowledge.

The research on regulation of cognition is of more direct interest and is largely based on the kinds of work on control in AI and IP psychology that was discussed above. Brown *et al.* (1983), Brown (1978), and Brown (in press) offer a fairly comprehensive overview.

> These processes include *planning* activities prior to undertaking a problem (predicting outcomes, scheduling strategies, various forms of vicarious trial and error, etc.), *monitoring* activities during learning (testing, revising, rescheduling one's strategies for learning), and *checking* outcomes (evaluating the outcome of any strategic actions against criteria of efficiency and effectiveness). It has been assumed that these activities are not necessarily statable, somewhat unstable, and relatively age independent, that is, task and situation dependent. (Brown *et al*, 1983, p. 107)

As the epigraph for this chapter indicates, these are the "basic characteristics of thinking efficiently in a wide range of circumstances." The research indicates that the presence of such behavior has a positive impact on intellectual performance. That its absence can have a strong negative effect—when access to the right knowledge is not automatic—was indicated in the discussions of Types A and B control in the first part of this chapter.

METACOGNITION AND MATHEMATICS EDUCATION

Metacognition has received a fair share of attention in the mathematics education literature, and there are some interesting treatments of it. Silver (1982a, b) and Silver, Branca, and Adams (1980) have paid specific attention to applications of psychological work to mathematics education. Lesh (1982, 1983a, b) has worked extensively in this area as part of his attempts to explain the mathematical performance of middle school students. Lesh argues that the key to an individual's problem-solving behavior is that individual's *conceptual model* of the situation (problem) at hand. Expert behavior, in which the appropriate resources are routinely accessed, is a result of the experts' possession of stable conceptual models. Conversely, many students' difficulties are due to the fact that their conceptual models are unstable.

> [T]he metacognitive functions that are hypothesized to be of greatest interest are associated with the following four characteristics of conceptual models and their use. First, the development of conceptual models is characterized by *both* incremental growth and discrete qualitative "jumps" (insights, stages, etc.). Second, the *stability* (i.e., degree of coordination) of a conceptual model is one of the most important variables influencing cognitive functioning. Third, the structures that make up conceptual models function as "forms" for lower order structures and as "content" for higher order structures; and, in this hierarchy, the influence of higher order on lower order models is at least as important as the converse. Fourth, parallel processing is typical for cognitive functioning in mathematical problem solving situations. (Lesh, 1982, pp. 9–10)

These comments echo many of the themes raised in this chapter. Lesh's first point is that, in the midst of working a problem, students will often come to a radically different reconceptualization of that problem. Indeed, they may pass through a number of such reconceptualizations while working a single problem. In consequence, models of cognition that take for granted stable views of the situation (e.g., production systems of condition–action pairs) may not provide an appropriate mechanism for representing such behaviors. Moreover, any discussion of strategies that are primarily appropriate for implementation in stable systems (e.g., means–ends analysis) may not be germane for characterizing unstable student behavior. The second point can be interpreted as a statement about the interactions between resources and control; weaknesses in conceptual systems (resources) result in problems at the control level. Lesh delineates two kinds of unstable subsystems: conceptual and procedural.

> (1) If information must be extracted (or interpreted) using a poorly coordinated *conceptual* system, then the student can be expected to (i) neglect to "read out"

some important facts or properties—focusing only on the more obvious features of the problem while ignoring less salient features, and (ii) "read in" subjective information or meanings—distorting interpretations of the situation to fit preconceptions.

(2) If a poorly coordinated *procedural* system must be executed, then the student can be expected to (i) lose track of overall goals (or the subsuming network of procedures) when attention is focused on the execution of individual steps, and (ii) execute individual steps incorrectly when attention is focused on overall goals (or the subsuming network of procedures). (Lesh, 1982, p. 13)

The same point is made by Matz (1982) in her discussion of students' errors in algebra:

> A failure while executing a procedure forces a departure from the simple execution of algorithms into a more strategic mode; a student is forced to decide what other goal (if any) to resume or pursue. When a local failure is wrongly assessed as a global failure, like the failure of a particular procedure to produce a desired result, some students quit instead of looking for another approach. Similarly, when perspectives are at too low a level, students can lose sight of the problem solving repertoire, locking themselves into one approach. (Matz, 1982, p. 45)

Other characteristics that Lesh associates with unstable procedural systems, and that plague people as they work problems in not-very-familiar domains, include (1) inflexibility, (2) rigidity in procedure execution, (3) the inability to anticipate the consequence of actions during their execution, and (4) cognitive "tunnel vision," where execution is so demanding that there is no energy for monitoring, and the like. Points 3 and 4 serve to reinforce the point that resources and control are deeply interconnected and that the qualitative separation suggested here should only be pushed so far.

SOCIAL INTERACTIONS AND THE DEVELOPMENT OF CONTROL STRATEGIES

One of the hallmarks of good problem solvers' control behavior is that, while they are in the midst of working problems, such individuals seem to maintain an internal dialogue regarding the way that their solutions evolve. Plans are not simply made, they are evaluated and contrasted with other possible plans. New pieces of information are sought but then challenged as to potential utility. Solutions are monitored and assessed "on line," and signs of trouble suggest that current approaches might be terminated and others considered. Complete solutions are checked both for correctness and for signs that something simpler, or more general, might have been missed. Colloquially speaking, one might say that part of competent problem solvers' control behavior is that they argue with themselves as they work.* How do people develop such skills?

Seminal work in this area, establishing the framework within which a substantial amount of recent work has been done was carried out by Vygotsky (1962, 1978). A major thesis of Vygotsky's (1978) *Mind in Society* is that social interactions play a fundamental role in shaping internal cognitive structures.

> A good example of this process may be found in the development of pointing. Initially this gesture is nothing more than an unsuccessful attempt to grasp something. . . . The child attempts to grasp an object placed beyond his reach; his hands, stretched toward that object, remain poised in the air. His fingers make grasping movements. At this initial stage pointing is represented by the child's movement, which seems to be pointing to an object—that and nothing more.
>
> When the mother comes to the child's aid and realizes that his movement indicates something, the situation changes fundamentally. Pointing becomes a gesture for others. The child's unsuccessful attempt engenders a reaction not from the object he seeks but *from another person.* Consequently, the primary meaning of that unsuccessful grasping movement is established by others. . . . [T]here occurs a change in that movement's function: from an object-oriented movement it becomes a movement aimed at another person, a means of establishing relations. *The grasping movement changes to the act of pointing.* . . . Its meaning and functions are created first by an objective situation and then by people who surround the child. (Vygotsky, 1978, p. 56.)

Vygotsky's strong position is stated as follows:

> Every function in the child's cultural development appears twice: first, on the social level, and later, on the individual level; first, *between* people *(interpsychological),* and then *inside* the child *(intrapsychological).* This applies equally to voluntary attention, to logical memory, and to the formation of concepts. All the higher functions originate as actual relationships between human individuals. (Vygotsky, 1978, p. 57.)

This transformation of an interpersonal process into an intrapersonal one is hypothesized to require a long and slow series of developmental events. In addition, the potential for change at any time is limited to what Vygotsky

* This phrase is intended as more than an idle metaphor. The notion has, for example, been taken quite seriously in the AI community. One view of control is that a single "executive" makes decisions. This top-down hierarchical approach is reflected in programs such as Sacerdoti's (1977) NOAH, even though that particular program is not rigidly top-down and has a fair degree of flexibility. A rather different view of control has such decisions made by what might be called an "executive board." Members of this board work somewhat independently, generating potential suggestions for consideration. These suggestions are recorded on a common data structure, and then a decision by the executive board (sometimes accomplished by fiat when one board member "shouts loudest") determines what gets done next. This approach was implemented in Selfridge's (1959) PANDEMONIUM, and is at the heart of the Hayes-Roths' (1979) opportunistic planning model.

calls the "zone of proximal development (ZPD)." Working as an individual, a child may perform up to a certain intellectual level. Working under adult guidance or in collaboration with more capable peers, the student may perform at a somewhat higher level. The range of skills that extends beyond what the student can currently perform, but that the student can perform with assistance, is the ZPD. Vygotsky's hypothesis is twofold: (1) immediate development is limited to the ZPD, and (2) development takes place as a result of social interactions.

While the evidence Vygotsky offered for his hypotheses is sometimes difficult to evaluate, there is some substantiation of these ideas from a variety of different sources. Working in the Piagetian tradition, Doise, Mugny, and Perret-Clermont (1975) reported on two experiments exploring students' collaborative work.

> The first experiment shows that two children, working together, can successfully perform a task involving spatial coordinations; children of the same age, working alone, are not capable of performing the task. The second experiment shows that students who did not possess certain cognitive operations involved in Piaget's conservation of liquids task acquire these operations after having actualized them in a social coordination task. (Doise *et al.*, 1975, p. 367)

A subsequent (1978) experiment by Mugny and Doise explores the causes of cognitive growth. It indicated that "more progress takes place when children with different cognitive strategies work together than when children with the same strategies do so, and that not only the less advanced but also the more advanced child makes progress when they interact with each other" (p. 181). Recent work by Petitto (1985) extends these results. Often the way that a pair of students approaches an estimation task differs qualitatively from the approach taken by either student alone. The new approach evolves during the solution as a result of interactions between the two students. Once it emerges, it can then become part of the individual students' repertoire. Thus social interactions spur individual cognitive development.

Returning directly to the issue of control, it is again the case that the bulk of the relevant literature deals with young children. The pattern of development has been summarized as follows.

> A great deal of this learning [by the internalization of social processes] involves the transfer of executive control from an expert to the child. Children first experience active problem-solving procedures in the presence of others, then gradually come to perform these functions for themselves. This process of internalization is gradual; first, the adult (parent, teacher, etc.) controls and guides the child's activity, but gradually the adult and the child come to share the problem-solving functions, with the child taking initiative and the adult correcting and guiding when she falters. Finally, the adult cedes control to the child and serves primarily as a supportive and sympathetic audience. (Brown *et al.*, 1983, p. 122)

Most learning, of course, is not nearly as well structured as the sequence described in the preceding paragraph. This is especially the case with regard to the broader aspects of control discussed in this chapter: looking at situations from multiple perspectives, planning, evaluating new ideas, monitoring and assessing solutions in the midst of working problems, and so forth. Where do such behaviors arise, and how does one learn to argue with oneself while solving problems? A Vygotskean hypothesis seems plausible here. It seems reasonable that involvement in cooperative problem solving—where one is forced to examine one's ideas when challenged by others, and in turn to keep an eye out for possible mistakes that are made by one's collaborators—is an environment in which one could begin to develop the skills that, internalized, are the essence of good control. It is interesting to note that in virtually all mathematics instruction beyond elementary school, learning and practicing mathematics is very much an individual and solitary endeavor. Our teaching practices deprive students of the opportunity for cooperative endeavors, and in doing so may inhibit the development of effective control strategies.

Summary

Control—resource allocation during problem-solving performance—is a major determinant of problem-solving success or failure. The first part of this chapter discussed the effects of two prescriptive control strategies on students' problem-solving performance and concluded with a discussion of the ways that good or bad control can affect problem-solving performance. The second part described some of the relevant literature.

The chapter began with a case study in a mathematical microcosm—techniques of integration. The work in integration yielded two results: (1) Even in this simple and straightforward domain, students do not develop efficient control strategies, and their problem-solving performance suffers because of it; and (2) a prescriptive control strategy can result in significant improvements in students' performance (with less total study time).

The work in integration was, in essence, a pilot study for an attempt to deal with similar issues regarding general mathematical problem solving via heuristic strategies. If bad control could dilute the effects of algorithmic strategies in integration, then it might wash out the effects of heuristic strategies altogether. As in the case of integration, a prescriptive "executive strategy" for problem solving was developed. That strategy served as the structural backbone for both upper- and lower-division undergraduate courses in mathematical problem solving. A variety of tests (detailed in Chapters 7, 8, and 9) indicated that students' problem-solving performance can be significantly improved by training in the use of heuristic techniques, if that training is done within the context of a prescriptive control strategy.

Prescriptive or otherwise, heuristic problem solving covers only part of the problem-solving spectrum. The third section of this chapter examined the range of effects that control can have on the way a problem solution evolves. Examples were described of poor control, in which students who had more than enough subject matter knowledge to easily solve a problem failed to do so, because they went off on wild goose chases and squandered the resources (including time) potentially at their disposal. At the other end of the spectrum, an example was described where an individual solved a problem precisely because he made superb use of the resources at his disposal—even though he began the problem with less subject matter knowledge than many others who failed to solve it. A rigorous framework for the macroscopic analysis of such behavior is given in Chapter 9.

The second part of the chapter summarized relevant work in related fields. *Control* is a term borrowed from AI. For both computational and epistemological reasons, many researchers in AI and in IP psychology have found it useful to make a qualitative distinction between resources and control. More broadly, there is a growing literature on metacognition documenting the role that competent executive processes play in contributing to competent intellectual performance. Finally, the role of social interactions in the development of control processes was discussed. A Vygotskean perspective suggests that the "internal dialogues" of competent problem solvers result from their having internalized aspects of the cooperative problem-solving sessions in which they had been engaged. Conversely, a paucity of cooperative problem-solving experience might (in depriving the student access to overt models of control behavior) hamper the development of individual control strategies.

5

Belief Systems

Chapter 1 gave a brief discussion of people's beliefs about mathematics and of the ways that such beliefs can affect people's behavior in mathematical situations. It described an attempt by the college students SH and BW to solve a construction problem, Problem 1.1, using straightedge and compass. The approach taken by SH and BW was entirely empirical. The students made conjectures and then tested them by construction, with the sole standard for accepting or rejecting a potential solution to the problem being the accuracy of the construction. (Indeed, a correct solution was rejected, because the construction did not look good enough.) From their behavior it appeared that the two students were completely unfamiliar with deductive geometry. Yet it was later shown that they were perfectly capable of making the deductive arguments that provide the answer to Problem 1.1. They simply had not thought to approach the problem that way.

The subjects SH and BW were not alone in this behavior. Of more than a hundred high school and college students asked to solve Problem 1.1, only one student derived a solution to the problem; only one other student confirmed the solution by proof after discovering it empirically. Most of these students, like SH and BW, had little trouble doing the proof problems when they were asked to. It simply had not occurred to them that doing so would be useful. Such behavior is the focus of this chapter. The chapter develops and elaborates one possible line of explanation, *naive empiricism,* for the students' separation of deductive and constructive arguments.

The first section of the chapter offers a description of related work in other disciplines. Research in naive physics, everyday reasoning, and decision theory illustrate some of the ways that people's cognition is shaped by their belief systems, which set the psychological context within which their more "purely cognitive" resources are accessed and controlled. Some of the literature on students' learning in geometry, and on affective issues in mathematical performance, is briefly discussed.

To establish a baseline of comparison for students' performance, an expert's solution to a geometric construction problem is given in full and analyzed in the second section. The third section returns to SH and BW's protocol, offering a model of empirical behavior that matches their performance quite closely. Despite the match at the behavioral level, that model is clearly a caricature—although the extreme version of empiricism it characterizes is not so far off the mark that it should not be taken seriously. The fourth and fifth sections introduce and expand on the notion of naive empiricism. In the fourth section a complex protocol is analyzed. In that protocol it appears that two students turn toward empiricism when other approaches fail. An alternate explanation is explored, in which their apparent tries at deduction early in the solution are seen to differ substantially in spirit from the expert's approach (given in the second section). The fifth section describes an extended problem set in which students' rejection of the results provided by geometric proof is quite clear.

Selections from the Relevant Literature

This section considers evidence from three domains, indicating that, even under relatively ideal cognitive circumstances (i.e., with as few complications as possible induced by negative affect, stress, etc.), humans are not the rational animals one would like to believe they are. Issues of students' geometrical resources and of affect are then briefly considered. The discussion begins with the research area that has the clearest and most direct relevance to mathematics learning.

NAIVE PHYSICS

Much of formal physics is counterintuitive. There is clearly documented evidence that

1. People develop strong sets of intuitions about the behavior of physical systems. These intuitions often stand in direct contradiction to the behavior of those systems—and, of course, to the formal characterizations of that behavior that are studied in physics courses.

2. The intuitions (preconceptions, misconceptions) that people develop about the world around them often interfere with the learning of the formal principles of physics in the classroom. They often persist after instruction. In some cases they supplant the formal knowledge. In some cases they coexist with it, with the individual simultaneously "believing" in mutually contradictory bodies of knowledge.

Two of the simplest and most frequently discussed problems used in the research on naive physics are the following:

Problem 5.1 Figure 5.1 shows a thin curved metal tube. In the diagram you are looking down on the tube. In other words, the tube is lying flat. A metal ball is put into the end of the tube indicated by the arrow and is shot out of the other end of the tube at high speed. Draw the path the ball follows after it emerges from the tube, ignoring air resistance and any spin the ball might have. (McCloskey, 1983a, p. 300)

Figure 5.1

Problem 5.2 Imagine that someone has a metal ball attached to a string and is twirling it in a circle above his or her head. In Figure 5.2 you are looking down on the ball. The circle shows the path followed by the ball and the arrows show the direction in which it is moving. The line from the center of the circle to the ball is the string. Assume that when the ball is at the point shown in the diagram the string breaks where it is attached to the ball. Draw the path that the ball will follow after the string breaks. Ignore air resistance.

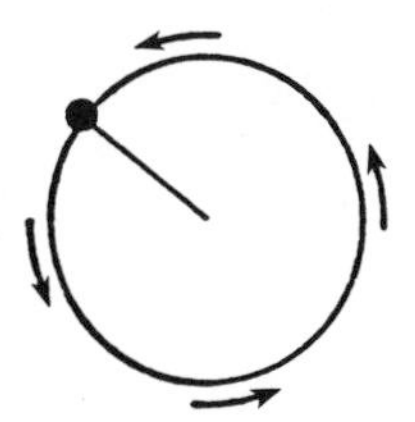

Figure 5.2

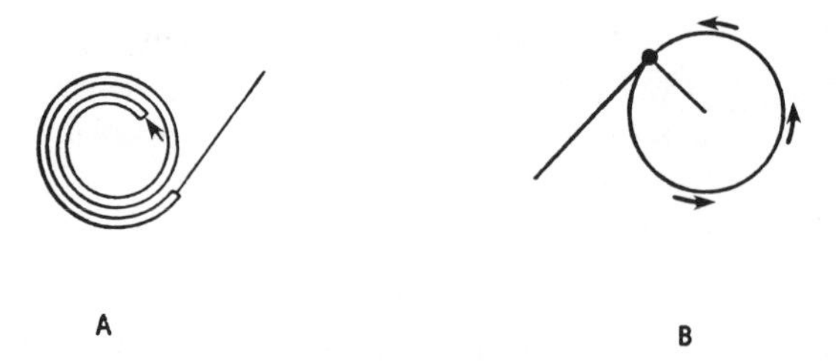

Figure 5.3

According to Newton's first law, an object in motion will, in the absence of net applied forces, travel in a straight line. In consequence, the correct answers to Problems 5.1 and 5.2 are those given in Figures 5.3A and 5.3B, respectively. A large number of people, including many who have had classroom instruction in physics, answer these problems incorrectly. Their incorrect answers apparently result from the belief that a trajectory that begins as a curve will remain curved. This belief often persists through instruction. A series of experiments on projectile motion (Caramazza *et al.*, 1981; McCloskey, Caramazza, & Green, 1980; McCloskey, 1983a, b) indicated that roughly half of the college students who had not had physics in high school, about a third of those who had a year of high school physics, and a seventh of those who had studied a year of college physics, predicted the curved path in Figure 5.4A in answer to Problem 5.1. In answer to Problem 5.2, roughly one naive subject in three predicted the kind of curved path in Figure 5.4B.

The error rate is much higher on more complex problems in mechanics (see, e.g., Clement, 1983). In some cases, people's false intuitions about physical phenomena are so strong that they distort perception and memory. Lochhead (1983) reported that some students describe the motion of a ball tossed upward as follows: The ball goes up for a while, stops *for a while,* and then goes downward. DiSessa (1983) reported that individuals will *mis*remember prior experiences in the light of their misconceptions.

Figure 5.4

DiSessa offered an explanation for misconceptions in physics in terms of *phenomenological primitives* (p-prims for short), a "collection of recognizable phenomena in terms of which they see the world and sometimes explain it" (p. 16). One such p-prim is that of force as a mover:

> Another relatively well-documented false intuition of naive and novice students is that a force causes motion in the direction of the force, ignoring the effect of previous motion [Figure 5.5A]. We propose to explain this by assuming that the most commonplace situation involving forces, pushing on objects from rest, becomes abstracted as the highest priority p-prim that one will use to predict motion in general circumstances. . . . More speculatively, we might suppose that the development to more expert understanding involves raising the priority of a competing primitive analysis. . . . Deflection as in [Figure 5.5B] offers such competition in the case of "force as a mover." (diSessa, 1983, pp. 30–31.)

DiSessa noted further:

> One of the surprising things about naive physics is that people will indeed recognize a phenomenon like deflection to exist, if one makes the circumstances compelling enough. It simply is not referenced for ordinary situations. (diSessa, 1983, p. 31.)

The theme raised in this last quotation needs to be emphasized and expanded upon. There has been some argument in the literature about which naive theory of motion students may have (i.e., What is *the* naive physics?), but such discussions miss the point. First, there are numerous conflicting interpretations of many physical phenomena. Second, and more important, people are capable of holding a number of such mutually contradictory theories in mind at once.

The idea suggested by diSessa is that phenomenological primitives are contextually bound. That is, they operate in a fashion similar to schemata; the conceptual framework that one invokes to explain a particular phenomenon may be cued by the perceived features of the phenomenon itself. A point of major interest is that a significant number of the people who produced a drawing like Figure 5.4A in answer to Problem 5.1 had completed a year of high school or university instruction in physics. One might consider

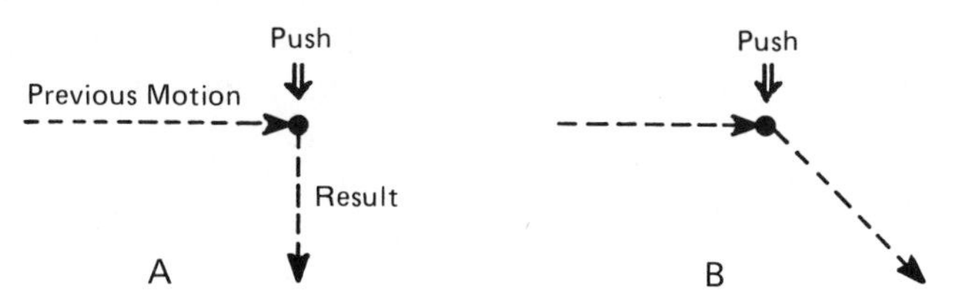

Figure 5.5 The "force as a mover" and "force as a deflector" phenomena. (From diSessa, 1983, p. 31.)

the following thought experiment. Suppose that students who had completed the relevant formal instruction in physics, but who had missed Problem 5.1, were asked to work Problem 5.3.

Problem 5.3 Figure 5.6 shows a thin curved metal tube seen from above. A metal ball is put into the end of the tube indicated by the arrow and is shot out of the other end of the tube at high speed. The point from which the ball emerges has coordinates $(2, -2)$. (Coordinates are measured in meters.) The ball emerges in the vector direction $(3i + 4j)$ with an initial velocity of 500 meters/sec. Give the coordinates of the ball one second after it leaves the tube.

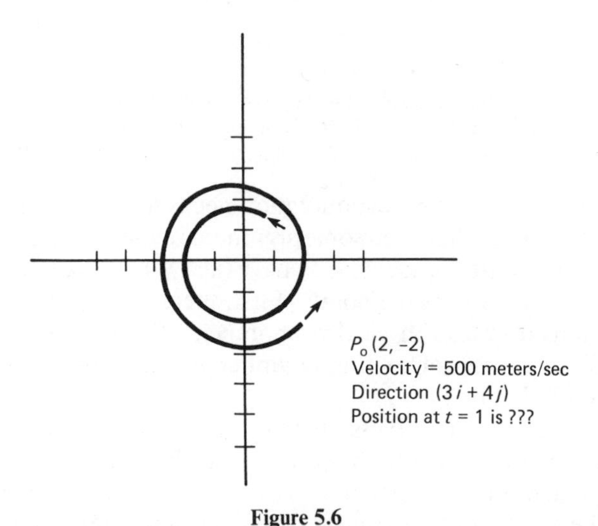

Figure 5.6

There is a good chance that a significant proportion of those students would get the correct answer to Problem 5.3. The contextual cues for Problem 5.3 are those of formal classroom physics. Those cues—the use of a coordinate system, of vector directions, and of the term *initial velocity*—should be enough to invoke the formal knowledge schemata of vector physics. This being the case, many of these students would complete the problem without even thinking about a curved trajectory for the ball. That is, the same person may invoke formal physics for one version of a problem and yet provide a contradictory qualitative interpretation of the phenomenon in another version.

In sum, individuals can hold in mind clearly contradictory bodies of information. The particular knowledge that the person brings to bear in a particular situation is likely to be a function both of that person's experience and of the problem context. This theme of knowledge tailored to circumstances appears again in the discussion of geometry.

EVERYDAY REASONING

Research on everyday reasoning also points to the contextually bound nature of thought processes. Perkins (1982) and Perkins *et al.*, (1983) explored people's attempts to present arguments about four issues, summarized briefly as follows:

> Would restoring the military draft significantly increase America's ability to influence world events?
>
> Does violence on television significantly increase the likelihood of violence in real life?
>
> Would a proposed law in Massachusetts requiring a five-cent deposit on bottles and cans significantly reduce litter?
>
> Is a controversial modern sculpture, the stack of bricks [by minimalist sculptor Carl Andre] in the Tate Gallery, London, really a work of art? (Perkins, 1982, p. 9)

The kinds of reasoning required for dealing competently with such issues are described below.

> One of the ironies in reasoning [about such issues] is that in many everyday situations there seems to be a wealth of information. The man or woman on the street has all sorts of practical experience with aspects of life such as raising children, buying cars, or voting for candidates for office. . . . In everyday circumstances there *seems* to be a richness of information, but there is not. The everyday reasoner must wring whatever truth can be gotten from the knowledge at hand by weaving a kind of web of plausible conjectures that hangs together well enough to be worth believing in. (Perkins *et al.*, 1983, p. 178)

Perkins and his colleagues conducted 320 interviews with individuals who were asked to make arguments about two of the four questions given above. The subjects were given one of the topics and asked to think about it for 5 minutes, reaching a position on the issue if possible. They were then asked to state the position and to provide the reasoning that justified it. Interviewers then asked follow-up questions, including "How does your reason support your conclusion?" and "Can you think of any objections to your reason supporting your conclusion?" Interviews were transcribed and analyzed, with a particular emphasis on pointing to deficiencies in the arguments that subjects provided. Only about a fourth of the flaws in those arguments reflected faulty reasoning, in that they produced results that would contradict those obtained by formal reasoning procedures. Most of the errors reflected what might be called "inadequate model building—various failures to use available knowledge in constructing a more elaborate and realistic analysis of the situation under consideration" (Perkins 1982, p. 4). A typical oversight in making an argument, for example, was for the reasoner to overlook a clear counterexample to a claim that was being advanced. What is of particular interest is that many of those who produced weak arguments could, when asked, find the flaws in the arguments and repair them. When asked to produce counterexamples to claims they had just made, most could do so. They could also produce more refined arguments than they had (rebutting their claims and then countering the rebuttals, e.g.) on demand. Perkins offers the following explanation of that somewhat paradoxical behavior.

> Naive reasoners might be said to have a "makes-sense epistemology." Of course, this does not mean that they have an explicit philosophy about what grounds are necessary for belief. But it does mean something in terms of manifested behavior: such reasoners act as though the test of truth is that a proposition makes intuitive sense, sounds right, rings true. They see no need to criticize or revise accounts that do make sense—the intuitive feel of fit suffices. (Perkins *et al.,* 1983, p. 186)

To sum things up briefly, people who believe* that making a convincing case does not require carefully reasoned arguments are not likely to produce carefully reasoned arguments—or even think of producing them—despite the fact that they could do so without difficulty.

DECISION THEORY

A major theme in discussions earlier in this book, especially in Chapter 4, is that the quality of one's decision-making processes during problem solving is a strong determining factor of success or failure. In a probabilistic sense at least, one would expect rational decision-making to pay off. Similarly, one would expect deviations from judgments suggested by (more or less) objective assessments of probabilities to hamper problem-solving performance. For this reason the general literature on decision-making, called decision theory, is of interest.

There is a large body of research indicating that (1) people's decision-making is often far from rational, even in fairly clean experimental circumstances, and (2) deviations from rationality are often systematic and can often be traced to systematic biases or beliefs. For a broad and interesting review of psychological decision theory see Einhorn and Hogarth (1981). For the richest collection of papers in the area see Kahneman *et al.* (1982).

* As with the quotation from Perkins *et al.* (1983) above, this does not mean that the individuals would explicitly avow having that belief. Their manifested behavior is entirely consistent with it, however.

A few examples are given here to indicate the kinds of issues with which the research has been concerned.

One major finding is that people's judgments in certain situations may depend less on the objective reality of the situations than on the way the situation is framed psychologically. The following two problems were given to students at Stanford University and the University of British Columbia by Tversky and Kahneman (1981).

Problem 5.4 Imagine that the U.S. is preparing for the outbreak of an unusual Asian disease, which is expected to kill 600 people. Two alternative programs to combat the disease have been proposed. Assume that the exact scientific estimate of the consequences of the programs are as follows:

If Program A is adopted, 200 people will be saved.
If Program B is adopted, there is a $\frac{1}{3}$ probability that 600 people will be saved, and $\frac{2}{3}$ probability that no people will be saved.

Which of the two programs would you favor?

Problem 5.5 Imagine that the U.S. is preparing for the outbreak of an unusual Asian disease, which is expected to kill 600 people. Two alternative programs to combat the disease have been proposed. Assume that the exact scientific estimate of the consequences of the programs are as follows:

If Program C is adopted, 400 people will die.
If Program D is adopted, there is a $\frac{1}{3}$ probability that nobody will die, and $\frac{2}{3}$ probability that 600 people will die.

Which of the two programs would you favor? (Tversky & Kahneman, 1981, p. 453)

To be perfectly rational about the two problems, the situations they describe —although framed differently—are identical. Problem 5.4 was framed in terms of saving people's lives. Of the 152 people who responded, 72% chose program A, and only 28% chose program B. Problem 5.5 focused on the number of people who would die. In answer to this question the percentages were reversed: only 22% of the 155 respondents chose program C (the same as program A), and 78% chose program D (the same as program B). Such dramatic shifts in behavior are not perfectly rational.

Nor, although it is consistent, is much of the behavior described in Kahneman *et al.* (1982). One example of consistently biased behavior is the use of what the authors called the "representativeness" heuristic. Many probabilistic questions provide information about the nature of an object X, and ask, What is the likelihood that X is a member of Class A? Research suggests that people will make their judgments based on the degree to which the characteristics of X match archetypal characteristics of Class A. Given a description of someone as "extremely bright, with an aptitude for the scientific and a passion to see what makes things tick," people will rate the likelihood of that person's being a nuclear physicist as fairly high, and of being (say) a farmer as fairly low—despite the fact that there are far more farmers than nuclear physicists in our society. Baseline statistics are swept aside by the use of representativeness. In fact, judgments by this heuristic defy standard probability. One of the basic laws of probability is that if B is a subset of A, then the probability of B must be less than or equal to that of A. In one study, Kahneman and Tversky (1982) presented subjects with the following personality sketch. Their description of people's reactions to it is given below.

> Linda is 31 years old, single, outspoken, and very bright. She majored in philosophy. As a student, she was deeply concerned with issues of discrimination and social justice, and also participated in anti-nuclear demonstrations. [R]espondents were asked which of two statements about Linda was more probable: (A) Linda is a bank teller; (B) Linda is a bank teller who is active in the feminist movement. In a large sample of statistically naive undergraduates, 86% judged the second statement to be more probable. In a sample of psychology graduate students, only 50% committed this error. However, the difference between statistically naive and sophisticated respondents vanished when the two critical items were embedded in a list of eight comparable statements about Linda. Over 80% of both groups [chose (B) over (A)].

DISCUSSION

Belief is an elusive issue. This discussion attempts to clarify the sense in which the term is used in this book and to establish a context for the balance of the chapter.

Issues of belief occupy a precarious middle ground between primarily cognitive and primarily affective determinants of mathematical behavior. To this point this book has focused on cognitive determinants of behavior: What one knows about a domain clearly sets the boundaries for what one can establish within it. Issues of affect are equally important, for they similarly establish boundaries and determine behavior. Perhaps the two clearest examples from within the affective domain deal with mathematics anxiety and fear of success—both of which in their stronger manifestations have debilitating effects on mathematical performance.

Mathematics anxiety is so well known that it hardly needs comment. Faced with mathematical situations, some people simply freeze; others do whatever they can to avoid situations that threaten to involve the use or discussion of mathematics. This problem is generally acknowledged as significant; *mathephobia* has entered our vocabulary, and there are mass-market books (Buxton, 1981; Tobias, 1978) for dealing with it. Much of the research literature is based on the use of the Mathematics Anxiety Rating Scale, or MARS (Suinn, Edie, Nicoletti, & Spinelli, 1972). Fear of success is less well known, but it provides another clear example of how social factors and matters of self-perception shape intellectual performance. To sum things up briefly, there are potential negative consequences for women who succeed in mathematics. The women (particularly young women in school settings) may fear social rejection because of success and may fear a loss of their own femininity for having succeeded in a typically male arena. The consequences of acting on such fears are obvious. Reviews of the relevant literature may be found in Tresemer (1976) and Leder (1980).

Major, and somewhat overlapping, foci of attention in the affective literature are as follows: Mathematics anxiety can be considered a subset of mathematics attitude. The general hypothesis in that area is that attitude and performance are correlated. Literature reviews are given in Aiken (1970, 1976) and Kulm (1980). The role of motivation (Atkinson & Raynor, 1974; Ball, 1977) in performance is obvious and important. So is the issue of perceived personal control (Weiner, 1974; Stipek & Weisz, 1981; Lefcourt, 1982), the degree to which individuals believe that their actions can shape events around them. People who believe their efforts will pay off, it is hypothesized, are more likely to make such efforts than people who do not. Another personal attribute that shapes behavior in mathematical situations is the degree of risk a person is willing to tolerate. The study of variations in such attributes, and of the impact of such variation, is the study of individual differences and cognitive style. Fennema and Behr (1980) offer a review of the relevant literature in mathematics. Broader discussions may be found in Witkin and Goodenough (1981) and Messick (1976).

Though they are of unquestioned importance, affective variables per se are beyond the scope of this volume. This does not mean that such issues are ignored. (In the discussion of protocol analyses in Chapter 9, e.g., it is stressed that the results of some "purely cognitive" laboratory studies have been misinterpreted precisely because the studies failed to take affective issues into account; the impact of affective variables was there, even if they were not designed for.) Rather, the intention here is to focus on behavior that may appear to be purely cognitive—simply a matter of "reasoning practices"—and to indicate that such behavior may well have an affective component. It is in that sense that the issue of *belief* straddles the affective and cognitive domains.

A nonmathematical example may serve to illustrate the way the term is used here. At a recent meeting of mathematics teachers, a sheet of paper folded in the shape suggested by Figure 5.7 was dropped from a point about 8 feet above the floor. As the paper figure descended very slowly toward the floor, the audience was asked to think of possible reasons for the slowness of its descent. More than 200 people were in the audience, and they were given some time to generate possible explanations. All of the explanations that were offered had to do with wind currents and the aerodynamics of the Y shape. Not surprisingly, the possibility that the paper figure's descent had

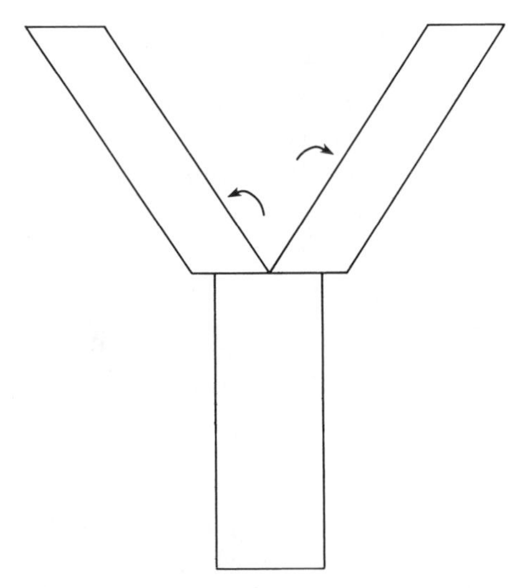

Figure 5.7 A paper "helicopter." The left part of the Y is folded away from the reader, the right part toward the reader.

been slowed by telekinesis was not raised by anyone—even though the audience had been asked to consider all plausible explanations. This was pursued further. Had anyone even raised the possibility of telekinesis to himself or herself and then rejected it as implausible? No, not a single member of the audience had even entertained the notion.

I argue here that the notion of belief is relevant to an explanation of this behavior. The issue is not as simple as asking the question, Do you accept the reality of psychic phenomena such as telekinesis?, although in that case the status of the issue as a matter of belief is clear. Nor, to phrase it somewhat differently, is the issue whether or not the members of the audience overtly believe in a Cartesian separation of mind and matter. (It may well be that some of the people in the audience would not be willing to take so strong a stand, although the vast majority might.)

One plausible explanation of their behavior is that the members of the audience have a world view that is entirely consistent with a Cartesian mind–matter dualism, and that this world view drives their behavior in much the same way that the overt belief would drive it. In the experience of the audience, the explanations of virtually all physical phenomena lie within the realm of physical causality. It is natural, then, to turn to that realm for the explanation of what appears to be an ordinary physical phenomenon. Routinized and generalized, this pattern becomes not only natural but exclusive. Other possible realms of explanation are not first considered and then rejected. Rather, it simply does not occur to the naive mind-matter dualist to consider anything besides physical causality as a means of explanation. In this way the individual's world view — not necessarily recognized or codified by the individual as part of an overt belief structure — determines the set of ideas the individual will bring to bear in a given situation. This kind of unarticulated world view can be thought of as the precursor of belief. Once recognized and codified by the individual, it becomes part of a clear belief structure.

The kind of behavior described in the previous paragraph is natural and, in general, highly functional. Humans function with efficiency precisely because they abstract from their experience as a matter of course and then depend on those abstractions as a guide to behavior. Much of this is done without conscious reflection. When, for example, a person crosses the street or tries to catch a fly ball, the person does not compute the trajectories of oncoming cars or of the ball de novo. Rather, past empirical experience structures present actions. The result is nearly automatic and highly efficient behavior, which suffices "for all practical purposes." Occasionally, however, this process causes problems. The research in naive physics indicates that people's abstractions of physical phenomena, though functional for everyday purposes, can contradict the formal laws of physics. When this happens one's "physical world view," or set of beliefs about the ways the physical world works, can cause difficulty. One can misperceive and misinterpret events; one can bring irrelevant or incorrect information to bear in certain situations; one can have difficulty learning formal physics because of interference from one's naive physics.

In a similar way, students abstract a "mathematical world view" both from their experiences with mathematical objects in the real world and from their classroom experiences with mathematics. Some aspects of their mathematical world view may be unarticulated and other aspects may be codified as overt beliefs. These perspectives affect the ways that students behave when confronted with a mathematical problem, both influencing what they perceive to be important in the problem and what sets of ideas, or cognitive resources, they use. Explorations of such perspectives, particularly with regard to geometry, are the focus of this chapter.

A Mathematician Works a Construction Problem

This section presents a mathematician's solution to Problem 5.6, a problem that has the same underlying structure as Problem 1.1. The approach taken to this problem is of interest because it indicates that, for the mathematician, mathematical argumentation serves as a form of discovery. In working the problem this individual uses mathematical proof procedures to help find a solution. This use of deduction as a tool for problem solving is typical of the way that mathematicians employ deduction. It stands in stark contrast to the empiricism demonstrated by SH and BW in Chapter 1 and establishes a baseline of performance with which the students' behavior discussed in this chapter can be compared.

When he is asked to work Problem 5.6, the mathematician returns to a once-familiar but no longer comfortable domain; he has not done geometry for about 10 years. He reads the problem and begins working on it. The full transcript of his attempt is given below.

Problem 5.6 Using a straightedge and compass, inscribe a circle in the triangle given below (Figure 5.8). The inscribed circle is a circle that lies inside the triangle and is tangent to all three sides of it.

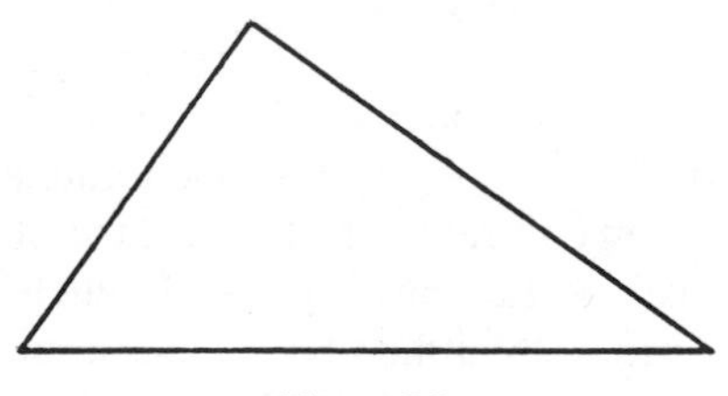

Figure 5.8

All right, so the picture's got to look like this [draws Figure 5.9], and the problem is obviously to find the center of the circle. . . . Now what do I

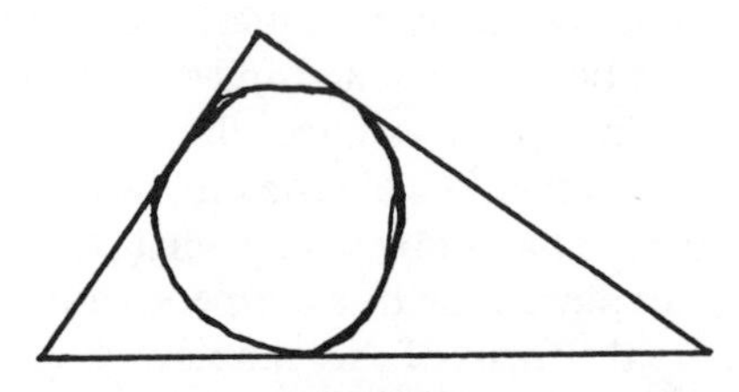

Figure 5.9

know about the center? We need some lines in here. Well, the radii are perpendicular at the points of tangency, so the picture's like this [draws Figure 5.10]. . . . That doesn't look right; there's something missing. What if I draw in the lines from the vertices to the center? [He draws Figure 5.11.] That's better. There've got to be congruent triangles in here. . . .

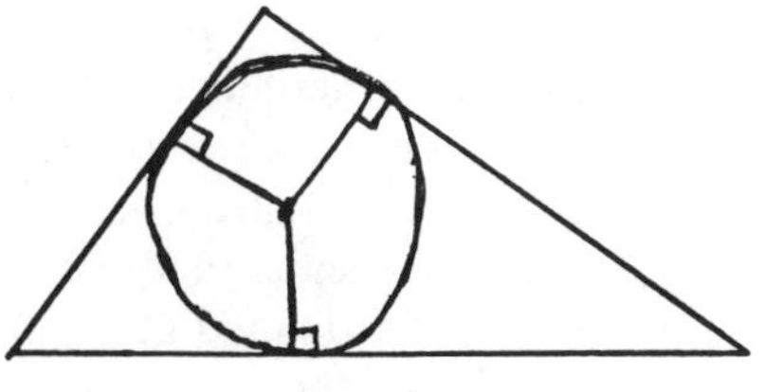

Figure 5.10

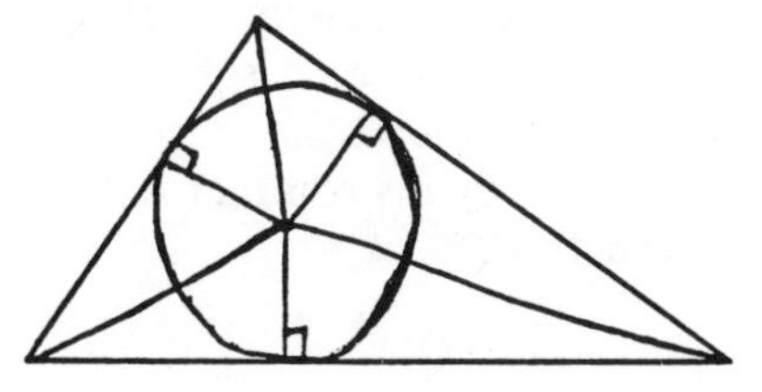

Figure 5.11

Let's see. All the radii are equal, and these are all right angles, and with this, of course, this line is equal to itself [marks the X on Figure 5.12], so these

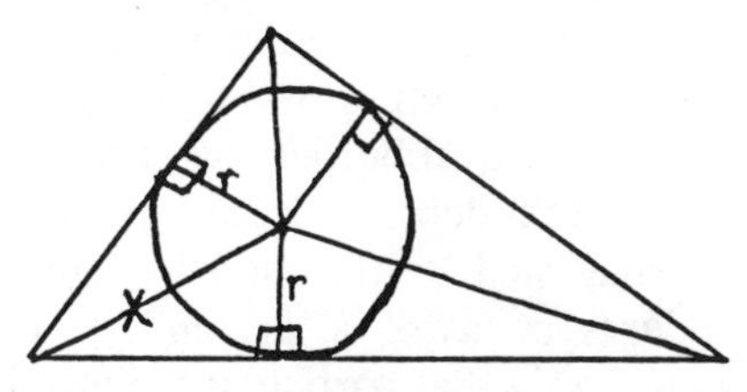

Figure 5.12

two triangles [the two at the lower left vertex] are congruent. Great. Oops, it's angle-side-side; oh no, it's a right triangle, and I can use Pythagoras or hypotenuse–leg or whatever it's called. I'm OK. So the center is on the angle bisectors.

[Turns to me] I've solved it. Do you want me to do the construction?

This approach to Problem 5.6 contrasts rather dramatically with the empiricism demonstrated by SH and BW as they worked Problem 1.1. Of course a number of factors contribute to the mathematician's successful work on this problem. Better control behavior, more reliable recall of relevant facts, and (not to be underestimated) more confidence all work to his advantage. But most important is the basic method that the mathematician adopts. Lacking relevant information about the desired circle, he derives the information that he needs through the use of geometric argumentation. Note that he looks for congruence—"There've got to be congruent triangles in here"—long before there is a conjecture to verify. Rather than being an afterthought or a means of after-the-fact verification, mathematical argumentation serves as a means of discovery for this problem solver. (It was Pólya, I believe, who defined geometry as the art of right reasoning on wrong figures.) This dependence on argument to provide information is very much a part of the mathematical world view, and it is evident from the very beginning of this solution attempt.* Also clear is the fact that the derived results guarantee a solution. The mathematician knows that it is not necessary to perform the construction in order to confirm its correctness. This is shown by his comments, "I've solved it. Do you want me to do the construction?"

The Student as Pure Empiricist: A Model of Empirical Behavior**

This section offers a characterization of pure empiricism based on a collection of "empirical axioms" and the rudiments of a model for predicting students' performance on construction problems like Problem 1.1, reproduced below. While this portrayal borders on caricature, it is remarkably good at predicting behavior: Students often behave in ways quite consistent with the performance suggested by the model. This, at least, raises the possibility that the students' behavior is "driven" by an empiricist perspective. Whether and how firmly that particular perspective (or any other) is held, and how such a perspective interacts with one's problem-solving performance, are delicate issues that are explored in the sequel.

Problem 1.1 You are given two intersecting straight lines and a point P marked on one of them, as in Figure 5.13 below. Show how to construct, using straightedge and compass, a circle that is tangent to both lines and that has the point P as its point of tangency to one of the lines.

The empiricist axioms that provide the "belief structure" underlying the model are as follows.

* This comment is not meant to suggest that mathematicians do not rely on intuition or empirical investigations as they work. Competent mathematical performance fully exploits intuition and empiricism—but only as two of many tools than can be used.

** This model of student behavior was first described in Schoenfeld 1983a.

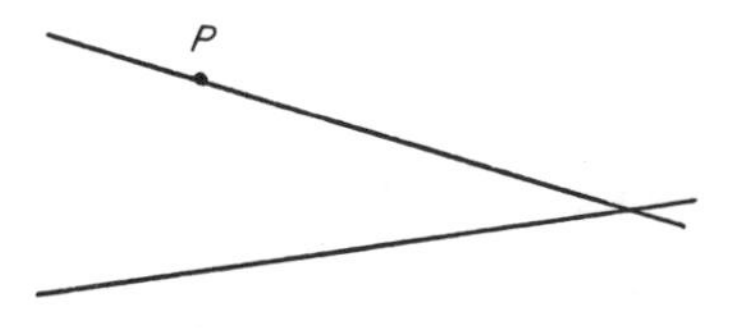

Figure 5.13

Axiom 1 Insight and intuition come from drawings. The more accurate the drawing, the more likely one is to discover useful information in it.

Axiom 2 Two factors dominate in generating and rank ordering hypotheses for solution. They are (1) the "intuitive apprehensibility" of a solution and (2) the perceptual salience of certain physical features of the problem. That is, (1) if you can "see your way" to the end of one plausible construction more clearly than to the end of another, the first will be ranked higher and tested first. This holds unless (2) some feature of the problem—say, the bisector of the vertex angle in Problem 1.1—dominates perceptually as an essential ingredient of a solution. If this is the case, the plausible solutions including the dominant perceptual feature are highest ranked (but those, in order of apprehensibility).

Axiom 3 Plausible hypotheses are tested seriatum: Hypothesis 1 is tested until it is accepted or rejected, then Hypothesis 2, and so forth.

Axiom 4 Hypothesis verification is purely empirical. Constructions are tested by implementing them. A construction is correct if and only if performing it provides the desired result (within some tolerance set by the individual).

Axiom 5 Mathematical proof is irrelevant to both the discovery and (personal, rather than formal) verification process. If absolutely necessary—that is, the teacher demands it—one can probably verify a result using proof techniques. But this is simply playing by the rules of the game, verifying under duress those things that one already knows to be correct.

Returning to Problem 1.1, we observe some of the perceptual features that may catch a student's attention. Note that, like the "facts" and "procedures" discussed in the chapter on resources, the perceptual features of a problem that a student takes to be true may or may not actually be true. Six common features noted by students are:

Feature F1 The radius of the desired circle is perpendicular to the top line at the point P (a recalled fact).

Feature F2 The radius of the desired circle is perpendicular to the bottom line at the point of tangency.

Feature F3 The point of tangency on the bottom line, P', is the same distance from the vertex as is the point of tangency on the top line (by some sort of perceived symmetry).

Feature F4 Any "reasonable-looking" line segment that originates at P and terminates at the bottom line is likely to be the diameter of the desired circle.

Feature F5 The center of the circle lies halfway between the two lines, and is thus on the angle bisector (again by a perception of symmetry).

Feature F6 The center of the circle lies on the arc that is swung from the vertex and that passes through P.

Of these six features, Features F4 and F5 tend to be perceptually dominant; the student who notes one of those will, in general, base further conjectures on that feature. Also, Feature F6 is generally noted only after F5, when a problem solver tries to identify which point on the angle bisector is the center of the desired circle. (See Figure 5.14 for illustrations of F4 and F5's dominance.) Feature F1 plays an interesting role. Students often recall Feature F1, but it is rarely perceptually dominant. Rather, it serves as a test condition; whatever construction the student arrives at, by whatever reasoning, must ultimately satisfy Feature F1.

Various combinations of the features listed above yield hypothetical solu-

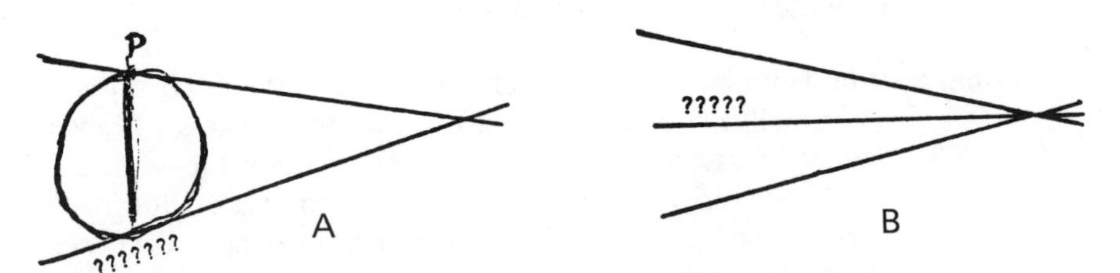

Figure 5.14 (A) Feature F4 dominates: Which point on the bottom line is the hypothetical endpoint of the diameter. (B) Feature F5 dominates: Which point on the angle bisector is most likely to be the center of the circle?

tions to the problem. Eight hypotheses can be generated as follows: Feature F4 combines with Features F1, F2, and F3, respectively, to generate Hypotheses H1–H3. The diameter of the desired circle is then . . .

Hypothesis H1 the line segment between the two given lines that is perpendicular to the top line at *P*.

Hypothesis H2 the line segment between the two given lines that is perpendicular to the bottom line and passes through *P*.

Hypothesis H3 the line segment from *P* to *P'*.

Feature F5 combines with Features F1, F2, F3, and F6 to yield the following. The center of the desired circle lies at the intersection of the line that bisects the vertex angle and

Hypothesis H4 the perpendicular to *P*.

Hypothesis H5 the perpendicular from *P* to the bottom line.

Hypothesis H6 the segment from *P* to *P'*.

Hypothesis H7 the arc from the vertex that passes through *P*.

Finally, the nondominant Features F1, F2, and F3 combine to yield

Hypothesis H8 the center of the circle lies on the intersection of the perpendiculars through *P* and *P'*.

The axioms above, along with some implementation rules, provide the basis for predicting behavior. Four such rules follow:

Rule R1 The features noted by students determine candidate sets of solutions. Thus Hypotheses H4–H7 can become candidates only when Feature F5 is noted, and so on.

Rule R2 Students' initial sketches determine the plausibility (and thus the ranking) of candidate solutions. If a rough sketch resembles Figure 5.15A, for example, Hypothesis H3 ranks high and is likely to be the first hypothesis tested by construction. An initial sketch like Figure 5.15B will rule Hypothesis H3 out of contention. (See the solution attempt below by AM and CS.)

Rule R3 Feature F3 is a default condition when F1 is not noted, Feature F4 when F5 is not noted.

Rule R4 Unless initial sketches suggest the contrary, Hypothesis H8—as the combination of three nondominant features—receives the lowest initial ranking of candidate hypotheses.

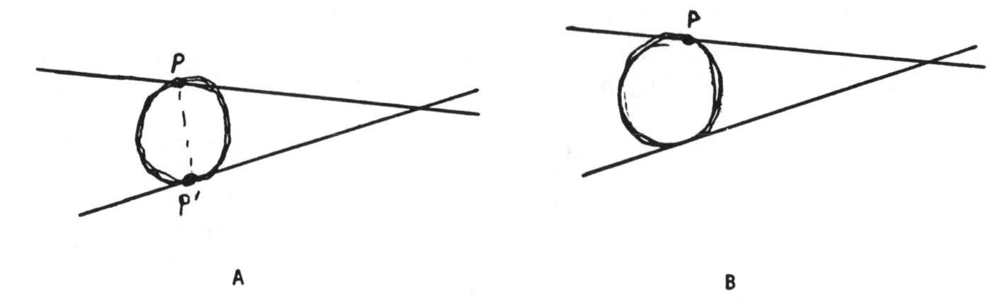

Figure 5.15

How the Model Corresponds to Performance

The reader may wish to review SH and BW's attempt to solve Problem 1.1 (described on pp. 36–44). The balance of this section compares their performance with the behavior that would be predicted by the model.

When they began working the problem, neither SH nor BW recalled Feature F1; nor did either student perceive Feature F5. According to the model (see Rule R3) the default conditions F3 and F4 should be triggered, and the students should conjecture Hypothesis H3. This is indeed what happened. (Note: This behavior is far from anomalous. Roughly half of the students who work Problem 1.1 try Hypothesis H3 as their first conjecture.)

Student SH did not say anything out loud as he drew Figure 1.20, so his intentions can only be inferred from his sketches. They appear rather clear, however. First SH drew the figure at the bottom of Figure 1.20. This was a test of Hypothesis H1. This rough sketch indicated that Hypothesis H1 would not work, so H1 was not tested by construction. Then SH drew the sketch at the top of Figure 1.20, this being a test of Hypothesis H2. Again, the students' behavior up to this point continued to be consistent with the model. With the implementation of Hypothesis H3, the "diameter feature," Feature F4 became perceptually dominant; and since Feature F1 takes precedence over F2 (for obvious reasons), Hypothesis H1 should have been tested before H2. All of the F4-related conjectures were tested, in appropriate order, before any others were entertained.

At this point SH made a tentative suggestion about Feature F5 ("might it involve angle bisectors at all?"). If pursued, Feature F5 might have become a dominant perceptual feature.* As it happened, BW shunted the sugges-

* The belated introduction by SH of Feature F5 into the solution calls for further comments about the nature of dominant perceptual features. A feature is dominant when a student perceives it, is convinced of it, and adopts it clearly in a solution. This is the role played by

tion aside. When SH proposed Feature F5, BW had already begun to trace Hypothesis H8; her response to SH's suggestion ("we can try") served to postpone the consideration of Feature F5 until her current hypothesis was evaluated. As a result SH and BW worked through the sequence H3 → H1 → H2 → H8, precisely as the model predicts.

It is emphatically not my intention to suggest that because the model of pure empiricism so closely matches SH and BW's performance at the behavioral level, that SH and BW must necessarily believe in Axioms 1–5 described above. (Indeed, a current difficulty with much AI work is the tacit assumption that matching performance at the behavioral level means that one has captured the essence of underlying psychological processes.) The axioms are deliberately overstated to make a point. As the analyses in the following two sections indicate, however, those axioms are not as much of a caricature as they might seem.

A Deeper Look at Empiricism: CS and AM Work Problem 1.1

This section considers an attempt to solve Problem 1.1 that is much more complex than either SH and BW's try at the same problem or the mathematician's solution of Problem 5.6. Both of those problem sessions were straightforward, in that the problem solvers' actions and perspectives were stable throughout the solution attempt. The students SH and BW took an empirical approach from the very beginning and retained it throughout the entire solution. In consequence, their behavior matched the model of pure empiricism given above fairly well—although their avoidance of geometric argumentation remains to be explained, in the light of their subsequently demonstrated ability to solve Problems 1.4 and 1.5. Although the mathematician's approach to Problem 5.6 differed substantially from the students', it was equally consistent: A deductive stance was taken from the very beginning and was adhered to throughout the solution.

As they begin to work on Problem 1.1, most students appear to be halfway between the two extreme cases described above. The students are clearly aware of mathematical argumentation, and not infrequently they begin their work on the problem with what seems to be an attempt to derive a solution.

Feature F4 in the solution attempt discussed here, was the case with SH and BW for Feature F4 and will be the case in DW and SP's adoption of F5 in the next section. However, Feature F5 did not enter into SH and BW's solution attempt even as a significant, much less as a dominant, feature. SH was not really convinced of Feature F5 when he suggested it. Just the opposite was the case; he was looking for something to try in the light of the failure of his other hypotheses.

When these attempts fail they appear to turn to empiricism, and their dependence on it grows during the solution. One explanation of these students' behavior is that the students, who begin the problem session with a rather shaky knowledge of geometry, are empiricists of last resort; predisposed toward geometric argumentation but unable to find a convincing argument using standard techniques, they turn to "proof by construction" as the only available way to convince themselves that their hypotheses are valid.*

The following problem session is typical of the protocols in which students appear to develop a dependence on empiricism. The denouement of the session is more spectacular than most, however.

Both AM and CS were second-semester freshmen at the time they worked Problem 1.1. Student AM was a very good student, and he had a great deal of confidence. He often plunged into problem-solving attempts with the faith that things would turn out well in the end, and they often did. Student CS had a shaky mathematical background and lacked confidence. As a result of these differences in background and personality, the two students' joint problem-solving effort sometimes took on the character of parallel conversation, with AM doing most of the talking. Occasionally CS served as a sounding board for AM or took on "cleaning up" tasks as AM moved on to something new.

In the beginning of the session, AM reads the problem out loud. He then says "so what we want to do is construct a circle like this . . ." and draws Figure 5.16. He then begins to make additions to the figure. As he does so it appears he is making a whole-hearted attempt to derive an answer to Problem 1.1. In some ways this attempt seems similar in spirit to the mathematician's. It is, unfortunately, unsuccessful.

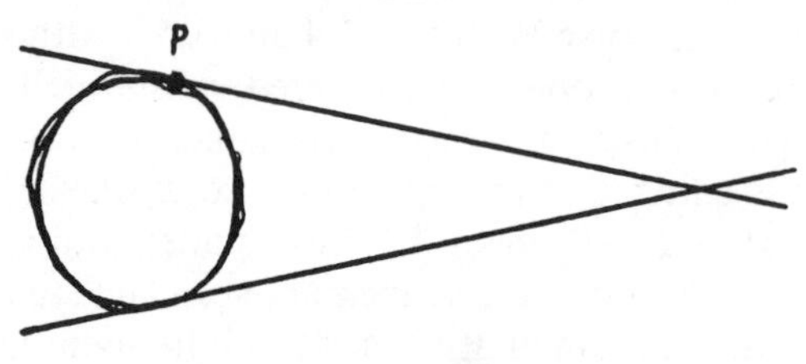

Figure 5.16

* This explanation of students' behavior is explored below. Certainly the large number of sessions where students vacillate between deduction and empiricism raises questions about the validity of the purely empirical model. Even if the model captures the sequence of actions produced by the students, it would appear to violate the "spirit" of their approach. I argue below that the discrepancy is not as large as it appears, and that a more subtle naive empiricism serves to unify all of the examples discussed here.

AM: What is a formula for constructing a circle inside a triangle? The radius of a circle times . . . times something gives you the area times the circumference. . . . All right, if we drew another line like this, say establish a triangle that we could work with, . . . we can draw a square here because it is a right angle. . . . We can always inscribe a square inside a triangle. [He draws Figure 5.17.] This *r* is equal to this which is equal to this, which is equal to this . . . and this is another right angle, because any, well, this angle [long pause].

While making these comments AM makes a number of modifications and additions to his diagram in the hope that the resulting figure will resemble a problem he knows how to solve. (If the match is perfect, of course, the problem is solved.) The line segment at the left is drawn, making a triangle. The problem has now been "reduced" to that of inscribing a circle in a triangle, which he knows can be solved. Unfortunately, he does not recall the procedure. Nor do the other additions to the diagram suggest anything directly useful. While he stares at Figure 5.17 in the hope that something else will come to mind, CS makes a suggestion we have seen before.

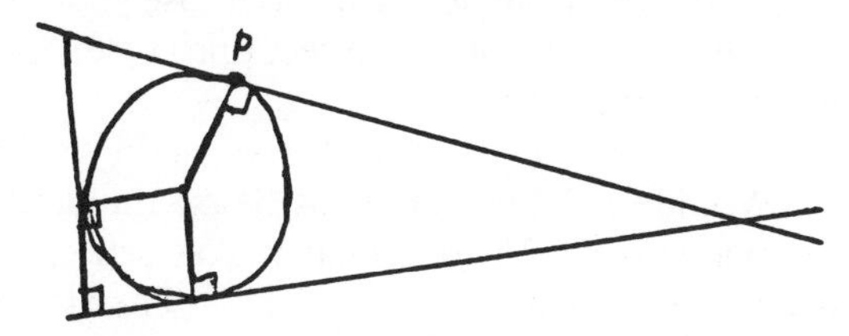

Figure 5.17

CS: Oh, Oh, I've got it. . . . You can make a straight line and bisect that [draws Figure 5.18].

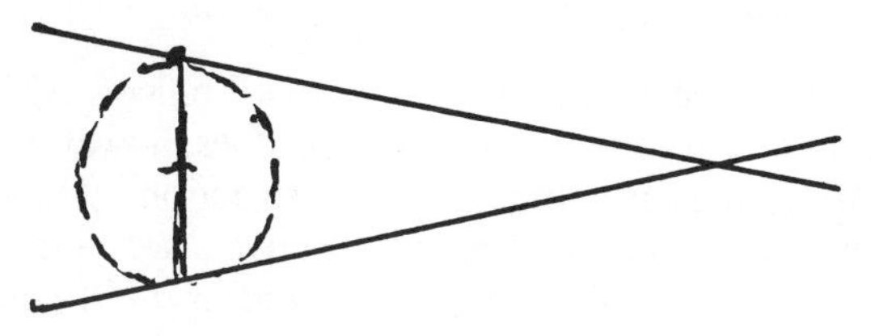

Figure 5.18

Student AM looks at the sketch he had drawn earlier (Figure 5.17) and rejects CS's suggestion, instead tracing Figure 5.19 with his finger.

AM: But you don't want a straight line, you want a curve.

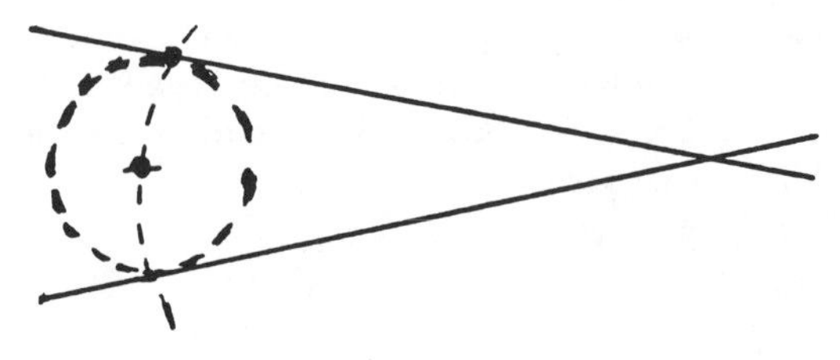

Figure 5.19

This suggestion is clearly grounded in empiricism. A look at Figure 5.16 indicates that the center of the circle they seek is to the left of the point that CS has suggested, so the students need a procedure that will yield a point in that direction. A curve (more specifically, an arc of a circle with center where the lines cross, and passing through *P*) will meet that constraint.

CS: But once you have a straight line between two points there, can't you just . . .

AM: Well, you can bisect the angle, right? Because you know the center point will be halfway, so you can bisect this angle. . . .

CS: You could do it either way.

AM: Yeah, bisect the angle. Then draw a curve, or do you? You draw an arc. . . . Wouldn't the center point be on the arc? [AM thinks about his suggestion.] Yeah. That makes sense. I don't know why. Now why does that make sense? . . . Then [looking at Figure 5.17] we also know that this is *r*. What does that tell us? It is a lopsided drawing. We didn't draw [sketch] too good of an arc. All right, whatever . . . *Proof,* yeah, how . . . Hmm [rereads problem] show how to construct . . . *well, we've shown how to construct, but . . . we cannot prove that it works. Of course we could construct it. . . .* [emphasis added]

The comment "we've shown how to construct" reflects AM's confidence that his proposed construction is correct. He does, however, indicate concern about verifying the correctness of his construction. Apparently unable to find rigorous substantiation for his perspective ("we cannot *prove* that it works"), AM now takes a step toward empiricism ("Of course we could construct it"). There is more exchange as CS reiterates the "diameter hypothesis" (Figure 5.18), but AM provides a strong argument that convinces CS to abandon it. Picking up the straightedge and compass, AM begins to implement his construction. He works quickly and somewhat inaccurately. Figure 5.20 results.

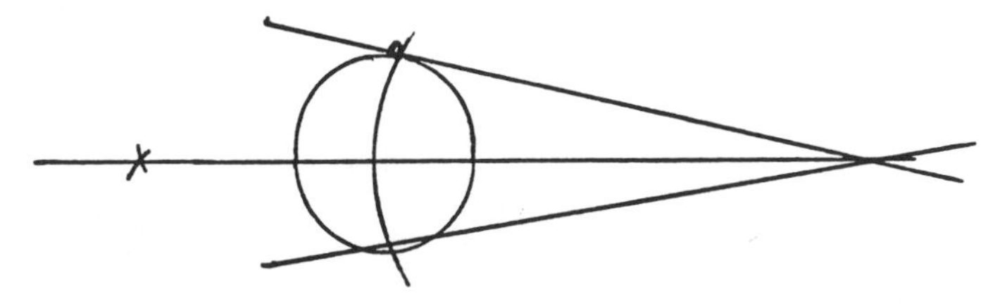

Figure 5.20

AM: Hmmm . . . things don't look so good. I don't think we should have that much error. . . .

CS points out that the construction is not accurate: The top half of the purported angle bisector appears clearly larger than the bottom half. As CS considers this, AM looks back at Figure 5.17. He worries that the radius of the circle should be perpendicular to the tangent at the point of tangency.

AM: I mean if we draw a perpendicular for the line of the tangent at *P*, right?

This gives rise to a new construction (the perpendicular through *P* and the angle bisector), which happens to be correct. However, sloppy work leads to a result as inaccurate as their previous attempt (Figure 5.21). Because it

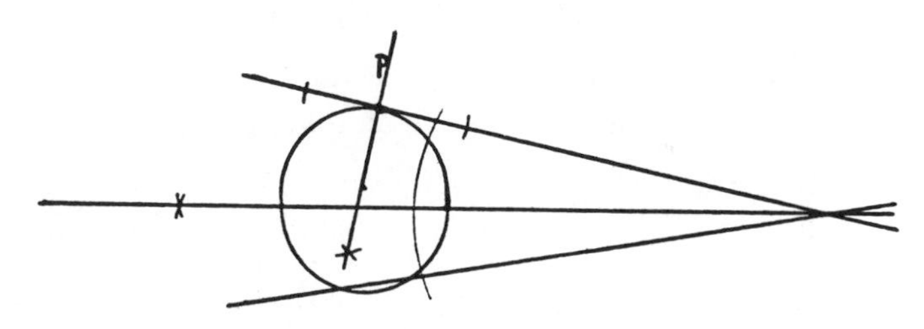

Figure 5.21

looks wrong, Figure 5.21 is also regarded as a failure. Student CS worrries about the inaccuracy of the angle bisectors in Figures 5.20 and 5.21. Since neither appears accurate, he considers making repairs to both constructions. In the meantime, AM traces over the perpendicular from *P*.

AM: Let's just assume that this is perpendicular . . . that would give us this line . . . [tracing down the perpendicular from *P*] . . . then you can draw a perpendicular like that [sketching a line perpendicular to the bottom line, Figure 5.22] . . . but is this [the distance from *P* to the purported center of the circle] the same distance as this [the

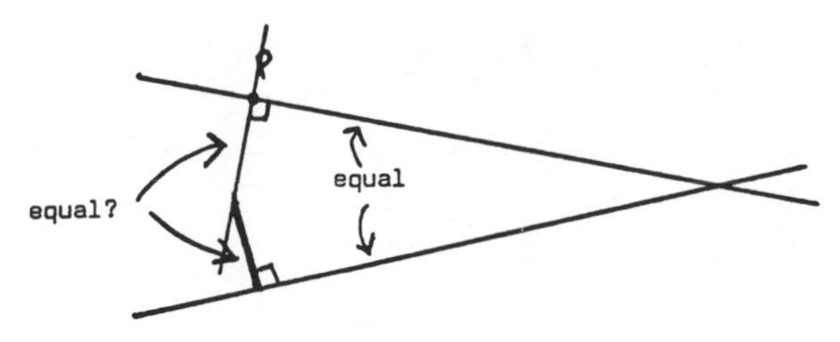

Figure 5.22

distance to the bottom line]? That's the question. Where they cross should be the midpoint. . . . *But can we prove that now? Yeah, by drawing a circle and seeing if it works.* [emphasis added]

As the result of purely empirical explorations, AM has come to conjecture H8: The center of the desired circle appears to lie at the intersection of the perpendiculars drawn from *P* and from its counterpart on the bottom line. More importantly, AM adopts empirical verification as his standard of proof; he will prove (his word) his conjecture by "seeing if it works." At this point CS and AM begin parallel constructions. AM works on the conjecture he has just made, while CS tries to repair the obviously inaccurate angle bisectors in Figures 5.20 and 5.21. All three conjectures now look equally good: AM's in Figure 5.23 and both of CS's in Figure 5.24.

CS: So . . . we proved it two different ways.
AM: Yeah, because it's tangent. All right, that's good.

I intervene. "From what I could hear, it sounds like you were going in the right direction. What did you finally decide upon?" Student AM says "We can do it two different ways," and tells me that the problem can be solved (1) with the construction suggested in Figure 5.23 or (2) "with the arc or the perpendicular" as in the two constructions in Figure 5.24.

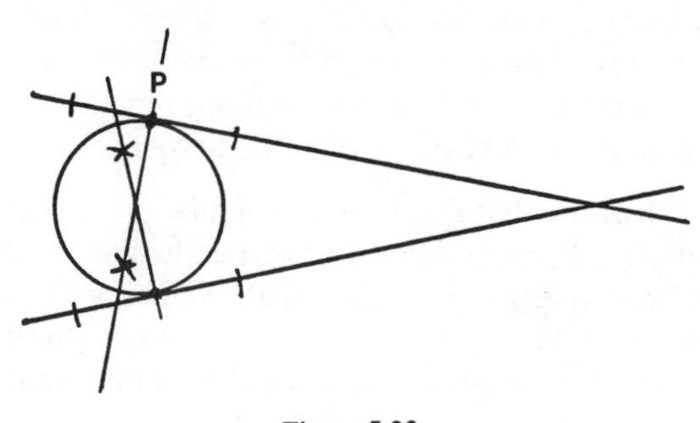

Figure 5.23

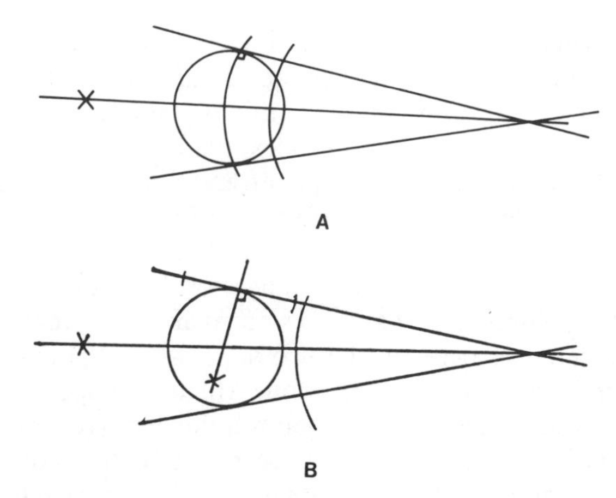

Figure 5.24 (A) The arc and the angle bisector repaired. (B) The perpendicular and angle bisector repaired.

In this session AM and CS appear to have evolved a dependence on a purely empirical standard for the correctness of a solution, and that dependence leads them into behavior that borders on the bizarre. Early in the solution attempt neither the arc nor the perpendicular (Figures 5.16 and 5.21) met the empirical standard, so both were rejected—although with some misgivings. As a result of those misgivings CS repaired the two constructions (Figure 5.24), and then they both looked equally good. When one plays by purely empirical rules, these two tentative solutions have to be accepted or rejected simultaneously. In consequence, CS and AM offer me the two *mutually contradictory* constructions as alternate versions of the same procedure.

For the record, AM knew more than enough deductive geometry to derive a solution to Problem 1.1. A few minutes after they worked the problem, I asked them "to work a couple of high school problems for me." Typically, AM took over the solution attempts. He solved Problems 1.4 and 1.5 in a total of less than 4 minutes. This is especially noteworthy in view of his earlier comments, "Well, we've shown how to construct, but . . . we cannot prove that it works."

DISCUSSION AND RESOLUTION

The key to understanding students' attempts to solve Problem 1.1 lies in reconciling their knowledge with their performance. The knowledge issue is clear-cut: Both pairs of students whose solutions we have discussed knew more than enough to produce a solution to Problem 1.1 without difficulty.

In a slightly different context and with no additional training, SH and BW solved Problems 1.4 and 1.5 in class the day after they were taped; in precisely the same context and only a few minutes after they worked Problem 1.1, AM solved the proof problems in less than 4 minutes. Moreover, the students (with the possible exception of CS) did understand the generic nature of proof arguments. The performance issue is not as clear-cut. Both pairs of students took an empirical approach to Problem 1.1. Each solution attempt lasted more than 15 minutes, and most of that time was spent with straightedge and compass in hand. As noted above, there did appear to be significant differences between the two approaches: SH and BW took an empirical approach from the outset, while AM and CS appeared to be forced into an empiricism of last resort as their solution evolved. I argue here that those differences are not as large as they seem. All of the above behavior can be explained coherently if the students are considered to be *naive empiricists.*

The fundamental distinction here is between a naive empiricism and the kind of overt empiricism suggested in the model. In the pure empiricism suggested by Axioms 4 and 5 of the model, the students use straightedge and compass as the tools of choice, actively and consciously rejecting mathematical argumentation as being of no value. Though SH and BW took an empirical approach, it is not clear that they eschewed deduction. It appears, rather, that they were unaware of it. In addition, AM and CS were clearly not antideductive; they simply seemed unsuccessful at deduction. As a point of departure, let us begin our discussion at the other end of the spectrum.

For the mathematician, dependence on argumentation as a form of discovery is learned behavior, a function of experience. This perspective is not "natural"; few B.A.s in mathematics possess it. Those who become mathematicians generally develop this perspective in graduate school, during their apprenticeship to the discipline. At first, "proof" is mandatory, an accepted standard. As one becomes acculturated to mathematics, it becomes natural to work in such terms. "Prove it to me" comes to mean "explain to me why it is true," and argumentation becomes a form of explanation, a means of conveying understanding. As the mathematician begins to work on new problems (perhaps on a dissertation), this progression continues. Mathematical argument becomes a way of convincing oneself that something ought to be true. Even unsuccessful attempts turn out to be valuable, because consistent failures point to weak spots in one's understanding. After numerous attempts to demonstrate a particular result, one can see a pattern in the failures and decide the result may not be true. To see if it is false, one may try to construct some examples that exploit the weak spots. If none of the examples one tries demonstrate that the result is false, one may again begin to believe the result is true—and a pattern in the failed examples may suggest the information that was missing from the original attempts. Thus mathematical argument becomes a tool in the dialectic between what the mathematician *suspects* to be true and what the mathematician *knows* to be true. In short, deduction becomes a tool of discovery.

This is hardly the experience of most students. The students whose performance was discussed above were typical of college freshmen who were (or had just been) enrolled in a calculus course. They had taken a year of geometry in high school and had spent most of the time in that course engaged in proof-related activities. From the perspective of such students, mathematical argumentation (or "proof") has a very specialized role. For the most part, these students have engaged in such argumentation in only one of two circumstances:

1. to confirm something that (as is often the case in geometry) is intuitively obvious, in which case proof seems redundant if not superfluous; or
2. to verify something that they are told is true (see, e.g., the statements of Problems 1.4 and 1.5), in which case the purpose of going through the exercise appears to be one of training; mathematical argument, like conjugating Latin verbs, is (supposedly) good for developing rigorous thinking skills.

Let us now return to the discussion that concluded the first section of this chapter. There it was suggested that in any domain, people's (not necessarily conscious) abstraction of experiential regularities can result in their having a world view that serves to shape their behavior in that domain. "Naive mind–matter dualism" was a case in point. For people accustomed to finding explanations of physical phenomena in the realm of physical causality, the idea of searching for explanations outside the customary realm did not even occur as a possibility. Whether or not the mind–matter separation was overtly declared did not make much of a difference insofar as their behavior was concerned.

It is quite likely that a similar process of abstraction takes place with regard to students' experience in geometry. For students who had abstracted the experiences with geometrical argumentation described just above, the idea of employing deduction as a means of discovering information would not even occur as a possibility. Whether or not these people overtly declared mathematical deduction to be irrelevant to the process of discovery would not make much of a difference insofar as their behavior was concerned. These naive empiricists would, ultimately, accept only empirical "proof" of their conjectures.

Once this perspective is adopted, the two problem sessions described above can be seen as entirely consistent. The attempt by SH and BW stands as described, save that the students' naive empiricism now explains why their geometric knowledge goes unused.* Since it does not occur to them that mathematical argumentation is useful, they do not think to use it in their approach to Problem 1.1; they (naturally) depend on intuition for insight and construction for verification. The attempt by Am and CS is more interesting. It can now be reinterpreted as follows.

Student AM's initial attempt can best be described as a "core dump." The sequence of actions that results in Figure 5.17 is not an attempt to *derive* useful information, but rather an attempt to *retrieve* it. His search for "a formula (!) for constructing a circle inside a triangle" is an attempt to remember something that will solve the problem directly. His modifications to the original diagram (drawing in radii, sketching in right angles, adding a line to the figure to make the desired circle an inscribed circle in an as-yet-undetermined triangle, etc.) are attempts to trigger the right memories. If something he remembers matches his diagram, the problem will be solved; if it almost matches, the problem is almost solved. This is quite sensible behavior, and it should not be demeaned. But neither should it be confused with an attempt to derive useful information by means of mathematical argumentation.

When this attempt to recall a relevant solved problem fails, AM and CS find themselves in the same position as SH and BW at the beginning of their attempt. The students have no easy way to deal with the problem, and have only intuition to guide them. The first suggestion CS gives is exactly what one expects from students in such circumstances, and it is, as one expects, rejected for empirical reasons (AM's initial sketch, Figure 5.16). What occurs in the balance of the solution is not at all the evolution of an "empiricism of last resort" in two students who began the problem with open minds about mathematical argumentation. It is, rather, the evolution of an avowed empiricism in two students who began the problem as naive empiricists.

Further Evidence Regarding Naive Empiricism: DW and SP Work Four Related Problems

To pursue the issues raised in the previous section, a series of more extensive interviews was begun in 1983. Students were videotaped as they worked four problems. The students and I then watched the videotapes

* To use the language of cognitive psychology, the students' initial formulation of the problem results in the formation of a *problem space* that excludes deductive knowledge. Until the problem space is changed, it is impossible for the students to make the relevant connections.

together and, in "clinical interview" fashion, we discussed what they had done. The four problems used in the study were the following:

Problem A The same as Problem 1.1.

Problem B Can you remember the procedure for bisecting an angle using straightedge and compass? If so, (1) use the procedure to bisect (a given angle), and (2) explain why this procedure works.

Problem C The circle in Figure 5.25 is tangent to the two given lines at the points P and Q. Using the techniques of high school geometry, prove that the length of the line segment PV is equal to the length of the line segment QV.

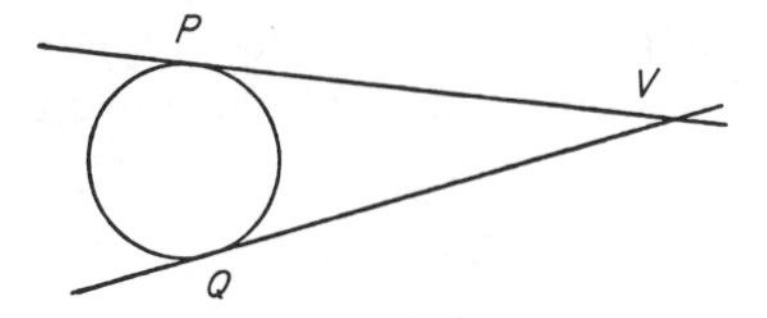

Figure 5.25

Problem D The circle in Figure 5.26 is tangent to the two given lines at the points P and Q. The center of the circle is C. Using the techniques of high school geometry, prove that the line segment CV bisects the angle PVQ.

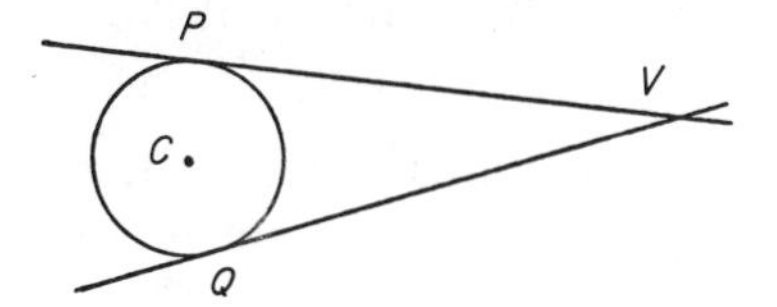

Figure 5.26

The students were told that Problem A has a straightforward solution but that many students seem to have difficulty with it; they should be careful to make certain of the correctness of their answer. To induce them to be careful (and to see if, with the promise of an extra reward, they would call upon mathematical argument to derive or substantiate their results), the students were offered an extra $5 each if they arrived at a correct solution.

At the time they were taped DW and SP were enrolled in first-semester calculus class. They had volunteered to be videotaped. In many ways they were nearly ideal subjects. They were extroverted and friendly, and they worked well together. They were comfortable working "out loud," and their conversations provide a good record of their progress. They were more lucid, and more competent, than most. They enjoyed the problems and worked hard at them, and happily spent more than an hour discussing them with me afterward. The transcript of their session follows.

Student DW reads Problem A out loud, and the two students reread it a few times to make certain that they understand it.

DW: Oh, so they want a circle right here, that's one circle, that touches here [at *P*] and then touches here [points to bottom line].
SP: The point *P* as its point of tangency. OK. So it's just in the middle.
DW: Right. So a circle would go that touches right here [top line] and then touches this [bottom] line too . . . Well that's a start. I think we should bisect these lines. You know how to do that with a compass. . . . And that way . . . then we know the center of the circle will be on that bisecting line [i.e., the angle bisector]. It has to be. Right?

The two students have agreed upon dominant Feature F5 of the model. Student DW suggests that they construct the angle bisector, and they do—even though (as confirmed by later discussions) they have not yet worried about which point on the angle bisector will be the center of the desired circle. The construction takes a few minutes and is fairly messy. During the construction DW recalls that the radius is perpendicular to the tangent and draws the appropriate conclusion:

DW: Now the next thing we have to do is draw a line that's perpendicular to here [point P] . . . a perpendicular here, and that will be the center of our circle, hopefully.

They redraw the figure. Like most students, they have trouble with the cheap compass. And, typically, they do not engage in any form of mathematical argument at all.

DW: Oh, I hope this works. Now we have to draw, get this [in the angle bisector] the same size as that. I don't like this compass.
SP: Then it should touch both here [at *P* and the bottom line].
DW: We're hoping that it does. If it doesn't, we've got a problem. [They fiddle with the compass.]
SP: Hey! Looks almost perfect.
DW: This, I think it's perfect, just the only reason was the compass wasn't big enough. I think any amount of error in here is just going to be from our drawing not perfectly. I think we did it, I think we had the right idea. [They work over the construction.]
SP: Yeah. Alright, perfect.

DW: Yeah, OK. Now we'll write this. Now we'll conclude: "It's right even if it looks crooked."

Confident of their work, they finish Figure 5.27.

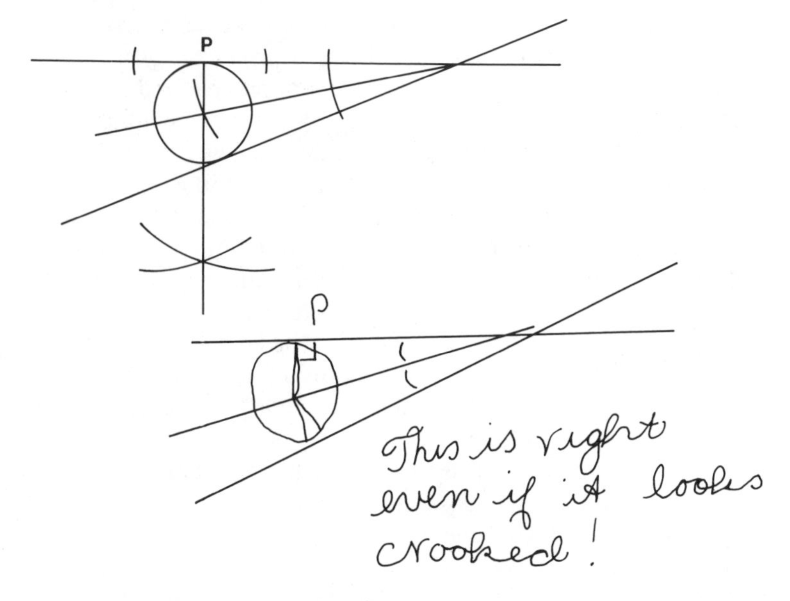

Figure 5.27

Problem B is dispatched in virtually no time. They laugh as they read the problem, since they have already done the construction.

DW: Now if you want to be real technical about this . . . this line goes from here to here and they're both the same length [*PV* and *QV* in Figure 5.28]. And this line is the same from here to here [*PX* and *QX*] and you can make triangles!

They make the congruence argument with no difficulty.

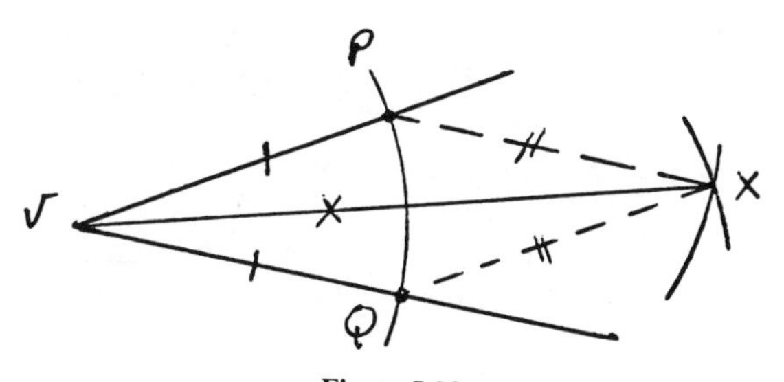

Figure 5.28

The students move confidently on to Problem C. They are disconcerted at first, since Figure 5.25 does not include the center of the circle, *C*. Since they believe that they need the point *C* for the proof, they decide to locate *C* by construction, using the angle bisector and the perpendicular (through the bottom line, "just to be different"). Having located *C* in that way, they then take less than 2 minutes to derive the appropriate congruence argument.*

In Problem D the students' constructive approach for finding the center of the circle catches up with them and causes them some difficulty. Asked to prove that *C* lies on the angle bisector, they realize that locating *C* by their construction would be circular ("It's like we're doing these problems backwards"). In consequence, they try to find a geometric argument using congruence that avoids anything vaguely resembling the picture that they had used for solving Problem C. This unfortunate decision sends them off on a wild goose chase that lasts for more than 10 minutes. It is finally terminated by a good control decision: "The best thing, I think, to do, is go over what we know, right from the beginning, and take it slow." Once DW and SP start over, they produce the correct congruence argument in about a minute and a half.

In sum, SP and DW finished the videotaping part of the interview session with correct solutions to Problems A, B, and D, and an almost-correct solution (the congruence argument was right) to Problem C. As expected, they had taken a purely empirical approach to Problem A; their hypothesis was generated intuitively and tested by construction. The students' work on Problems C and D demonstrates that they had the deductive knowledge to verify their construction, and their work on Problem B indicates their understanding of the generic role of geometrical argumentation. Moreover (as the comment about "doing these problems backwards" indicates), they were fully aware that the four problems were related.

I ask SP and DW to tell me about their solutions to the problems. They begin talking about Problem A simultaneously, in exclusively procedural and intuitive terms.

SP: We first bisected that angle [at the vertex], using here, then we connected—

DW: Well, we knew . . . wait. Tell him, yeah we knew that when we bisected that [at the vertex], that had to be the center of the circle since—

SP: Since it would be equidistant from both these lines.

* Locating *C* by construction is not legitimate. This confusion was caused by the fact that *C* was not given in the diagram. Had *C* been given or had they simply begun their argument with the phrase "let *C* be the center of the given circle," their argument would be completely correct.

DW: And then we made . . . [W]e just drew a perpendicular and where they crossed had to be the center of the circle. Because that [pointing to the perpendicular] was the radius, and we knew that that [pointing to the angle bisector], right, that that had to go—Like I said, this [angle bisector] was on the center, and this was the line that was the radius, and where they crossed was the center.

SP: We needed two lines to make sure we had the proper point. If we only had one line we couldn't be sure it was the center. So we did the other line also. We just drew the line.

AHS: Well, the picture looks good. Not only that, it turns out to be correct.

This statement is greeted with an enthusiastic choral "Yay!" from both students. It is absolutely clear from their response that this is their first confirmation that their approach to Problem A is correct and that up to this point (despite the assertion scrawled on Figure 5.17) they were not really certain that their construction was correct.

Later in the retrospective we return to Problem A.

AHS: How did you know that the center was going to lie on the angle bisector? You said over here [pointing to their work sheet] that it would. And I said "Oh, yeah?" and you said "Yeah," but I just wonder. How do you know that?

DW: Because if you draw, if it's going to touch both circles, if each line is going to touch both circles. Let me see if I can say this without getting confused. . . .

I repeat the question, pointing to the bisector in her figure. Student DW returns to the figure. Saying "Because the distance from here to here has to be . . . ," she traces out, dynamically, the set of points equidistant from the top and bottom lines. This produces the angle bisector as an intuitive locus (Figure 5.29) but provides no "hard evidence." She persists:

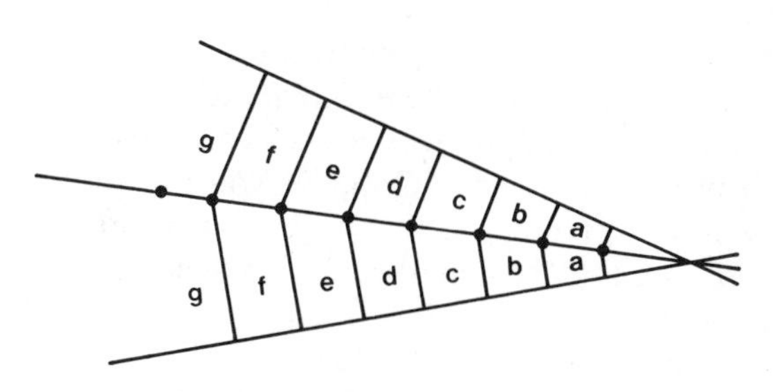

Figure 5.29

DW: It has to. It has to. I'll say it really loud and you'll believe it, [I'll say it] with a lot of authority. . . . OK, umm, give me a minute and I will tell you the exact reason.

AHS: OK. I'll give you a minute.

DW: OK. [Looking at the figure and enumerating what she can take as being certain.] This is tangent to this. It's given that the circle is tangent to this line and this line at this point and this point. OK. So we know that beyond a doubt. And—

Student DW again makes a qualitative attempt to justify her intuition, exhibiting behavior typical of the entire problem session. Not once during the half hour of "postmortem" in which their work on the four problems was discussed, or in the half hour of more general conversation that followed, did either SP or DW refer to any of what they had proved in Problems C or D in order to substantiate the assertions they made. Any time they were asked about why something was true, the students' response was empirical and intuitive; proof was never mentioned. The discussion that followed provided additional evidence regarding the students' views about the utility of proof. Toward the end of the interview the students were told what the purpose of the problem set had been. To provide some background and to explain how mathematical argument can be used as a means of discovery, the students were told about experts' solution to Problem 5.6 (which was discussed in the second section of this chapter).

AHS: [Describes the expert solution.] Now what he was doing was using mathematical argument the way a mathematician does. It told him what the answer was going to be and guaranteed that it worked.

DW: Then you go back and prove it.

This comment indicated that DW did not grasp the point of the example. It was explained further and then contrasted with the students' behavior.

AHS: That's the way the mathematician uses geometry. The way I think most students think of it is that "proof" is the thing the teacher forces you to do when you already know the answer. So when you try to solve a problem, like over here [Problem A], you don't even think of it. Even when I ask you why it's correct, and you've just done the proofs over here (Problems C and D), you don't think of it. . . . [In high school] you didn't learn that proof can actually help you. It's up there in a little box marked "High School Geometry. Do not open unless"—

SP: [Interrupting AHS in mid-sentence] I think we're sort of learning about that in [calculus] class right now, because we get the equation, and then we prove it. So that seems to follow.

Since the statement about the role of proof in geometry came from the experimenter and not from the students, one can hardly take the students' accord as a strong statement of belief on their part. On the whole, however, the discussions supported that statement. And the fact that SP drew the parallel between her previous geometry instruction and her current calculus instruction indicates that, at minimum, the hypothetical description of their perceptions struck a resonant chord. From the students' perspective, the role of "proof" is to confirm, for the record, what one knows for certain to be true.

DISCUSSION

The problem-solving session with SP and DW provides a complex and sometimes contradictory illustration of many of the points raised in this chapter. Before elaborating on their behavior, it should be stressed that these two students were quite bright and that their grasp of the geometric subject matter was a good deal better than average. They were also highly motivated. In the retrospective session SP said that, having done so well on Problems A and B, they had decided to solve all four problems in record time, and be "world champions."

Let us begin with a discussion of the students' work on Problem B, which provides clear documentation of the students' understanding of both the confirmatory and explanatory roles played by geometrical argumentation. There had been no mention of proof before the students worked Problem B, either by the students (in working Problem A) or when the stage had been set for the problem sessions. Moreover, the statement of Problem B does not mention proof; it asks the students to explain why their construction works. It is significant, then, that the students took the approach that they did. With virtually no hesitation they embarked on a congruence argument demonstrating that angles PVX and QVX in Figure 5.28 are equal, so that the segment VX bisects angle PVQ. From this behavior it appears that the students fully comprehended the role of "proof arguments" in not only confirming that geometrical constructions are guaranteed to work, but also in explaining why they work. There is certainly no question about their understanding of the generic nature of proof.

Consider now the students' behavior on Problems A, C, and D. If, for the time being, one ignores Problem B, SP and DW's work on these three problems and in the subsequent discussions points strongly to their being naive empiricists. The students' solutions to Problems C and D indicate that they possessed more than enough subject matter knowledge to enable them to derive the answer to Problem A. Yet their approach to Problem A was empirical from start to finish, from the intuitive way that they generated their hypothesis to the constructive way that they tested it. At minimum one can say that intuition took precedence over deductive knowledge in that approach. A much stronger case can be made, however. After having solved Problems C and D—which guarantee that their solution to Problem A is correct—the students were still not certain that they had the correct answer to Problem A. When they were told that their construction was correct, they expressed genuine relief and surprise. Their subsequent behavior indicated that they did not accept the results of their proofs as being compelling arguments. Asked why the center of the circle lies on the angle bisector, for example, they could have said "because we proved it does." Instead, they turned to an intuitive justification based on an apparent perception of symmetry. This behavior was entirely consistent through the problem session. The students appeared to see no connection between Problem A and Problems C and D.

This description raises two sets of questions. The first has to do with the accuracy of the assessments in the previous paragraph. Alternative hypotheses can be raised for almost all of the students' behaviors. Even if one ignores the students' behavior on Problem B, is the description of SP and DW as naive empiricists really plausible? The second has to do with reconciling their ostensible naive empiricism on Problems A, C, and D with their behavior on Problem B. In working that problem the students were fully aware of the explanatory nature of deductive arguments. How could they employ such knowledge there and ignore it elsewhere, especially since they demonstrated clear awareness that all four problems were related?

There is no definitive answer to the first set of questions. What can be said is that a broad range of alternative hypotheses regarding the students' actions has been considered, and those explanations have been found wanting. For example, consider the students' relief when they were told that their answer to Problem A was correct. Might it have been a matter of academic training; that is, the teacher is the ultimate arbiter of correctness, whether or not one is convinced of the answer (or has proved it correct)? The students were asked, and they said that, no, they were not really sure that their construction was right; the proof statements had not really convinced them. Similarly, one can find alternate explanations for the fact that the students, having proved that the center of the circle lies on the angle bisector, sought to justify that claim by intuitive means. Here are two such arguments.

First, the mathematical proof produced by the students provides a guarantee that the center of the circle lies on the angle bisector, something that can be accepted as fact. However, the (somewhat mysterious) argument by congruence does not provide a causal mechanism; it merely says what must be true. The attempt by DW to trace out the locus with her finger can be seen as an attempt at elaborating that causal mechanism, trying to demonstrate why the center lies on the angle bisector. If so, her argument should not be perceived as being contradictory to the mathematical argument, but as supplementary to it.

The second argument is that perhaps DW did indeed believe the formal proof she had produced. She may have produced the locus argument because of the way she interpreted the request that she justify her construction. Since she had already produced a formal proof, she may have thought that something additional was being requested.

These hypotheses are plausible, but they are unlikely when one views them in the context of the whole solution. Both hypotheses are based on the notion that the students accepted as valid the proofs in Problems C and D, but then felt that more was being asked for in the follow-up questions. But if the students really accepted the validity of the proofs in Problems C and D, they would have been certain of their answer to Problem A—which they were not. In fact, they only needed to accept the statements of Problems C and D to know for certain that their construction in Problem A was correct. They did not make that connection, so it is difficult to accept either argument above. Similar objections can be found to other plausible explanations of the students' behavior. In sum, the characterization of DW and SP's behavior as naive empiricism seems to fit fairly well, and, as a whole, other explanations do not stand up to close scrutiny.

What then of the apparent contradiction between the students' rather sophisticated behavior on Problem B and their naive empiricism on the others? This difference in behavior can, I think, be traced to a difference in context. As indicated by the literature on naive physics, people are capable of believing, at the same time, in theories that—once brought out into the open and compared—are mutually contradictory. People who invoke "force as mover" in some circumstances will invoke "deflection" in others, if the circumstances are compelling enough (diSessa, 1983, pp. 30–31). Context determines the explanatory framework one brings to bear in a situation, and contradictions between explanatory frameworks may not be perceived even though individuals invoke radically different knowledge in closely parallel situations. Something similar is taking place here. Problem B asked the students about a procedure they had used in Problem A, and about which they were absolutely confident. They knew that their procedure for bisecting an angle was correct. Moreover, the problem was trivial; it was an exercise rather than a problem. Given all this, SP and DW interpreted the term *explain* in the problem statement as they would interpret the term in a schoolbook exercise: One is being asked to explain for the teacher, and such explanations call for employing the formal procedures of geometry. The initial comment by DW, "Now if you want to be real technical about this . . . you can make triangles!" makes it quite clear that she understands that a particular game is being played. Good student that she is, she knows how to play by the rules.

The students' behavior on Problem B can now be seen as further confirmation of their naive empiricism. Both DW and SP were indeed capable of invoking proof as explanation. They could do so whenever the teacher demanded, or when they saw such demands being made (implicitly, but clearly) in a problem statement. These, however, are precisely the circumstances in which the students will freely invoke such knowledge: when they perceive themselves to be playing an academic game according to the rules established by the teacher. When the context is different—in particular when the students see themselves as needing to discover information in a real problem-solving situation—then they simply do not think to invoke that kind of knowledge. In some cases, such behavior results from an explicit and overtly manifested belief structure; and in others, because the students have a particular kind of mathematical world view in which it does not occur to them that engaging in mathematical argumentation might be useful. Either way, the results are similar. When placed in problematic situations, the students act as though engaging in mathematical argumentation has nothing at all to do with discovery or problem solving.

POSTSCRIPT

The focus of this chapter has been fairly narrow, concentrating solely on geometry after the opening literature review. The effects of belief systems or mathematical world views on performance are hardly limited to geometry, however; they pervade virtually all mathematical behavior. The discussion of geometry should simply be taken as a case in point. As indicated by the brief discussion of belief in Chapter 1, people's beliefs about the nature of mathematics help to determine whether or not they invoke formal mathematics to solve "real-world" problems; beliefs shape how people try to learn mathematics (memorizing vs. trying to understand); they shape how long people are willing to work on mathematical problems, and so on. This chapter just points to the tip of the tip of the iceberg, and much more work needs to be done. Also, this chapter paused only briefly to suggest the ways in which mathematical world views are developed as abstractions of individual's experience with mathematics. The implications for pedagogy are serious and deep, and much more work needs to be done in this area as well. I return to this issue at some length, in Chapter 10.

Summary

This chapter focused on the ways that people's mathematical world views shape the ways that those people behave in mathematical situations. The first section provided a brief literature review, in which similar phenomena were explored in other fields. Work on naive physics, everyday reasoning, and decision theory was described, and the borderline status that belief occupies between the cognitive and affective domains was discussed. The second section described an expert's approach to a construction problem, in which the subject (a professional mathematician) took a deductive approach to a problem that apparently called for discovery. The fact that deduction and discovery are flip sides of the same coin is part of the mathematician's world view. As indicated by the examples described in the sequel, students do not (for the most part) see things the same way.

The balance of the chapter was devoted to an exploration of the nature of students' geometrical empiricism. The third section offered a characterization of *pure empiricism* based on a collection of "empirical axioms" and a model for predicting the sequence of hypotheses students would test on simple straightedge-and-compass construction problems. Though the model captured the students' actions fairly well at the behavioral level, its description of the causes for the students' behavior was far too simplistic; the model was based on an *avowed empiricism,* and there is not compelling evidence to believe that students openly and actively reject the value of mathematical argumentation. The last two sections of the chapter elaborated a much richer explanation of such behavior, called *naive empiricism.*

In principle, the development of naive empiricism in geometry is similar to the development of, for example, naive physics. The idea in the latter case is that, abstracting from their experience, people develop models of "real-world" phenomena that enable them to function efficiently in the real world. Such models enable them to cross the street and not be hit by oncoming vehicles, to catch fly balls, and so on. Unfortunately such models—which are not necessarily overt or formally codified by the individual—may also contradict the formal laws of physics. When that happens, this natural and highly functional mechanism of abstraction from experience can cause substantial difficulties.

The situation in mathematics is similar but even more problematic, for much of mathematics deals with formalizations of abstract structures with which individuals have little real-world experience. Here the nature of the students' classroom experiences becomes absolutely critical. If the bulk of students' experience with particular mathematical ideas occurs in the classroom, then the students' mathematical world views—their abstraction of their experiences with those mathematical ideas—will be based on those experiences. In geometry, students' perspectives regarding the utility of mathematical argumentation (loosely speaking, "proof") will be based on the way proof has been used in the classroom. In the experience of most students, mathematical derivation is used only to verify propositions put forth by the teacher, propositions already known (by authority) to be true.

Such experience is abstracted as part of the students' mathematical world view as follows: Mathematical argumentation only serves to verify established knowledge, and argumentation (proof) has nothing to do with the processes of discovery or understanding. As a result, students who are perfectly capable of deriving the answers to given problems do not do so, because it does not occur to them that this kind of approach would be of value. Such students may fail to see that "proof problems" that they have already solved provide the answers to related "discovery problems" that they are now trying to solve. In fact (see Chapter 10), the answers they propose to current discovery problems may contradict the answers they have just derived in closely related proof problems.

Finally, it should be noted that naive empiricism is just one example of the ways that people's behavior is shaped by their beliefs about the nature of mathematics. This is an issue of some generality, with broad implications.

Part Two

Experimental and Observational Studies, Issues of Methodology, and Questions of Where We Go Next

Overview

Part One, consisting of Chapters 1 through 5, outlined and elaborated a broad theoretical framework for investigations of mathematical thinking. It defined four major categories of mathematical knowledge and behavior: resources, heuristics, control, and belief systems. The definitions were fleshed out with observational data, examples of mathematical behavior chosen as cases in point to clarify the nature of the behavior being discussed.

Part Two describes a series of research studies that, in part, provide the rigorous underpinnings for the discussions in Part One. Each of Chapters 6 through 9 introduces a set of tools for analyzing some aspect of mathematical problem-solving performance. These chapters are all dual-purpose. On the one hand, each provides a detailed elaboration of some behavior broadly characterized in Part One. As such, it serves to clarify some aspect of the framework. On the other hand, each is a methodological study in itself; the tools described in each chapter may in the long run prove to be at least as useful as the particular results discussed there. In consequence, there are extended discussions of the strengths and weaknesses of some approaches (particularly those for analyzing protocol data) and of methodological issues such as consistency and reliability.

Chapter 6 describes a small-scale laboratory study undertaken in the late 1970s. The literature on heuristic training at that time was suggestive but ambiguous. Simply put, there was no clear evidence that students could learn to master heuristic strategies. One hypothetical obstacle to such mastery could have been the lack of attention in most studies to control or executive behavior. This experiment took place within a format that minimized the difficulties that might be caused by poor control and that allowed for explorations of the ways that students used individual heuristics. It demonstrated that students can master individual heuristic strategies but that in general a variety of conditions necessary for such mastery have been underestimated. The experimental format presented here can be modified for further explorations of heuristic strategies.

Chapters 7 and 8 move from the laboratory to the classroom. They focus on the development of paper-and-pencil tests to capture various aspects of problem-solving performance. Chapter 7 describes measures of three aspects of problem-solving performance: (1) the frequency with which students generate a variety of problem-solving heuristics, how far they pursue them, and with what success; (2) the students' subjective assessments of their own problem-solving behavior, and (3) the students' transfer of heuristic behavior to problems related in various degrees (from quite similar to completely unrelated) to training problems in heuristics. Chapter 8 describes a fourth measure, an adaptation of a card-sorting task from cognitive psychology to explore people's perceptions of the structure of mathematics problems. All of these tests were given to students in a problem-solving course and to students in a control group, both before and after the courses were given. A comparison of the results indicates the kinds of learning that can take place in a problem-solving course.

Chapters 9 and 10 take as their raw data the transcripts of classroom sessions and the protocols of people working problems out loud. Chapter 9 deals with methodological issues related to protocol analysis. It presents a framework for parsing protocols at a macroscopic level, allowing for rigorous characterizations of control behavior and for tracing the effects of control decisions on the evolution of problem solutions. Protocols from before and after a problem-solving course are again discussed, providing another perspective on the effects of problem-solving instruction. Chapter 10 also examines such protocols, this time with an eye toward belief systems. A classroom session that unintentionally promotes the growth of counterproductive "mathematical world views" is described, and some issues for further consideration are raised.

6

Explicit Heuristic Training as a Variable in Problem-Solving Performance*

This chapter presents the results of a small-scale research study conducted at Berkeley in 1977–1978. It describes, in some detail, the problem-solving processes of seven upper-division college students working a series of problems that can be solved by the application of one or more heuristic strategies. In the study, two small groups of students were given practice on, and shown the solutions to, a set of problems that could be solved by heuristic methods. The amount of time working the problems was identical for the two groups, and the solutions they were shown were nearly identical. The primary difference in the treatments the two groups received was that the solutions shown to the experimental group explicitly included mention of the heuristic strategies used to solve the problems. The control group saw the same solutions but without an elaboration of the heuristic method. The major purposes of the experiment were as follows.

1. To see, first, whether students who received explicit training in the use of five particular heuristic strategies would be able to use those strategies to solve posttest problems comparable to, but not isomorphic to,

* Chapter 6 is an expanded and revised version of Schoenfeld (1979b). Permission from the *Journal for Research in Mathematics Education* to reproduce parts of the article is gratefully acknowledged.

the instructional problems. More generally, to see which aspects of the heuristic training would transfer to the postinstruction problems and to explore the conditions necessary (and possibly sufficient) for such transfer.

2. To test the hypothesis that problem-solving experience suffices for the apprehension of heuristic strategies. If so, the control students, who worked the same problems for the same amount of time and who saw the same solutions—but without the explicit heuristic instruction—should intuit the strategies as a result of their problem-solving experience and use them on the posttest problems.
3. To see, by comparing the two groups, if explicit instruction in heuristics makes a difference—both as measured by pretest-to-posttest gains and as indicated by an examination of the problem-solving procedures used by the students.

A Brief Discussion of Relevant Literature

About the time this study was conducted, E. G. Begle (March 1977, personal communication) wrote that "problem solving, in my opinion the most important outcome of mathematics education, at the upperclass undergraduate level, has not been seriously studied before." In consequence, the body of literature relevant to this study (those studies of mathematical behavior that focused on the processes rather than the products of mathematical problem solving) dealt for the most part with problem-solving tasks that were easier than, and called for less mathematical background than, the tasks discussed in this chapter. A brief summary of the relevant work within mathematics education follows.

The starting point for work in mathematical heuristics is, of course, Pólya. The source of much inspiration about problem solving, Pólya's books (1945, 1954, 1962, 1965) lay the groundwork for explorations in heuristics. They did not themselves deal with research issues about the ways that students could learn to use such strategies. Nor did college-level textbooks on problem solving (Rubinstein, 1975; Wickelgren, 1974). Lipson (1972) studied senior mathematics majors enrolled in a course on heuristics, but the focus of the research was on the carry-over effects of heuristic instruction on those students' students (her subjects were student teachers); her study does not bear directly on the work discussed here.

Goldberg's (1974) dissertation, which examines the behavior of nonmathematics majors taking a course in number theory, provides some evidence that explicit instruction in heuristics can enhance students' ability to construct proofs. That research is suggestive, although it should be noted that the study differed from this one along a number of dimensions. The level of mathematics required for solving the problems used here, and the level of mathematical sophistication of the subjects in this study, are a good deal more advanced than those in Goldberg's. That study was a "teaching experiment," this one a laboratory exploration.

In another teaching experiment, Smith's (1973) dissertation focuses on the transfer of heuristic learning. Unfortunately it produced disappointing results: The transfer of general heuristics was far less than was hoped for. Given the discussion of heuristic strategies in Chapter 3, that finding should come as no great surprise. Heuristics are far more complex and subtle than they appear to be on the surface, and mastering them even in fairly ideal situations is nontrivial. Transfer is much more difficult than mastery, for even if a student masters a strategy in one context it may be quite difficult for the student to see how to use it in another. Such problems are exacerbated in real-world instruction, where a broad range of things takes place at the same time and where it may be difficult for students to sort out what is essential. In sum, transfer is not to be expected unless the foundations for it have been carefully established.

Lucas (1972) studied students' use of heuristics in calculus classes. The results, although suggestive (decreased study time and more checking over solutions for the heuristics group, but, alas, no clear improvements on performance measures), are difficult to interpret clearly—a problem, unfortunately, with much real-world instruction and with much of the literature on heuristic strategies.

There is a substantial literature of problem-solving studies in mathematics education. Perhaps the best sense of the state of the art in the mid- to late 1970s can be found in Harvey and Romberg's (1980) monograph. That volume offers a literature review and the detailed summaries of nine dissertations on problem solving, most of which (like Lucas's, which is one of the studies included) offered interesting but often ambiguous or flawed results. Since problem solving has become a popular topic, one has many literature reviews to choose from. There are relevant surveys in the National Council of Teachers of Mathematics' (NCTM) 1980 Yearbook, *Problem Solving in School Mathematics;* the NCTM's (1980) research handbook, *Research in Mathematics Education;* Mary Lindquist's (1980) *Selected Issues in Mathematics Education;* Frank Lester's (1982) *Mathematical Problem Solving; Issues in Research;* and the Association for Supervision and Curriculum Development's (1981) *Mathematics Education Research: Implications for the 80's.* An extensive review of the mathematics education literature on problem solving prior to 1970 may be found in Kilpatrick (1969). Virtually all of this work, however, can be summed up in Begle's (1979, pp. 145–146) lament:

> A substantial amount of effort has gone into attempts to find out what strategies students use when solving mathematics problems. . . . No clearcut directions for mathematics education are provided by the results of these studies. . . . This brief review of what we know about mathematical problem solving is rather discouraging.

In part, the reason for such discouraging results was that the complexity of heuristic strategies had been consistently underestimated; with hindsight it appears that adequate groundwork was not established for heuristic mastery. In part, it was due to the fact that most studies of heuristic behavior took place in the context of treatment comparison experiments, where the comparison was between ordinary and modified classroom instruction. In real-world instruction, whose primary attention is to a particular body of subject matter (which may or may not be amenable to a heuristic approach), a large number of things are taking place at once. The instruction mixes attention to the subject matter and to the heuristics, and the students are trying to master both at the same time. It is not easy in these circumstances to say exactly what the students have been taught. If they do not perform at the desired or expected level (almost always the case with heuristic instruction), it is not easy to identify which aspects of the subject matter they did learn, and what might have kept them from learning to employ the heuristic techniques. Difficulties in this regard were compounded by the methodologies used for treatment comparisons. While a statistical comparison on a given measure will indicate whether or not there were significant differences in performance between two groups, it will not say much about what the subjects actually did. Thus in the case of negative results (no significant performance differences between treatment groups), the evaluation measures were, in general, uninformative. So, for somewhat similar reasons, were the results obtained from the use of "process coding schemes" to analyze problem-solving protocols.*

* It became increasingly apparent through the 1960s and 1970s that statistical treatment comparison measures were not yielding insights into problem-solving processes. Kilpatrick's (1967) dissertation represents the first significant attempt to code protocols of problem-solving performance. Kilpatrick's coding scheme was modified for Lucas's (1972) study, and it has been used in modified form in a number of studies since then. These and other such measures will be discussed at length in Chapter 9, so they are mentioned here only briefly. In this context the important thing to know is how such measures were used. In general, heuristic behaviors were coded for in problem-solving transcripts. Then the analyses undertaken were statistical; correlations were sought between problem-solving success and the use of certain heuristic strategies. Some results did emerge, such as Kantowski's (1977) obervation that success was generally correlated with the use of goal-oriented heuristics. This correlational approach did not, however, reveal information about how the heuristics were used. As above, the statistical approach proved generally uninformative when there were no significant differences in performance between "heuristic" and "control" treatment groups.

For these reasons the study described in this chapter was deliberately narrow and deliberately controlled. It took place in a laboratory setting, where as much control as possible over the details of instruction and training was maintained. It dealt with a small number of heuristic strategies, so that issues of executive or control behavior (as defined in Chapter 4) would not have a substantial impact on performance. It specified in detail the differences in training between the experimental and control groups, so that the differences (if any) in their performance after instruction could be traced to the differences in their training. In assessing the students' work after instruction, it focused on the processes that the students used as they worked the posttest problems. Thus if any differences were found, they could be traced to differences in behavior.

Experimental Design

The participants in the study were seven upper-division science and mathematics majors recruited as volunteers from advanced undergraduate courses in mathematics at the University of California, Berkeley. When they were recruited the students were told that the purpose of the experiment was to have them "work a bunch of specially chosen problems designed to improve your problem-solving ability." Four of the seven students were randomly chosen for the experimental treatment. Two of these four were mathematics majors, as were two of the three students who served as the control group. The mathematical backgrounds of all the students were comparable.

All of the students were trained and tested individually in order to obtain as much detailed information as possible about the problem-solving processes that they used. Before the experiment, the students were trained to talk out loud as they solved problems. Then each took a pretest, consisting of the five problems given in Table 6.1. They worked out loud on each problem for 20 minutes, or until they were confident that they had solved it. (This process was repeated for the posttest, given in Table 6.9.) The "data" produced by each student consisted of written work on the problems plus the transcripts of his or her out-loud protocols.

The subject matter for the experiment consisted of the five heuristic strategies listed in Table 6.2. Each strategy is particularly useful in the solution of one of the pretest problems and for a corresponding posttest problem as well. It should be noted that almost all of the test problems can be solved two different ways; many of them can be solved four or five different ways. Any solution to a problem was given credit, of course. The problems are

Table 6.1
Test A: Pretest

A1. Let a and b be given real numbers. Suppose that for all positive values of c the roots of the equation

$$ax^2 + bx + c = 0$$

are both real positive numbers. Present an argument to show that a must equal zero.

A2. Ten people are seated around a table. The average income of these 10 people is $10,000. Each person's income is the average of the incomes of the people sitting immediately to his left and right. What is the possible range of incomes for each person? (Incomes are given in whole dollar amounts.)

A3. Let n be a given whole number. Prove that if the number $(2^n - 1)$ is a prime, then n is also a prime number.

A4. You are given the real numbers a, b, c, and d, each of which lies between 0 and 1. Prove the inequality

$$(1-a)(1-b)(1-c)(1-d) > 1-a-b-c-d.$$

A5. What is the sum of the series

$$\frac{1}{1\cdot 2}+\frac{1}{2\cdot 3}+\frac{1}{3\cdot 4}+\cdots+\frac{1}{n\cdot(n+1)}?$$

Prove your answer if you can.

fairly difficult, however. Thus one would expect that being able to use the heuristic approaches, which are generally among the more direct and efficient ways to solve the problems, would contribute significantly to improved performance on the posttest.

The instructional sequence consisted of 20 problems, which included the five pretest problems. There were four training problems for each heuristic strategy. As an example, the complete set of problems that illustrate the use of induction is given in Table 6.3.

The training sessions took place over a period of two weeks, with each student coming for five individual sessions. Every attempt was made to control for the significant instructional variables that could be controlled during the training and to monitor the rest. Virtually all of the instruction was carried out through written materials supplemented by tape recordings. In each of their training sessions, the students were given a tape recorder and

Table 6.2
The Five Problem-Solving Strategies

1. Draw a diagram if at all possible.
   Even if you finally solve the problem by algebraic or other means, a diagram can help give you a "feel" for the problem. It may suggest ideas or plausible answers. You may even solve a problem graphically.

2. If there is an integer parameter, look for an inductive argument.
   Is there an "n" or other parameter in the problem that takes on integer values? If you need to find a formula for $f(n)$, you might try one of these:
   a. Calculate $f(1), f(2), f(3), f(4), f(5)$; list them in order, and see if there's a pattern. If there is, you might verify it by induction.
   b. See what happens as you pass from n objects to $n + 1$. If you can tell how to pass from $f(n)$ to $f(n + 1)$, you may build up $f(n)$ inductively.

3. Consider arguing by contradiction or contrapositive.
   Contrapositive: Instead of proving the statement "If X is true then Y is true," you can prove the equivalent statement "If Y is false then X must be false."
   Contradiction: Assume, for the sake of argument, that the statement you would like to prove is false. Using this assumption, go on to prove either that one of the given conditions in the problem is false, that something you know to be true is false, or that what you wish to prove is true. If you can do any of these, you have proved what you want.
   Both of these techniques are especially useful when you find it difficult to begin a direct argument because you have little to work with. If negating a statement gives you something solid to manipulate, this may be the technique to use.

4. Consider a similar problem with fewer variables.
   If the problem has a large number of variables and is too confusing to deal with comfortably, construct and solve a similar problem with fewer variables. You may then be able to
   a. adapt the method of solution to the more complex problem,
   b. take the result of the simpler problem and build up from there.

5. Try to establish subgoals.
   Can you obtain part of the answer, and perhaps go on from there? Can you decompose the problem so that a number of easier results can be combined to give the total result you want?

Table 6.3
Problems Amenable to an Inductive Approach

1. (pretest) What is the sum of the series

$$\frac{1}{1 \cdot 2} + \frac{1}{2 \cdot 3} + \frac{1}{3 \cdot 4} + \cdots + \frac{1}{n \cdot (n+1)}?$$

   Prove your answer if you can.
2. You are given n points in the plane, none of which lie on a straight line. How many straight lines can you draw, if each straight line must pass through two of the n points?
3. Let x be any odd integer. Show that x^2 leaves a remainder of 1 when divided by 8.
4. Determine a formula for the product

$$(1 - \tfrac{1}{4})(1 - \tfrac{1}{9})(1 - \tfrac{1}{16}) \cdots (1 - \tfrac{1}{n^2}).$$

   Prove it if you can.
5. (posttest) Let S be a set which contains n elements. How many different subsets of S are there, including the null set?

a booklet with four practice problems. The students worked on each problem for up to 15 minutes or until it was solved. At that point they turned to the next page of the booklet, which presented a solution to the problem. The students also turned on the tape recorder to listen to an explanation of the solution. (These explanations ran parallel to, but were not identical to, the written solutions. The purpose of the tapes was to allow students to hear the solutions, which often helps in comprehension.) If, after thinking about a problem solution for 10 minutes, a student had a question, he or she could ask me. If the question could be answered by the instructional materials, the student was referred back to them. Other questions were generally limited to technical matters such as "the converse isn't always true but the contrapositive is, right?"

A number of my colleagues listened to the complete set of instructional tapes before they were used, comparing the explanations that would be heard by the experimental and control groups. Their judgment was that there was no discernible difference between the two instructional treatments in either enthusiasm or clarity of presentation, so that any differences in scores between the two groups should not be attributable to "experimenter bias." During the training sessions, the amount of time allotted (and spent) for solving problems in both the control and experimental groups was kept the same. With one slight variation described below, the students in both groups saw (and heard) the same solutions. The differences in treatment between the two groups were as follows:

1. The students in the experimental group were informed (on tape) at the beginning of their first practice session that the experiment would try to show that five specific strategies would help them to solve problems. They were then given the list of strategies (Table 6.2). The list was placed conspicuously in front of them during all practice sessions and during the posttest. At the first session the students in the heuristics group listened to a 10-minute tape providing a general introduction to the strategies. The students in the control group were also welcomed to the experiments on tape. They were told that the problems and the instruction had been carefully chosen to develop their problem-solving skills, and they were told how to use the materials.

2. Although the solutions to each problem seen by both groups were identical, the students in the heuristics group saw in addition an overlay to each solution. That overlay indicated which heuristic strategy had been used to solve the problem and how it had been used. Tables 6.4 – 6.8 provide the complete solutions to each of the pretest problems, each of which illustrates the use of a different one of the five strategies. The experimental group saw each table in its entirety. The control group saw everything on the page with the exception of the boxed description of the heuristic strategy at the far left of each table. This was the only difference in the written materials.

The tape recordings heard by each group recapitulated the essential ideas in the written solutions, although not word for word. In the tape corresponding to Table 6.8, for example, the nonheuristics group listened to a tape that said "Let's calculate a few of the sums and see what happens. The first term is $\frac{1}{2}$." The heuristics group heard "Notice that there is an *n* in the problem statement. When we see an integer parameter, we should calculate the values and see what happens. The first term is $\frac{1}{2}$; . . ." It should be stressed that colleagues had listened to all of the tapes and determined that there was no difference in clarity, level of exposition, or enthusiasm of presentation between the tapes that the two groups of students heard.

3. The order of the problems was different. In each session the experimental group focused on practicing one particular strategy. The first three of the four problems they worked provided training in the use of that strategy. The fourth was different, both for variety and to keep them from thoughtlessly falling into a pattern. For the nonheuristics group the order of the problems was scrambled. The reason for scrambling the order of the problems worked by the control group is discussed below.

Table 6.4 Pretest Question 1

THIS IS A SIMPLE EQUATION: DRAW A DIAGRAM.
HOW CAN YOU SHOW SOMETHING IS 0? ASSUME IT IS NOT, AND CONTRADICT A GIVEN STATEMENT. (YOU HAVE MORE TO WORK WITH WHEN YOU ASSUME $A \neq 0$.)
A PICTURE OF WHAT IS HAPPENING CAN OFTEN SUGGEST THE ARGUMENT.

Let a and b be given real numbers. Suppose that for all positive values of c, the roots of the equation

$$ax^2 + bx + c = 0$$

are both real, positive numbers. Present an argument to show that a must equal 0.

SOLUTION

What if the number a is not 0? It's either positive (Case I) or negative (Case II). Examine the two cases.

Case I: What if a *is positive?*

The graph of $y = ax^2 + bx + c$ represents a "right-side-up" parabola. When we say the roots of the equation ($ax^2 + bx + c = 0$) are both real numbers, we are saying that the parabola crosses the x axis. But what about the number c? The larger c is, the further *up* we shift the parabola (see Figure 1).

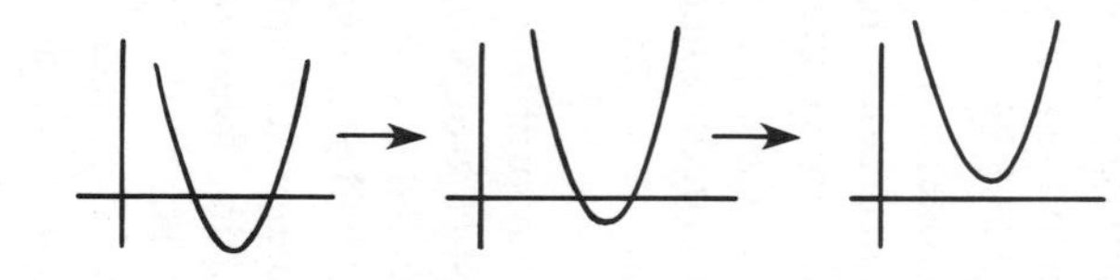

Figure 1: The parabola $y = ax^2 + bx + c$, for increasing values of c when a is positive.

For very large positive c, the curve is shifted so far up that it does not cross the x axis. But then its roots are not real, a contradiction. So a cannot be positive.

Case II: What if a *is negative?*

The graph of $y = ax^2 + bx + c$ is now an "upside-down" parabola. If we make c larger, we shift the parabola upward and raise the point where the parabola crosses the y axis (see Figure 2). For large c, that point is shifted above the x axis, and the parabola will have a negative root. This is a contradiction, so a cannot be negative. Since a cannot be either positive or negative, it must be 0.

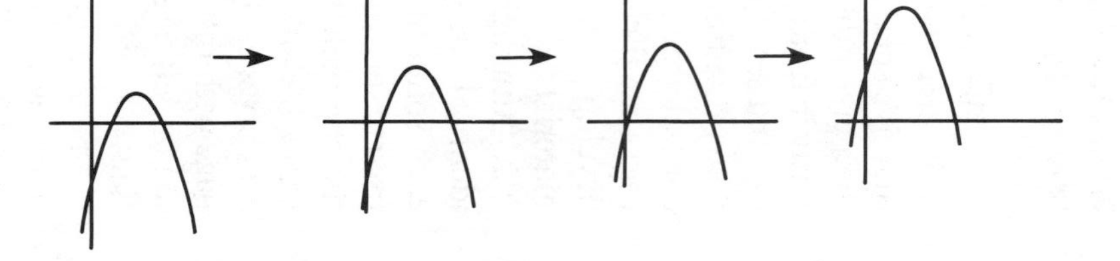

Figure 2: When a is negative, the parabola $y = ax^2 + bx + c$, for increasing values of c.

Table 6.5
Pretest Question 2

Ten people are seated around a table. The average income of these 10 people is $10,000. Each person's income is the average of the incomes of the people sitting immediately to his left and right. What is the possible range of incomes for each person? (Incomes are given in whole dollar amounts.)

SOLUTION

What about the person, say *A*, who makes the most money? We are told that his (or her) income is the average of the incomes of the people to his right and left. Now if one of these people makes *less* money than *A* does, the other person must make *more.* That is impossible, since *A* makes the most. That says that all three earn the same. Suppose we call the person to the right of *A*, person *B*. What about *C*, the person to the right of *B*? Person *B* makes the same salary as *A*, but *B* is also the average of *A* and *C*. So *C* makes the same as *A* and *B*. We can make the same argument all the way around the table. Everybody makes exactly the same amount of money: $10,000.

WHERE CAN YOU START?
ESTABLISH A SUBGOAL.
CHOOSE A PARTICULAR PERSON, AND DETERMINE THAT PERSON'S INCOME.

THE GOAL IS: PROVE n IS PRIME.
HOW DO YOU PROVE THAT? THERE ISN'T MUCH TO WORK WITH.
IF WE START WITH CONTRADICTION OR CONTRAPOSITIVE, THE ARGUMENT BEGINS WITH . . .
n IS NOT PRIME, SO

$$n = a \times b.$$

THIS GIVES US MUCH MORE TO WORK WITH.

Table 6.6
Pretest Question 3

Let n be a given whole number. Prove that if the number $(2^n - 1)$ is a prime, then n is also a prime number.

SOLUTION

Suppose n is not prime. Then $n = a \cdot b$, where a and b are both integers greater than 1. We can write

$$2^n - 1 \quad \text{as} \quad 2^{ab} - 1, \quad \text{or} \quad (2^a)^b - 1.$$

Now any number of the form $(X^b - 1)$ has a factor of $(X - 1)$, so that $2^n - 1 = (2^a)^b - 1$ has a factor of $(2^a - 1)$. Note that this factor, $2^a - 1$, is larger than 1 but smaller than $2^n - 1$. That tells us that $(2^n - 1)$ is not a prime, contradicting our given fact. Thus n must be a prime.

Table 6.7
Pretest Question 4

You are given the real numbers a, b, c, and d, each of which lies between 0 and 1. Prove the inequality

$$(1-a)(1-b)(1-c)(1-d) > 1-a-b-c-d.$$

SOLUTION

Suppose we start by proving the equation

$$(1-a)(1-b) > 1-a-b \qquad \text{(Equation 1)}$$

If we multiply out the left, Equation 1 is true if and only if

$$1-a-b+ab > 1-a-b, \text{ which is true if and only if } ab > 0.$$

But $ab > 0$, since we were given that a and b are both positive. This proves Equation 1. Now let us build on this. The number c is between 0 and 1, so $(1-c)$ is positive. Multiplying both sides of Equation 1 by $(1-c)$, we get

$$(1-a)(1-b)(1-c) > (1-a-b)(1-c)$$

or

$$(1-a)(1-b)(1-c) > 1-a-b-c+ac+bc.$$

Since ac and bc are both positive, we obtain

$$(1-a)(1-b)(1-c) > 1-a-b-c. \qquad \text{(Equation 2)}$$

Continuing in the same vein, we notice that $(1-d)$ is positive; multiplying both sides of Equation 2 by $(1-d)$, we obtain

$$(1-a)(1-b)(1-c)(1-d) > (1-a-b-c)(1-d),$$

or

$$(1-a)(1-b)(1-c)(1-d) > 1-a-b-c-d+ad+bd+cd.$$

As before, we see that ad, bd, and cd are all positive. Thus

$$(1-a)(1-b)(1-c)(1-d) > 1-a-b-c-d,$$

which is what we wanted to prove.

THE FOUR-VARIABLE PROBLEM IS TOO COMPLICATED.
CAN WE LEARN SOMETHING FROM A SIMILAR ONE-VARIABLE PROBLEM? NOPE.
HOW ABOUT THE COMPARABLE TWO-VARIABLE PROBLEM? IT'S EASY TO SOLVE.
CAN WE USE THE RESULT? YES
. . . BUILD UP TO THREE VARIABLES
. . . THEN BUILD UP TO FOUR.
REMEMBER, WHEN A PROBLEM IS COMPLICATED . . .
CONSIDER A SIMILAR PROBLEM WITH FEWER VARIABLES. THEN TRY TO USE EITHER THE METHOD OR RESULT TO SOLVE THE ORIGINAL PROBLEM.

Table 6.8

Pretest Question 5

What is the sum of the series

$$\frac{1}{1 \cdot 2} + \frac{1}{2 \cdot 3} + \frac{1}{3 \cdot 4} + \cdots + \frac{1}{(n) \cdot (n+1)}?$$

Prove your answer if you can.

NOTICE THE INTEGER PARAMETER . . .

Solution

CALCULATE THE VALUE FOR $N = 1, 2, 3, 4, \ldots$

Let us examine a few of the sums.
The first term is $\frac{1}{2}$.
The sum of the first two terms is $\frac{1}{2} + \frac{1}{6} = \frac{2}{3}$.
The sum of the first three terms is $\frac{2}{3} + \frac{1}{12} = \frac{3}{4}$.
The sum of the first four terms is $\frac{3}{4} + \frac{1}{20} = \frac{4}{5}$.

GUESS THE FORMULA FROM THE PATTERN ⟶ We can be pretty certain at this point that the sum of the first n terms is $\frac{n}{(n+1)}$. As usual, we

VERIFY IT BY INDUCTION

verify a formula like this by induction. For $n = 1$, the sum of the first n terms is $\frac{1}{2}$, so the formula checks. Suppose the sum of the first k terms is $\frac{k}{(k+1)}$; that is,

$$\frac{1}{1 \cdot 2} + \frac{1}{2 \cdot 3} + \frac{1}{3 \cdot 4} + \cdots \frac{1}{(k) \cdot (k+1)} = \frac{k}{k+1}.$$

Then the sum of the first $(k+1)$ terms is

$$\frac{1}{1 \cdot 2} + \frac{1}{2 \cdot 3} + \cdots + \frac{1}{(k) \cdot (k+1)} + \frac{1}{(k+1) \cdot (k+2)} =$$

$$\frac{k}{k+1} + \frac{1}{(k+1)(k+2)} = \frac{k^2 + 2k + 1}{(k+1)(k+2)} = \frac{(k+1)^2}{(k+1)(k+2)}$$

$= \frac{k+1}{k+2}$, which is the desired formula for $n = k + 1$.

This completes the argument.

Table 6.9
Test B: Posttest

B1. Suppose p, q, r, and s are positive real numbers. Prove the inequality

$$\frac{(p^2+1)(q^2+1)(r^2+1)(s^2+1)}{pqrs} \geq 16.$$

B2. For what values of "a" does the system of equations

$$\begin{Bmatrix} x^2 - y^2 = 0 \\ (x+a)^2 + y^2 = 1 \end{Bmatrix} \quad \text{have}$$

a) no solutions?
b) 1 solution?
c) 2 solutions?
d) 3 solutions?
e) 4 solutions?

B3. Let S be a set which contains n elements. How many different subsets of S are there, including the null set?

B4. Prove that the product of any three consecutive whole numbers is divisible by 6.

B5. Let A and B be two given whole numbers. The *greatest common divisor of A and B* is defined to be the *largest* whole number C that is a factor of both A and B. For example, the GCD of 12 and 39 is 3, and the GCD of 30 and 42 is 6. PROVE that the greatest common divisor of A and B is unique.

4. Finally, there was one slight difference in the posttest. Both groups of students were interrupted every 5 minutes as they worked the posttest problems. At these periodic interruptions the control students were told the following: "You've been working for 5 minutes now. You may be on the right track, you may not. But stop, take a deep breath, and look over your work. Then decide whether you want to continue in that direction." The experimental group was told the following: "You've been working for 5 minutes now. You may be on the right track, you may not. But stop, take a deep breath, and look over the list of strategies. Then decide whether you want to continue in that direction." The reason for such systematic intrusions is also given below.

Results

The posttest used in this study is given in Table 6.9. For the record, the heuristic approaches most relevant to solving the posttest problems are (B1) the "fewer variables" technique, (B2) drawing a diagram, (B3) induction, (B4) subgoals, and (B5) contradiction. Table 6.10 summarizes the test results. Two measures of success were used. The first, which is stringent but unambiguous, was an all-or-nothing evaluation of the students' papers. A score of 1 was awarded for a problem solved completely, a score of 0 otherwise. The results are given in the first two columns of Table 6.10.

Table 6.10
Paper-and-Pencil Scores Earned by the Students

	Completely solved problems			Almost completely solved problems		
	Pretest	Posttest	Difference	Pretest	Posttest	Difference
Nonheuristics students						
S_1	2	2	0	3	4	1
S_2	1	2	1	2	3	1
S_3	2	1	−1	2	1	−1
Heuristics students						
S_4	2	5	3	3	5	2
S_5	0	2	2	0	2	2
S_6	2	4	2	2	5	3
S_7	0	2	2	0	2	2

Since the problem-solving abilities of students can vary substantially at the onset of an experiment, the most appropriate measure of the instructional materials' effect is the difference in each student's scores from pretest to posttest. These differences are recorded in the third column.

Observe that all four of the heuristics group improved from pretest to posttest, while only one of the nonheuristics group did. In fact, the average net gain for the nonheuristics group is 0, and for the heuristics group it is more than 2. More importantly, all four of the students who received heuristics training outscored all three who did not. The probability of this happening randomly is $\frac{1}{35}$, so the differences in the students' performance are statistically significant at $p < .05$.*

A second gross measure of success is to look at "almost completely solved" problems. In some cases the students had made minor arithmetic mistakes in working the posttest problems. In the all-or-nothing scheme they were given a 0, a rather harsh evaluation. A slightly more lenient grading scheme offered a chance to compensate for that harshness. Also, the problem-solving process was cut off at 20 minutes for each posttest problem. We can ask if it is clear that the student would obtain a solution to the problem if given another 5–10 minutes. In these circumstances as well the student is now assigned a score of 1; otherwise the student receives a score of 0. The most lenient awarding of "almost" credit for a student's work is described below in the discussion of Subject 1's approach to Problem B1, so the reader can see how the standard was applied. A second scorer blind-graded the papers

* One hardly sets out to do statistical analyses with sample sizes of four and three. I had intended initially only to examine the transcripts of the students' solutions, looking for differences in their problem-solving performance. The strength of these results came as somewhat of a surprise, but I am not averse to listing statistically significant results when they fall in my lap.

independently, and his ratings agreed with mine on all but one problem (where he would have given "almost" credit for a partial solution to a student in the experimental group and I had not). The results obtained with this measure are given in the last three columns of Table 6.10. The comments made about the data in the first three columns apply once again. The results are again significant at $p < .05$.

The data provide a clear indication that the heuristics group's performance improved in comparison with the nonheuristics group's performance. They do not, however, indicate how or why that improvement took place. For that we turn to the protocols of the solutions themselves. We shall examine how all seven students worked posttest Problem B1:

Suppose p, q, r, and s are positive real numbers. Prove the inequality

$$\frac{(p^2+1)(q^2+1)(r^2+1)(s^2+1)}{pqrs} \geq 16.$$

The problem is most easily dispatched by employing the fourth strategy on the list, "consider a similar problem with fewer variables." In the problem statement the variables p, q, r, and s all play identical roles. The left-hand side of the inequality is (in essence) the product of four identical terms, each of the form $(x^2 + 1)/x$; the right-hand side is 2^4. Thus one need only prove the one-variable inequality

$$\frac{(x^2+1)}{x} \geq 2.$$

One can then substitute p, q, r, and s for x in turn and take the product of the four resulting inequalities.

None of the three students in the nonheuristics group solved the problem, although Subject 1 was in the right ballpark and was given an "almost" for her efforts. The general outline of Subject 1's approach was much like the solution I have just described. Within a few seconds of reading the statement of the problem, she said "OK, I've got to show that $(p^2 + 1)/p$ is always greater than or equal to 2." Then she got sidetracked. She noticed that the quotient equals 2—the hypothetical minimum value for the function—when p is 1. After noting that the value of $(p^2 + 1)/p$ increased for the next few integer values of p, she tried to prove the result by induction! At the third "5-minute warning" she gave this up and made a sketch of the graph of $f(p) = (p + 1/p)$ that was accurate but was not analytically justified. She finished this just within the allotted 20 minutes and said "yes" when asked if she were satisfied with the solution. When asked "What if we weren't satisfied?", she said "I guess I could always find the minimum of $(p + 1/p)$ by calculus." She proceeded to do so in short order, which earned her an "almost" for the problem. As noted above, this was the most lenient granting of partial credit in all the grading.

After Subject 1 showed me her solution, I complimented her on it. Her response was the following. "I noticed that whenever a practice problem had lots of variables in it, you tried to do the one- or two-variable problem. So I tried to do that here." We return to this observation in the discussion below.

While Subject 1's behavior changed radically from pretest to posttest, the posttest behavior of both Subject 2 and Subject 3 very much resembled their pretest behavior. These two students had each jumped into messy algebraic computations on pretest Problem A4, and they acted the same way on posttest Problem B1. Each persisted, through the third 5-minute warning, in trying to show by some sort of clever factorization that

$$(p^2 + 1)(q^2 + 1)(r^2 + 1)(s^2 + 1) - 16pqrs \geq 0.$$

With less than 5 minutes left, Subject 2 decided to set $p = q$ and $r = s$, but that led nowhere and Subject 2 ran out of time. Subject 3 observed that substituting the value of 1 for each of p, q, r, and s led to the minimum value of 16 and then ran out of time. Neither student had come anywhere near a solution.

The performance of the heuristics group was quite different. Subject 4 read the problem statement, said "that's a fewer variables problem," and tried to analyze the two-variable inequality

$$\frac{(p^2 + 1)(q^2 + 1)}{pq} \geq 4.$$

She was unsuccessful with this. After about 8 minutes she stopped, said "What about the one-variable problem?", and went on to solve the problem within 4 minutes.

Throughout the entire posttest, Subject 5 consciously relied on the list of strategies. After reading the statement of Problem B1, he turned to the list and checked off the strategies one by one: "Let's see, I don't think I can really draw a diagram for this one, and there isn't an integer parameter so I can't do an induction, and let's see, what would the contrapositive be? No, that doesn't make sense; Ok, how about fewer variables? Yeah. Let me see if the one-variable problem makes any sense. . . ." He solved the problem by the first warning.

Much like Subjects 2 and 3, Subject 6 jumped right into a complicated algebraic morass; she was happily calculating away when she was given the 5-minute warning. At that point she stopped, looked at the strategies, and chose the appropriate one. After spending a few minutes on the two-variable problem, she solved the problem correctly.

Subject 7 first tried to disprove the problem statement by showing that the product of the four terms must be less than 16 and took about 8 minutes to realize that this logic was flawed. After that he wrote out the left-hand side of the inequality as the product of four similar factors. He then decided to explore the nature of $(p^2 + 1)/p$ by plugging in a few values of p and then seeing what happened as p grew infinitely large. He was nowhere near a solution when he ran out of time.

This brief look at the students' solutions to Problem B1 supports the results suggested by the statistics: The conscious application of the problem-solving strategy does indeed have a positive impact on their problem-solving performance. Three of the four students in the experimental group solved the problem, all by the deliberate application of the heuristic strategy. Only one of the three control students solved the problem, and she did so by intuiting and applying the strategy that the others had been taught. None of the students had used this strategy on pretest Problem A4.

The results were more dramatic for the induction Problems A5 and B3. On the pretest two of the students (one from each group) solved Problem A5 by recalling from their calculus days that the given series can be reexpressed as a "telescoping series." The rest got nowhere. All four students in the experimental group solved the posttest problem, but none of the others did. Each student who observed that the integer parameter n in the problem statement suggests induction as a method of solution was able to solve the problem; each student who did not make that observation failed to solve it. The results were similar but not as strong for the "diagram" problems, A1 and B2. The pretest problem was quite difficult, and none of the students did well on it. Those who drew a diagram on the posttest were able to make some progress on Problem B2, whether or not they were given credit for it. Those who did not do so soon got lost in a maze of computations.

The results were inconclusive for the other two strategies. Posttest Problem B5 disturbed some of the students, who either saw nothing to prove or who got jumbled in the mechanics of a proof by contradiction. There was comparable, and disappointing, performance for both groups. More instruction, or perhaps a different posttest problem, might have given different results.

The two groups of students also performed comparably on the two "subgoals" problems, A2 and B4. In the light of what I now know, this should have been expected. Using subgoals is a very complex strategy, and it was naive to expect that, as a result of having worked the five practice problems, the students learned enough about the strategy to be able to use it to solve posttest Problem B4.

Two Methodological Questions

Before the results of the study and their implications are discussed, the two methodological issues raised in the section on experimental design should be addressed: (1) Why was there a 5-minute warning? (2) Why was the problem order scrambled for the control group but not for the experimental group?

The blunt answer to question 1 is that, without periodic reminders to sit back and take stock, the students might well go off on wild goose chases. If so, the effects of bad control might well wash out any positive effects of heuristic learning. One major purpose of the experiment was to see if, and under what circumstances, students could learn to use the five heuristic strategies. It was essential to make certain that the students had the opportunity to use them. Otherwise, one might never find out whether they were capable of doing so.

As Subject 6's performance on Posttest Problem B1 indicates, this concern was justified. Even though Subject 6 was taking part in an experiment designed to teach the five strategies, had the list of strategies in front of her at all times, and was fully competent (as her protocols show) at using the "fewer variables" strategy, she ignored it completely when given Problem B1. Right after she read the problem she jumped into the kind of complex computations that could easily occupy her for 20 minutes—just as similar computations had occupied her for 20 minutes on the corresponding pretest problem. Only at the 5-minute warning did she stop to reconsider. Changing her approach, she soon solved the problem. This kind of behavior occurred more than once during the posttest. On posttest Problem B2, for example, Subject 4 immediately began working on an algebraic solution to the two given simultaneous equations. She was deeply involved in an incorrect solution when asked to "take a deep breath" and reconsider. That was enough. With an "Oh, sure; draw a diagram," she went on to solve the problem easily.

It should be recalled that the control group also received 5-minute warnings. As with the experimental group, reactions to the warnings varied with individuals. Some continued with a brief pause, some sat back and reflected. The major difference in behavior is that the warnings did not trigger access to the strategies for the control group, as they did with the experimental group.

In answer to question 2, the problems were ordered differently for the following reason. In the case of the heuristics group, the problems were explicitly ordered for optimal training. Again, the idea was to see whether the students could learn to use the strategies. This kind of ordering would be stacking the deck for the control group, since one purpose of the experiment was to see whether those students would use or intuit the strategies as a result of their problem-solving experience. Even with the scrambled order, however, the deck was fairly well stacked in favor of the control group's picking up the heuristics. Within a period of 2 weeks the control students worked and were shown the solutions to a total of 20 problems. Four were solved by induction, four by the use of diagrams, four by contradiction, four by fewer variables, and four by subgoals. If one learns to solve problems by solving problems—an argument colleagues have often advanced to justify the claim that there is no need to provide explicit instruction in problem-solving techniques—then a concentrated dose of problems like those used in the training should have served as an enriched breeding ground for heuristic strategies.

Discussion

We consider, in order, the questions listed in the first section of this chapter. Capsule versions of the questions are given below.

Question 1A Could students master the five heuristic strategies taught in the experiment?

The data indicate that, with explicit instruction, students can learn to use some heuristic strategies. Despite the limited amount of training (working on four practice problems is hardly overkill) the students in the heuristics group did learn to use three of the five strategies (fewer variables, induction, and drawing diagrams) by the time they took the posttest.

Well, perhaps. Deciding what these students really learned as a result of their heuristic training is a delicate issue. All of these students had fairly extensive backgrounds in mathematics when they began the experiment. All of them had already mastered the skills required to apply the heuristic techniques, and in all likelihood they had already solved many problems using those skills. The most probable interpretation of what took place during the practice sessions is that the explicit mention of the heuristic techniques served to bring those skills to the students' conscious attention and to help them codify and reorganize their existing knowledge in such a way that those skills could now be accessed and used more readily. It is not necessarily the case that students with weaker backgrounds, who would need both to master the mathematical techniques and to learn to recognize when they were useful, would fare as well. This is an empirical question.

Question 1B More generally, what can be said about the conditions necessary to support the learning and transfer of these and other heuristic strategies?

In a sense, the failures of this experiment—the places where students ran into difficulty—were more interesting and enlightening than the successes. Students did not do well on either the subgoals or contradiction problems.

"Establishing and using subgoals" is one of the broader and more complex heuristic strategies. With hindsight, it is clear that practice on four problems that do not appear closely related (in form) to the posttest problem would not be adequate to ensure transfer. The training did not provide enough practice to make it likely that students would master the mechanics of using the subgoals procedure. Nor was it sufficiently explicit about how to decompose problems and to select subgoals appropriate for their solution. What would be adequate is again an empirical question. A revised version of this study might contain more subgoals problems, or a more cleanly delineated subclass of such problems, or more detailed instructions on how to implement the strategy. Subsequent work of this type would provide greater insight into what it would take for students to master it.

In the case of contradiction, the difficulty appeared to be with the notion of uniqueness, which had not been singled out for special treatment in the practice problems. On the posttest the students (despite their experience, which surely included uniqueness arguments) seemed troubled by the very notion. They seemed to feel that the answer to Problem B5 was obvious and that there was "nothing to prove." The lesson from the students' reaction to this problem appears to be that, when students are taught to make uniqueness arguments, more attention will have to be given to the underlying rationale that supports their use. Simply including uniqueness arguments as a subclass of contradiction arguments, without giving them separate attention, may be insufficient if we expect students to master them.

The comments in the preceding paragraphs are tentative, because it is clear that we need to know a good deal more about heuristic strategies before anything resembling definitive statements can be made. The data offered here do, however, underscore one of the points emphasized in Chapter 3: Heuristics are complex and subtle strategies, and it is dangerous to underestimate the amount of knowledge and training required to implement them. The data do other things as well. They begin to suggest what, in ideal circumstances, might be required to master three of the strategies; they suggest what else must be considered in order to have students master (again, in ideal circumstances) two other strategies. As noted above, the resolution of these issues is an empirical question. Now, however, there is a firm experimental basis for making such empirical inquiries.

The performance data also point to the importance of control decisions in problem solving. To put things simply, the fact that students *can* master a particular problem-solving technique is no guarantee that they *will* use it—even in a very constrained environment that is conducive to its use. Despite having the list of strategies in front of them at all times, and despite being reminded periodically of its existence, the students in the experimental group still jumped impulsively into (sometimes relevant, sometimes irrelevant) solution attempts on a number of the posttest problems. Despite all the environmental support for heuristic usage, it was only the periodic warnings to stop and reconsider that kept the students from spending their allotted 20 minutes (on particular problems) on wild goose chases. This issue will be pursued below.

Question 2 Isn't problem-solving practice all you really need?

This is the issue of learning to solve problems by solving problems. One does, of course. That assumption was explicitly acknowledged in Chapter 3 when we considered the ways that individuals develop their own problem-solving styles and strategies. It was argued that over a period of years individuals are "trained" by their experiences in mathematics and come (whether consciously or not) to rely upon the problem-solving techniques that bring them results. The data from this experiment make one point absolutely clear, however: The acquisition of heuristic strategies from one's own problem-solving experience is a remarkably inefficient, hit-or-miss process.

The control group in this experiment practiced on 20 problems, four for each heuristic strategy, in an intensive series of problem sessions over a period of 2 weeks. If "experience" is all it takes to apprehend the strategies, these students should certainly have figured out the strategies and used them with some effectiveness. For the most part they did not.

The pattern of behavior demonstrated by the students on the "fewer variables" problems was typical. As we saw above, Subject 1 did note the fewer variables strategy and did use it on the posttest. Moreover, she learned this behavior during the experiment. On pretest Problem A4, which asked the students to demonstrate that if

$$0 \leq a, b, c, d, \leq 1,$$

then

$$(1-a)(1-b)(1-c)(1-d) \geq 1-a-b-c-d,$$

Subject 1 had multiplied out the terms on the left-hand side and had spent 20 minutes in algebraic manipulations. In working that problem she showed no signs of awareness of the relevant heuristic strategy. As her comment after the posttest indicates, she had perceived the way the strategy was used, and learned to apply it, as a consequence of working the practice problems. Yet this happened infrequently. With the identical work on the practice problems, neither Subject 2 nor Subject 3 made the connection—and in all honesty that connection does not seem a terribly hard one to make, given the nature of the experimental environment. In contrast, the experimental group used the strategy on the posttest with some success.

The students' performance on the induction problems is even more dramatic. The control students were all volunteers from advanced mathematics courses, and all of them were familiar with the mechanics of induction arguments. Thus they did not have to master a new technique in working the practice problems; they only needed to recognize that one can "discover" the answers to certain kinds of well-specified problems by looking for patterns. In three of the four practice problems (recall Table 6.3), there was an explicit integer parameter n in the problem statement, and values for $n = 1, 2, 3, 4, \ldots$ were calculated right at the beginning of the solutions to those problems. Although n was implicit in induction Problem A3, the solution process was the same. If, a priori, one were forced to choose which of the heuristic strategies used in this experiment had the greatest potential for transfer, surely one would choose induction as the most likely candidate. Yet there was no transfer at all! During their work on Problem B3 not one of the control students tried to discover a pattern by making calculations for small integer values of n. It should be stressed that this lack of transfer was not due to lack of memory. After the experiment was over the students were all told what the purpose of the experiment had been. When I showed the control students the solutions to the posttest problems, they often spontaneously recalled (almost verbatim, and with chagrin) one or more of the practice problems that had been solved in similar fashion. In a word, the answer to Question B is "no."

Question C Does explicit instruction in heuristics make a difference?

In a word, "yes"—as long as the environment supports the use of the heuristic strategies once they have been mastered.

Implications and Directions for Extension

Despite its diminutive size, this experiment does have some clear implications for teaching. The study placed students in an enriched environment that contained a small number of strategies and that gave them a concentrated dose of relevant practice problems. Even so, the degree to which the students in the control group were able to abstract for themselves the heuristics appropriate for working the posttest problems was minimal. In addition, the experimental group, having just mastered some of the strategies, might well have ignored these newly developed skills on the posttest; on a number of posttest problems they seemed ready to spend all of their allotted time chasing wild mathematical geese.

These problems, already significant in an enriched microcosm, become even more significant when one considers "real-life" instruction in problem solving. This study suggests that instruction in problem-solving strategies must be quite detailed and explicit. Heuristic strategies should be explicitly identified when they are used, the reasons for selecting the strategies (e.g., the integer parameter in the case of induction arguments) should be indicated wherever possible, and the mechanics of implementing the strategies should be carefully explicated. Issues of executive behavior, or control, will have to be explicitly dealt with as well. It will do the students no good to master the heuristic strategies if they do not give themselves the opportunity to use them.

This study has indicated that students can master certain heuristic strategies, at least in an enriched microcosm. Classroom experience is not as concentrated, but it is far more extensive. We now have an "existence proof" for heuristic learning—with regard to three strategies. We know that more training is required for two others, although how much and of what precise nature is open to question. As noted above, these are empirical issues.

My hope is that the experimental format used in this study, and variants of it, will be used to help resolve such issues. In its present form the format can be used to examine any of the large number of heuristic problem-solving approaches that have been identified in the literature. By systematically varying the strategies to be studied and the amount of training in them, we can begin to accumulate a large body of information about what students need to master them in relatively "ideal" circumstances. That information has been lacking to date and would prove extremely useful.

The "learning environment" in this study was deliberately artificial. Teaching was done via tape recordings to minimize the chance that differences in performance between experimental and control groups could be attributable to enthusiasm or bias on the part of the experimenter in favor of the heuristics group. In the future, when more heuristics, and the kinds of training it takes to master them in the laboratory, have been explored, instruction via tape recordings will no longer be necessary. The format can be changed to more closely approximate what takes place in a classroom setting. Instruction might be given by a teacher to small groups of students, say, half a dozen at a time. The experiment might be rerun in that format until the conditions are specified that allow the students, under those circumstances, to learn to use the strategies. The instruction might then be moved to a regular classroom setting. With (comparatively) large numbers of students, better validation of the results suggested here should be attainable.

Summary

This chapter described a small-scale laboratory experiment designed to explore students' mastery of five heuristic strategies. Two groups of students took part in the study. The instructional materials used in the study were tape recordings and written booklets. These materials were checked for bias by external reviewers and were found to be comparable. The control group worked 20 practice problems. After working the problems they were given written solutions to the problems, and they listened to tape-recorded explanations of those solutions. The experimental group was given a list of the strategies being studied and that list was kept in front of them at all times during the experiment. The experimental students worked the same problems for the same amount of time as the control group. The solution write-ups they were given contained the same text as the solutions for the control group. In addition, however, they contained "heuristic overlays" that clearly identified the strategies being used and that pointed out how they were being used. In postinstructional testing the control group was reminded every 5 minutes to "stop, take a deep breath, and look over your work. Then decide whether you want to continue in that direction." The experimental group was reminded every 5 minutes to "stop, take a deep breath, and look over the list of strategies. Then decide whether you want to continue in that direction." Pretests and posttests were given individually, and the students were recorded as they worked the problems out loud. Data for analysis consisted both of the written work and the tape recordings of the problem solutions.

Two comparisons of pretest-to-posttest gains, with regard to two different scoring procedures, indicated that the experimental group significantly outperformed the control group. More important, however, are the results provided by the protocol data. They suggest the following.

1. Problem-solving practice, by itself, is not enough. For the most part the control students failed to apprehend the heuristic strategies being used to solve the problems, despite the enriched nature of the experimental environment. This was not due to poor memory. When, after the experiment, it was explained what its purpose had been, these students often spontaneously recalled (nearly verbatim) the training problems that were used for each particular strategy. Thus explicit training is required.

2. At least in a fairly ideal environment—where problems of executive or control behavior are kept to a minimum, the heuristic strategies are explic-

itly labeled and explicated in some detail, and practice is given in concentrated doses—students can master certain heuristic strategies well enough to use them on related but not isomorphic problems.

Explicit heuristic instruction does (or can) make a difference with regard to problem-solving performance. This study, however, is simply an existence proof in a laboratory setting. Much more work needs to be done in elaborating both the training and the conditions necessary for students to learn a broad range of heuristic techniques—even in the laboratory. Work needs to be done in adapting the study done here so that the conditions of the study will more closely approximate the conditions of classroom instruction; in that way we can learn more about the conditions necessary for heuristic mastery in ordinary teaching situations. I hope that the experimental format described here will serve as a useful vehicle for such explorations.

7

Measures of Problem-Solving Performance and Problem-Solving Instruction*

The primary purpose of this chapter is to introduce and discuss three paper-and-pencil tests of problem-solving performance. As indicated by the discussion below, there has been a scarcity of reliable and informative testing procedures that focus on problem-solving *processes.* In order to examine the effects of a college-level course in problem solving, three pairs of such tests (and associated grading procedures) were developed. In brief, these tests can be characterized as follows.

Measure 1 consists of a pair of matched tests (pretest and posttest). These examinations are matched not according to problem type, but according to solution methods. Although not necessarily similar in form or from the same mathematical domains as their counterparts, corresponding problems on the two examinations will yield to the same general problem-solving approaches. Two different grading schemes are given for Measure 1. These grading schemes are used to evaluate (1) the frequency and variety of the problem-solving approaches generated by the students, (2) the degree to

* Chapter 7 is an expanded and revised version of Schoenfeld (1982a). Permission from the *Journal for Research in Mathematics Education* to reproduce parts of the article is gratefully acknowledged.

which the students pursue those approaches, and (3) the success that they have in pursuing them.

Measure 2 is a qualitative companion to Measure 1, examining students' subjective assessments of their problem-solving behavior. It records students' perceptions of how well they planned and organized their work on the problems in Measure 1 and their perceptions of the difficulty of those problems.

Measure 3 is a test of heuristic transfer. Keyed to specific problem-solving instruction, it too consists of a pair of tests (pretest and posttest) that are matched according to solution methods. These two examinations contain subsets of problems that are (1) closely related, (2) somewhat related, and (3) completely unrelated to the problem-solving instruction. Performance on these three categories of problems indicates the degree to which students have generalized their newly learned skills and the degree to which they can transfer them to new situations.

The second purpose of this chapter is to begin the detailed discussion and evaluation of a course in mathematical problem solving. The "experimental treatment" described in this chapter consists of a month-long intensive course in problem solving via heuristic strategies that was given at Hamilton College (New York) in 1980–1981. The three measures described in this chapter, which focus primarily on the use of heuristic strategies, are among the tools that were used to examine the students' behavior before and after the problem-solving instruction. Other examinations of that behavior are given in the sequel. Chapter 8 focuses on the knowledge organization that provides the support structure for problem solving. Chapter 9 focuses on control behavior, Chapter 10 on belief systems. Though the evaluation of the students' performance in each of these chapters is secondary to the discussion of more general theoretical, experimental, and methodological issues, the four chapters combine to provide fairly comprehensive documentation of the kinds of skills that students can be expected to master in such a course.

A Brief Discussion of Relevant Work

Consonant with the dual theme of this chapter, there are two relevant bodies of literature. The literature on empirical studies of problem solving was discussed in Chapter 6, and the discussion is not repeated here. To sum things up in a sentence, the mathematics education community has been enthusiastic about the notion of teaching problem solving via heuristic strategies, but the literature has offered little in the way of solid support for such enthusiasm.

Perhaps more importantly, the testing literature has offered few methods, whether for purposes of research or for use by teachers, of directly examining the procedures used by individuals as they attempt to solve problems. Virtually all available examinations of problem-solving performance have used *product* measures rather than *process* measures. That is, the tests they employ focus for the most part on assessing the correctness of the answers that students produce to problems, rather than focusing on the procedures that the students use in trying to solve them.

There are a fairly large number of tests of mathematical ability and of mathematical performance, many of which are commercially available. Virtually all of these tests are normative. They are generally multiple-choice examinations designed for large-scale implementation and for easy grading, for comparisons of whatever group is being measured (a student, a class, a school, or a school district) against a reference population. On none of the major commercially used tests are the methods that students use to solve individual problems examined. Rather, scorings and rankings (on parts or the whole of the examinations) are based on the number of correct responses given by a student. The research methodology used on such tests is most often statistical; "mathematical abilities" are determined by factor analyses of performance data, and correlations are sought between possession of such abilities and various kinds of mathematical performance. This type of research does indeed have its value, particularly in large-scale assessments of the utility of various instructional programs. It is, however, of little assistance in investigations of problem solving. Quite simply, it tells us nothing about the actual procedures employed by students in solving problems. For a general discussion of the limitations of the statistical approach to measurement and a rather severe critique of the perspectives implicit in such testing, see Chapter 2 of Krutetskii (1976). For an extended discussion of readily available commercial instruments, see Zalewski (1980) and Wearne (1980). Zalewski's summary evaluation of the commercially available tests is as follows.

> The examination of commercial tests as mathematical problem solving measures revealed several reasons to doubt their validity: (a) the "problems" were usually simple written items; (b) the scoring only focused on the correct response without considering the processes used; (c) tests set limits which gave students little opportunity to practice problem solving techniques; (d) the test writers provided no validity measures except the usual content validity statements. Thus, the commercial tests were judged not to be valid problem-solving measures and other procedures were examined. (Zalewski, 1980, p. 121)

Zalewski found those other procedures (mostly from research studies in psychology and mathematics education) equally wanting, as did Wearne in her review of the literature. An update of those reviews undertaken for this chapter does not indicate much of a change in the state of the art. For the most part, the paper-and-pencil tests that are available focus on answers and not on processes. (Zalewski, 1980 and Wearne, 1980, both tried to develop alternate approaches, with mixed success.) One notable exception to this "answer orientation" is a measure offered by Malone, Douglas, Kissane, and Mortlock (1980). In fact, the scoring procedure they introduced is similar to one of the scoring procedures for Measure 1 (see below). Save for that approach, I am not familiar with any paper-and-pencil measures that deal with heuristic processes in a meaningful way.

As noted in Chapter 6, there has been a fair body of research in recent years focusing on problem-solving processes. That literature will be reviewed at some length in Chapter 9. For the most part the relevant mathmatics education research (e.g., Kantowski, 1977; Kilpatrick, 1967; Lucas, 1972; Lucas, Branca, Goldberg, Kantowski, Kellogg, and Smith, 1979) has been based on extensive protocol analyses. Performing such analyses is an incredibly time-consuming task, and (at least insofar as such measures have been used) often results in data that consist of "frequency counts" of students' use of various heuristic strategies. Two of the tests described here, Measures 1 and 3, provide straightforward and easy-to-gather assessments of heuristic fluency and transfer. They can be used instead of protocol analysis for pretest versus posttest comparisons over a broad range of heuristic behaviors. As a result, researchers can reserve the more costly and time-consuming procedures of protocol analysis for more specialized use. In addition, the measures offered here can be adapted by classroom teachers both as a means of charting their students' problem-solving performance and for measuring the impact of their teaching. It is hoped that these measures, as well as being a research tool, will be of some practical use for teachers.

The Experimental and Control Treatments

This chapter describes the use of Measures 1, 2, and 3 to evaluate the problem-solving performance of two groups of students at the beginning and end of the 1980 winter term at Hamilton College, a small liberal arts institution in upstate New York. Unlike regular semester offerings, courses given during the winter term were brief and intensive. The term lasted a month, including time scheduled for examinations. Students were allowed to enroll for only one course during the winter term, and they were expected to devote all of their attention to that course. Most courses, including the courses that served as the experimental and control treatments described here, met for 18 days. There were $2\frac{1}{2}$ hours of class meetings each day, and each day's assignment kept the students at work for another 4 – 5 hours. This is roughly the equivalent of a semester's work, although condensing the work in this way has advantages and disadvantages. On the one hand, the students who are enrolled in a particular course study one subject, and one subject only, for nearly a month. Since immersing oneself in a discipline can often serve as a catalyst for deep understanding, there can be clear benefits to this type of intensive experience. On the other hand, so brief and intensive a course offers little time for assimilation and reflection. It may not allow for the development of skills that need to take shape over a longer period of time.

The "experimental group" for this study consisted of the 11 students who had enrolled in Mathematics 161, "Techniques of Problem Solving." All of the students in the course were freshmen and sophomores. Eight of these students had just completed their first semester of college calculus. The other three had just completed the third term of the calculus sequence. The "control group" consisted of eight students with comparable backgrounds (five with one semester of calculus, three with three semesters) who were recruited from a concurrent course in structured programming. Grade distributions were roughly the same for the two groups.

THE EXPERIMENTAL TREATMENT

Mathematics 161 can best be described as a workshop course in problem solving. The primary emphasis was on heuristic techniques, but a substantial amount of time was also spent in discussions of control behavior. The reader may recall a discussion of the first course handout, which provided a brief introduction to the course. (See "Modeling a Control Strategy for Heuristic Problem Solving," Chapter 4, pp. 106–114). An extended description of the structure and organization of my problem-solving courses may be found in my (1983c) *Problem Solving in the Mathematics Curriculum*. A few relevant aspects of the 1980 winter term course are touched on here.

As might be expected, virtually all of our class time was devoted to solving problems and to discussing problem solutions. The outline of a typical day's work was as follows. Class usually began with a discussion of homework problems, the solutions to which were often presented by students. After the homework had been discussed, a sheet of problems would be distributed. The class then broke into small groups (three or four students in each) to work on the problems. I traveled around the room as a "roving consultant" while they worked. When they had made a reasonable amount of progress, or had run out of steam, we reconvened to see what they had accomplished and to tackle the hard problems as a group.

There were, of course, lectures on the problem-solving strategies in which sample solutions were given. There were discussions of how to use particular heuristics and of the control decisions that should be considered while one is working on the problems. Decision-making processes were explicitly "role modeled" when sample solutions were presented. When I arrived at a critical point in a solution, I raised three or four possible options and evaluated them. On the basis of those evaluations I chose to pursue one option (perhaps the wrong one). After a few minutes I evaluated progress and considered whether to take another approach to the problem, and so on. The idea in these verbal solutions was to unravel some of the covert thought processes that students do not generally see when subject matter is presented to them.

When the class convened as a whole to work problems (40–50% of class time), I served as orchestrator of the students' suggestions. My role was not to lead the students to a predetermined solution, although there were obviously times when it was appropriate to demonstrate particular mathematical ideas or techniques. On problems for which the students had adequate mathematical backgrounds, my task was to role model competent control behavior—to raise the questions and model the decision-making processes that would help them to make the most of what they knew. Discussions started with "What do you think we should do?," to which some student usually suggested "Let's do X." Often the suggestion came too rapidly, indicating that the student had not adequately thought through what the problem called for or how the suggestion might be useful. The class was then asked, "Are you all sure you understand the problem, before we proceed with X?" A negative response from some students would result in our taking a closer look at the problem. After doing so, we returned to X as a possible solution approach. Did X still seem reasonable? Not infrequently the answer was "no." When it was, this provided the opportunity to remind students about the importance of making sure that one has understood a problem before jumping into its solution.

Even when the first suggestion made by a student could have been exploited to provide a straightforward solution to the problem, the class was asked for other possible ideas. There might be three or four suggestions, at which point a discussion was called for: Which suggestion should we try to implement, and why? The suggestions were implemented at the board. After a few minutes of working on the problem—whether or not we were on a track that would lead to a solution—the process would be halted for an assessment of how things were going. The class was asked, "We've been doing this for 5 minutes. Is it useful, or should we switch to something else? (and why?)" Depending on the evaluation, we might or might not decide to continue in that direction; we might decide to give it a few more minutes before trying something else. Once we had arrived at a solution, I did a post-mortem on the solution. The purpose of that discussion was to summarize what the class had done and to point out where it could have done

Table 7.1
Questions Posted during Problem-Solving Course

What (exactly) are you doing?
(Can you describe it precisely?)
Why are you doing it?
(How does it fit into the solution?)
How does it help you?
(What will you do with the outcome when you obtain it?)

something more efficiently, or perhaps to show how an idea that the class had given up on could have been exploited to solve the problems. When the class had succeeded in solving a problem using one of the suggestions it had generated, we often went back to the other suggestions to see if they could provide different solutions. The same problem was often solved three or four different ways before we were done with it.

Solving problems in small groups (also 40 – 50% of class time) provided the opportunity to work with the students individually as they grappled with the problems. During this part of the instruction, my role, in essence, was that of a problem-solving coach. There were matters of both technique and strategy. Technique (mostly questions regarding the use of particular heuristics, or other mathematical tools) is fairly straightforward. Issues of strategy (control) are more complex. Students' decision-making processes are usually covert and receive little attention. When students fail to solve a problem, it may be hard to convince them that their failures may be due to bad decision-making rather than to a lack of knowledge. That issue was dealt with in the following way. During the term, Table 7.1 was posted conspicuously in the classroom. The instructor had the right to stop students at any time while they were working on the problems and to ask them to answer the three questions on Table 7.1. At the beginning of the course the students were unable to answer the questions, and they were embarrassed by that fact. They began to discuss the questions in order to protect themselves against further embarrassment. By the middle of the term they could generally answer the questions when asked to do so. By the end of the term, asking the questions of themselves (not formally, of course) had become habitual behavior for some of the students.

THE CONTROL TREATMENT

Eight students with backgrounds comparable to those of the students in the problem-solving course (both in terms of course experience and grades) were recruited from a winter term course in structured programming. That course was given concurrently with the problem-solving course and made roughly the same work demands on its students. It met for the same amount of time and had extensive programming assignments that kept the students in the computer lab for 4 to 5 hours a day. The students in the control group did not study the mathematics problems studied in the problem-solving course. Its instructor, however, considered the computer programming course to be a vehicle for teaching problem-solving skills. The course was explicitly designed "to teach a structured, orderly way to approach problems."

Since the students in the control group did not work the same problems as the students in the experimental group, the role the control group plays is obviously limited. Despite these limitations, the control group serves two very important purposes. First, it provides some validation of the measures themselves, as a check that the pretests and the posttests are of comparable difficulty. (The data indicate that the control group did no better on the posttests than on the pretests, so that improvements in the experimental group's score are not attributable to a difference in test difficulty.) Second, it provides a baseline of performance against which the treatment group's scores can be compared.

A NOTE ON SAMPLE SIZE

The fact that the sample sizes were only $n = 11$ for the experimental group and $n = 8$ for the control group in the present study may be of concern. These are hardly large samples. This was not the measures' first use, however. Measures 1 and 3 were developed, pilot tested ($n = 8$), and revised at Berkeley in 1976–1977. The full complement of measures was used at Hamilton in 1979 on an experimental group only ($n = 20$). The results obtained in both of those testing situations were quite similar to the results presented below. It seemed appropriate, however, to postpone disclosure of the results until further replication had been obtained and until testing with a control group could help to validate the measures.

Measure 1: A Plausible-Approach Analysis of Fully Solved Questions

Measure 1 consists of matched pretests and posttests containing five problems each. Students took the tests, given in Figures 7.1 and 7.2, on the first and last days of the term, respectively.

The instructions for both examinations were identical. They told the students that the examiners "are interested in everything you think about as you work on these problems, including (a) things you try which don't work,

Measure 1: Pretest

1. If S is any set, we define $O(S)$ to be the number of subsets of S that contain an odd number of elements. For example, the "odd" subsets of $\{A, B, C\}$ are $\{A\}$, $\{B\}$, $\{C\}$, and $\{A, B, C\}$; thus $O(\{A, B, C\}) = 4$. Determine $O(S)$ if S is a set of 26 objects.
2. Suppose you are given the positive numbers p, q, r, and s. Prove that

$$\frac{(p^2+1)(q^2+1)(r^2+1)(s^2+1)}{pqrs} \geq 16.$$

3. Suppose T is the triangle given in Figure 1. Give a mathematical argument to demonstrate that there is a square S such that the four corners of S lie on the sides of T, as in Figure 2.

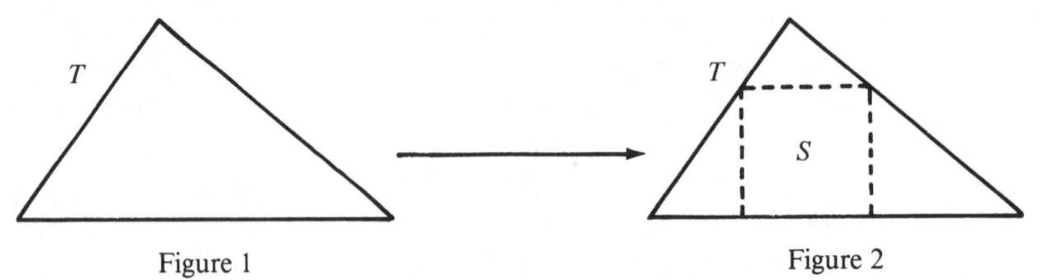

Figure 1 Figure 2

4. Consider the set of equations

$$\begin{Bmatrix} ax + y = a^2 \\ x + ay = 1 \end{Bmatrix}$$

For what values of a does this system fail to have solutions, and for what values of a are there infinitely many solutions?

5. Let G be a (9×12) rectangular grid, as illustrated to the right. How many different rectangles can be drawn on G, if the sides of the rectangles must be grid lines? (Squares are included, as are rectangles whose sides are on the boundaries of G.)

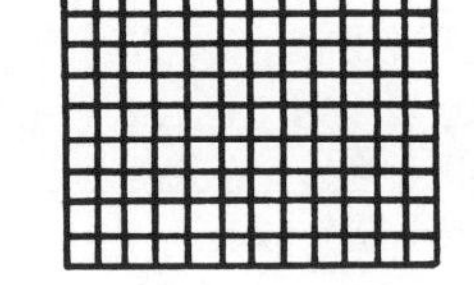

Figure 7.1

(b) approaches to the problems you think might work but don't have the time to try, and (c) the reasons why you did try what you did." Students were told to take the test with a pen. So that the graders could see everything they had written, students were asked not to erase their work or to black it out, but simply to put a large X through anything that they decided was wrong. They were given 20 minutes for each question.

Both the pretest and posttest versions of Measure 1 were related to the problem-solving instruction received by the students in much the same way that the test problems discussed in Chapter 6 were related to the heuristic training problems. The problems on the two examinations are "matched" in that the strategies that help to solve a particular pretest problem also help to solve the corresponding posttest problem. An inductive approach is useful for each Problem 1, the "fewer variables" approach for each Problem 2, exploiting "an easier, related problem" or "relaxing a condition and reimposing it" for each Problem 3, drawing a diagram for each Problem 4,

Measure 1: Posttest

1. What is the sum of the coefficients of $(x + 1)^{37}$? Indicate, as well as you can, why you think the answer is what you say it is.
2. Suppose you are given the positive numbers x, y, and z. Prove that

$$\frac{(x^2 + x + 1)(y^2 + y + 1)(z^2 + z + 1)}{xyz} \geq 27.$$

3. Suppose T is the triangle given in Figure 1, and P any point on the perimeter of T. Give a mathematical argument to demonstrate that there is an equilateral triangle, with its three vertices on the perimeter of T—and one of its vertices at P.

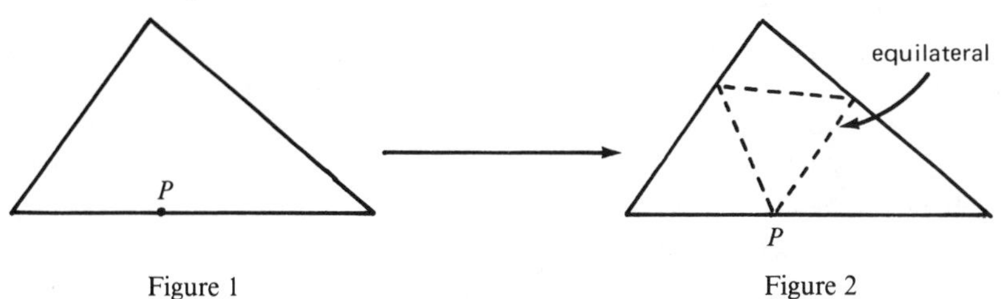

Figure 1 Figure 2

4. Consider the set of equations

$$\left\{\begin{matrix} x^2 - y^2 = 0 \\ (x - a)^2 + y^2 = 1 \end{matrix}\right\}.$$

For what values of a does this system have no solutions, one solution, two solutions, three solutions, or four solutions?

5. The student council at Gecko College consists of 2 men and 2 women picked at random from the student body. If there are 12 men and 9 women enrolled at Gecko, how many different student councils might be chosen?

Figure 7.2

and establishing subgoals for each Problem 5. Each of these techniques was explicitly studied in the course, though it was not exemplified on problems isomorphic to those on the exams.

The purpose of Measure 1 is to focus on problem-solving processes. It was designed examine students' ability to generate, select, and pursue plausible approaches to mathematical problems. In the scoring, a "plausible" approach need not yield a solution to a problem, but it must be germane. Thus the suggestion "try values of $n = 1, 2, 3, 4, \ldots$ and look for a pattern" would be considered a plausible approach to the problems

$$\text{Determine } \sum_{n=1}^{\infty} \frac{1}{n(n+1)}$$

and

$$\text{Determine } \sum_{n=1}^{\infty} \frac{1}{n^2}$$

even though the suggestion yields a solution in the former case but not in the latter. However, it would not be considered a plausible approach to the problem:

Given positive numbers a and b, determine

$$\lim_{n \to \infty} (a^n + b^n)^{1/n}.$$

Each of the problems in Measure 1 can be solved a number of different ways. For example, there are two straightforward inductive solutions and one combinatoric solution to Question 1 on both the pretest and the posttest; there are perhaps other solutions as well. Pretest and posttest Questions 3 can each be solved three or more ways, as can most of the other problems. This point is important to keep in mind with regard to grading procedures for Measure 1. Although one major goal of the problem-solving course was to teach problem solving via heuristics, and although one would expect the heuristics to make a difference in the students' problem-solving performance, credit is not keyed to the use of heuristics; any plausible approach to a problem is given credit.

Two scoring schemes are given below for Measure 1. In the first, Multiple Count Scoring, students are given credit for each plausible approach to a problem with which they give evidence of familiarity. The measure is scored this way for two reasons. First, fluency in generating reasonable approaches to problems is valuable. As one begins work on a problem one does not necessarily know which approaches to the problem may prove profitable. The more ways one can generate to attack problems, the more likely one is to solve them. Second, a student who solves a problem in two different ways has indeed done more than a student who produces only one solution to the problem. In this first scoring scheme, a student who solves a problem using two different methods is given credit for two correct solutions.

MULTIPLE COUNT SCORING

Suppose P is a specific problem on one of the examinations. First, a list is compiled of all the plausible approaches to Problem P that were taken by at least one student. Then, each student's attempted solution to Problem P is graded as follows. The following questions are asked about each plausible approach to Problem P on the list.

1. *Evidence.* Does the student show any evidence of being aware of this particular approach to Problem P? The student may describe the strategy overtly: "I tried to look for an inductive pattern." Or, evidence may appear in the student's work. For example, he or she may draw a diagram for either

pretest or posttest question 4, or pursue the $(1 \times n)$ special cases for pretest question 5.

2. *Pursuit.* Does the student pursue the approach? Since the instructions asked for "everything you think about as you work on these problems," a student might mention an approach without pursuing it.

3. *Progress.* If the student does pursue an approach, how much progress does he or she make toward a solution? There are four levels of progress:

a. Little or none *(little).* For pretest question 1, the student might compute $O(S)$ for sets of one, two, and three elements but make no conjecture; perhaps one of the computations would be incorrect.

b. A reasonable amount but not enough to claim the solution is almost in hand *(some).* For example, a student working posttest question 4 thought the solution of $(x^2 - y^2 = 0)$ was the line $y = x$ but graphed the second equation correctly.

c. Something close to a solution, marred by an incorrect calculation *(almost).* On posttest Problem B1, a student reported "I thought the answer was 2^n but the sum of the coefficients for $(x+1)^4$ was 15. I'm stuck."

d. A complete solution *(solved).*

Scores in each category are either 0 or 1, and under "progress" a 1 is given for the highest-ranking solution achieved. Thus a student who almost solves a problem with a particular approach receives scores of 1 in the categories evidence, pursuit, and almost, and 0's elsewhere. The student's final set of scores for a problem is a listing, for each category, of the sum of his or her scores over all of the plausible approaches to the problem. Thus a student who solves a problem with one approach, makes some progress on it with a second, and mentions a third but does not pursue it receives cumulative scores of 3 (evidence), 2 (pursuit), 0 (little), 1 (some), 0 (almost), and 1 (solved).

While such a scale may at first seem highly subjective, it turns out surprisingly that classifying behavior at this coarse a level of detail can be done with a high degree of consistency. The reliability between my scoring of the problems and my assistants' consensus scores exceeded 90%. We used the following procedures. Three advanced undergraduates assisted in the grading. The nature of plausible and correct solutions to each problem was discussed with me before the students graded the papers. Then the assistants graded papers separately from me, assigning a consensus grade to each solution. Experimental and control papers were intermixed and graded blindly: the assistants saw only a code letter for each. Tables 7.2 (control group) and 7.3 (experimental group) provide the summary data for multiple count scoring, giving for each problem the sums over all plausible approaches for all students.

Table 7.2
Multiple Count Scoring of Measure 1: Control Group ($n = 8$)[a]

			Progress			
Problem	Evidence	Pursuit	Little	Some	Almost	Solved
1	6 (10)	3 (5)	1 (3)	0 (1)	1 (0)	1 (1)
2	8 (6)	1 (2)	1 (2)	0 (0)	0 (0)	0 (0)
3	0 (0)	0 (0)	0 (0)	0 (0)	0 (0)	0 (0)
4	7 (5)	3 (0)	3 (0)	0 (0)	0 (0)	0 (0)
5	8 (8)	0 (7)	0 (0)	0 (1)	0 (4)	0 (2)
Total	29 (29)	7 (14)	5 (5)	0 (2)	1 (4)	1 (3)
Average per student	3.63 (3.63)	0.88 (1.75)	0.63 (0.63)	0 (0.25)	0.13 (0.5)	0.13 (0.38)

[a] Pretest scores are followed by posttest scores in parentheses.

Table 7.3
Multiple Count Scoring of Measure 1: Experimental Group ($n = 11$)[a]

			Progress			
Problem	Evidence	Pursuit	Little	Some	Almost	Solved
1	6 (14)	6 (11)	3 (1)	0 (0)	2 (1)	1 (9)
2	9 (11)	3 (10)	2 (0)	1 (1)	0 (2)	0 (7)
3	2 (11)	2 (11)	0 (1)	0 (4)	1 (5)	1 (1)
4	13 (15)	4 (10)	3 (4)	0 (1)	0 (1)	1 (4)
5	12 (12)	1 (11)	1 (0)	0 (2)	0 (1)	0 (8)
Total	42 (63)	16 (53)	9 (6)	1 (8)	3 (10)	3 (29)
Average per student	3.82 (5.73)	1.45 (4.82)	0.82 (0.55)	0.09 (0.73)	0.27 (0.91)	0.27 (2.64)

[a] Pretest scores are followed by posttest scores in parentheses.

"BEST APPROACH" SCORING

In some regards the multiple-count scoring method described above offers a more comprehensive and accurate picture than standard grading schemes of students' problem-solving performance. On either examination Question 1, for example, the student who pursued both the inductive and combinatoric approaches to Question 1 had indeed done more than the student who pursued (to the same degree) only one of those arguments. Multiple scoring provides a quick and easy-to-score tally of a class's overall ability to generate and pursue plausible approaches to problems. It provides useful information for the teacher and to the students; it also has the side benefit of rewarding students for trying more than one approach to a problem. For the researcher it is an inexpensive supplement to protocol analysis. However, this kind of scheme has two potentially serious drawbacks: (1) A few prolific students might contribute a large number of tallies, making it difficult to interpret the meaning of a class's average score; (2) because the scoring is nonordinal, individual scores are hard to rank or interpret. The following is an alternative, "standard" scoring of the tests. It rewards each student's best effort (rather than multiple efforts) on each problem.

First each approach the student takes to a problem is scored separately, as follows. An approach that is not pursued is given 0 points. Those approaches that were considered as making "little" progress under multiple count scoring will earn from 1 to 5 points, depending on how much progress was made. For example, the student who, on pretest Problem A1, computed O(*S*) successfully and in organized fashion for sets of one, two, and three elements, but did not make a conjecture, would receive the full 5 points. The student who computed one of them incorrectly would receive 2 points. Similarly, papers on which "some" progress was made would receive from 6 to 10 points; "almost" solutions from 11 to 15, and "solved" from 16 to 20. A score of 16 indicates that the correct answer was obtained but that the work and the justification were sloppy; a 20 means that all work was coherently and clearly justified.

For each of the problems, the grade assigned to the student is the highest grade that the student earned on any of the approaches he or she took to that problem. (More simply, the student's best effort on the problem is graded.) This grading scheme, modulo a scaling factor, is the scheme described by Malone *et al.* (1980). As with multiple count scoring, my assistants graded experimental and control papers blindly, reaching consensus grades. Reliability with my grading was better than 90%. With a 20-point scale for each question, the maximum score for each of the five-problem examinations is 100 points. The results are given in Table 7.4.

Table 7.4
"Best Approach" Scoring of Measure 1[a]

	Average score for problem					
Group	#1	#2	#3	#4	#5	Total
Control ($n = 8$)	9.0 (7.6)	2.9 (2.4)	0.2 (0.0)	0.8 (1.5)	1.1 (12.5)	14.0 (24.0)
Experimental ($n = 11$)	8.4 (17.9)	2.7 (14.3)	3.9 (11.3)	3.9 (11.3)	1.8 (17.4)	20.8 (72.2)

[a] Posttest scores appear in parentheses.

Discussion of Testing Results

With the exception of their scores on question 5, the control group's performance on the posttest version of Measure 1 was nearly identical to its performance on the pretest. This held for both the multiple scoring and best scoring approaches, as seen in Tables 7.2 and 7.4, respectively. As it happens, the control group's improved performance on posttest Question 5 is due to the nature of the problem rather than increased knowledge on the part of the students. That problem is more amenable to a brute force counting approach than its pretest counterpart, and the scores reflect this. (Recall that all plausible approaches are given credit.) In addition, the experimental and control groups' performance on the pretests were similar. Thus the control group fulfilled its limited purposes. The pretest and posttest versions of Measure 1 can be considered roughly equivalent, and the control group's scores provide a baseline against which the experimental group's performance can be compared.

The control group's performance on the tests indicates two things. First, taking the pretest did not condition the students for better performance on the posttest. This is entirely consistent with the results obtained in the previous chapter. Second, learning a "structured, hierarchical, and orderly way to approach problems" does not necessarily enhance problem-solving performance. This may have been the case because the control students did not have at their disposal the specific (in this case, heuristic) tools they need; it may be because more work needs to be done to ensure transfer from one domain (programming) to another (mathematics).

The experimental group's performance indicates clearly that those students learned something; the issue is to figure out what they learned, and why. I think it is reasonable to conclude that much of the students' improvement is due to their having learned to use certain problem-solving heuristics with some efficiency, and not simply to their having worked a lot of related problems. A few different lines of argument support this conclusion. First, the study reported in Chapter 6 suggests that problem-solving experience alone would not suffice to generate the improved performance seen here. A second line of substantiation comes from Measure 3, discussed below. Some of the test problems in Measure 3 were completely unrelated to the instructional problems. The students from the problem-solving class showed marked improvement on those problems while the control group did not. Clearly practice did not help there. Third, changes in heuristic and control behavior of students in the experimental group are apparent in the videotapes of their problem solving before and after the course. See Chapters 9 and 10, respectively, for some sample "before and after" transcripts.

Measure 2: Students' Qualitative Assessments of Their Problem Solving

Measure 2 was designed to assess students' qualitative reactions to their performance on Measure 1. After they had spent 20 minutes working on each question of Measure 1 (pretest and posttest), the students were given 4 minutes to answer the questions given in Figure 7.3.

Table 7.5 summarizes the percentage of "yes" responses to Questions 1, 2, and 3 of Measure 2.

For what they are worth, one can draw the following conclusions from the data. At the time of the pretest the two groups' perceptions of their knowledge were roughly comparable, with the control group thinking it had seen more "closely-related" problems than the experimental group. (I have no idea why this is the case.) The increase in the control group's positive responses to Question 1 on the posttest was due to some students' claims (in response to the "if so, where?") that the posttest problems were the same as the pretest problems. This is a plausible claim, given that a month had

Measure 2: Students' Qualitative Assessments of Their Problem Solving. Asked after Each Problem of Measure 1.

1. Have you seen this problem before? Yes ___ No ___
2. Have you seen a closely related problem before?
   Yes ___ No ___
   If "yes," where?
3. Did you have an idea of how to start the problem?
   Yes ___ No ___
4. Did you plan your solution or "plunge" into it?
   Check one of the boxes: | A | B | C | D |
   I jumped in (A) — I planned a bit (C) — I thought it out first (D)
5. Did you feel your work on the problem was organized, or disorganized?
   Check one of the boxes: | A | B | C | D |
   More or less random (A) — Somewhat structured (C) — Well organized (D)
6. Please rate the difficulty of the problem.
   Check one of the boxes: | A | B | C | D |
   Easy (A) — Approachable and solvable (C) — Hard (D)

Figure 7.3

Table 7.5
"Yes" Responses to Questions 1, 2, and 3 of Measure 2, Taken Over the Five Questions of Measure 1[a]

Group	% Responding "yes"		
	1: Seen before?	2: Seen related?	3: Know how to start?
Control ($n = 8$)	2 (13)	56 (53)	53 (53)
Experimental ($n = 11$)	2 (0)	25 (75)	48 (92)

[a] Posttest responses are given in parentheses.

elapsed between the two tests and the problems on them were quite similar. It is interesting, however, that the students in the experimental group were able to make finer distinctions and did not report having seen those problems before. The percentage of "related problems" they reported having worked increased dramatically, but none of the students claimed that any of the pretest problems had been used on the posttest. It is also interesting that the percentage of "related problem" responses only rose to 75%, since the fraction of "know how to start" responses rose to 92%. Thus the students used fairly stringent criteria for relatedness, and they did not perceive relatedness in some situations where I did. Yet they felt that they did know how to approach many of the other problems. Perhaps more important, however, are the students' responses to Questions 4, 5, and 6 of Measure 2. These are given in Table 7.6.

The conventional wisdom is that students will jump into the implementation of any plausible approach that they might have for a problem; if something in a problem statement strikes a student as familiar, the student is likely to be "off and running" in whatever direction the recollection suggests. As the data in Table 7.5 indicate, the percentage of the time that the experimental students had some idea of how to start the problems nearly doubled from pretest to posttest. In consequence, one might expect that the percentage of jumps into a solution would increase. As the answers to Questions 4 and 5 summarized in Table 7.6 reveal, just the opposite happened. Prior to instruction only 29% of the responses from the experimental group indicated some reasonable degree of planning (categories C and D combined); after the course the figure reported was 65%. The corresponding figures for the control group were 33% and 36%, respectively. Before the course only 36% of the solution attempts were from the experimental group were called "somewhat structured" or better, and the posttest figure was 70%. The corresponding figures for the control group were 45% and 55%, the increase most likely due to the students' increased attention to structure as a result of their programming course. This increase of only 10% should not be taken as

Table 7.6
Responses to Questions 4, 5, and 6 of Measure 2, Taken Over the Five Questions of Measure 1

		Percentages responding			
Question	Treatment	A	B	C	D
4	Control	45 (22)	21 (42)	21 (17)	12 (19)
	Experimental	35 (15)	37 (20)	21 (30)	8 (35)
5	Control	33 (18)	23 (26)	33 (26)	12 (29)
	Experimental	32 (11)	32 (18)	25 (45)	11 (25)
6	Control	0 (13)	20 (18)	25 (33)	55 (38)
	Experimental	4 (25)	20 (24)	30 (35)	46 (16)

an indication that they did not learn structure in programming. Rather, it points to difficulties in transfer. Behavior learned in a programming context (where there are specific guidelines for putting together programs in hierarchical, orderly fashion) may not carry over easily to another, messier domain where one's resources are more shaky. If one expects transfer at the control level, one had better lay the foundations for it.

Recall that the instruction had a strong emphasis on developing the appropriate "managerial" behavior. I stressed that the students should read and analyze the problems before they started implementing a solution. Their own subjective perceptions are that they did. These perceptions are substantiated by the videotapes of the students solving problems before and after the problem-solving course. Not only was there much more "plunging in" to a solution on preinstruction problems, but much of it was directed at the implementation of something minor—and often irrelevant or misleading. (The tapes of KW and AM and of DK and BM discussed in Chapters 1 and 4 were both examples of pretest behavior.) The students, in essence grasping at straws, were looking for something to focus on; when they found it they jumped into an exploration of it, without any sense of how counterproductive such behavior can be. After the instruction, there was clear evidence of more deliberation on the videotapes. There was slower problem reading and more rereading, more and longer problem analysis, and so forth. I believe this change in the students' behavior to be a strong contributing factor to their success.

Measure 3: Heuristic Fluency and Transfer

In novel problem situations, good problem solvers can often identify and modify relevant techniques from tasks that appear only marginally related to the ones they are presently considering. Thus good problem solvers are often able to generate plausible approaches (in the sense discussed above) to problems that are unfamiliar. One goal of problem-solving instruction should help to develop these skills. Since Measure 1 tested students' performance on five heuristic strategies that had been taught in some detail in the problem-solving course, Measure 1 left unaddressed left this aspect of the students' problem-solving behavior. Moreover, the scope of Measure 1 was far too narrow. Since the strategies examined there were discussed at length in the course, it could be argued that Measure 1 was simply a subject matter test. Perhaps the students did not learn to be better problem solvers. Perhaps they simply learned some new subject matter ("induction," "special cases," etc.) that was tested for in a new way.

Measure 3 was designed to deal with this possibility and to explore the boundaries of what the students might have learned in the course. It is a test of transfer and of students' ability to generate plausible approaches to new and unfamiliar problems. For this kind of testing it is appropriate to examine as broad a spectrum of problems as possible. However, the time required for such testing can quickly get out of hand. The students required a full 2 hours for each version (pretest and posttest) of Measures 1 and 2, which only asked the students to work on and react to five problems. An alternative that allowed me to obtain data on nine problems in 1 hour of testing was to ask students simply to plan solutions rather than to try to solve the problems. This was the approach taken in Measure 3. The instructions for the test read in part as follows.

Please read each problem, think about it for a few minutes, and write the plan of a solution. In other words, tell me what you would do if you had two or three hours to work on the problem. What would you look at. What would you consider a reasonable approach to the problem? How would you start solving it and in what direction would your argument go?

The pretest and posttest versions of Measure 3 are given in Figures 7.4 and 7.5. The problem order was scrambled on the examinations, and the problems have been grouped together in coherent categories for ease of presentation. The nine problems on each test were broken into three categories of three problems each: those that were closely related to, somewhat related to, and unrelated to the instruction. Problems were placed in the first category if they could be solved by a method that closely paralleled a method used to solve a problem in the course. For example, the problem

Let A be an arc that winds through the interior of an sphere S of radius 1. Suppose that the endpoints of A are on the boundary of S. If the length of A is less than 2, show that there is a hemisphere of S which does not intersect A.

Measure 3: Pretest

Closely Related Problems

1. Given the triangle ABC (to the right), determine the point P such that the areas of the three triangles it forms are all equal.

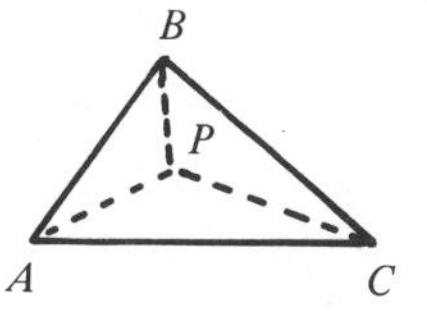

2. Show that the expression $n^2(n^2 - 1)(n^2 - 4)$ is divisible by 360 for all positive integers n.
3. Let P be a parallelpiped in space (as to the right). Prove that the four diagonals of P (the lines from one corner of the base to the opposite corner of the top) intersect in a common point.

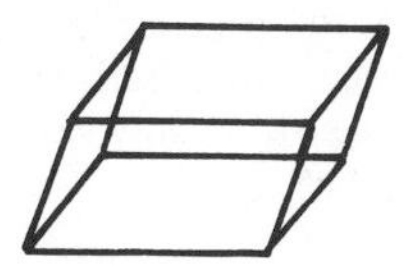

Somewhat Related Problems

4. Suppose C_1 and C_2 are two smooth, nonintersecting curves in the plane. Prove that the shortest straight line that connects some point of C_1 with some point of C_2 is perpendicular to both C_1 and C_2.
5. Prove that if the number $2^n - 1$ is a prime (has no factors other than 1 and itself), then the number n is a prime.
6. You are given the frustrum of a right pyramid with a square base, as in the figure to the right. If the upper base is b, the lower base B, the height H, find the volume of the solid.

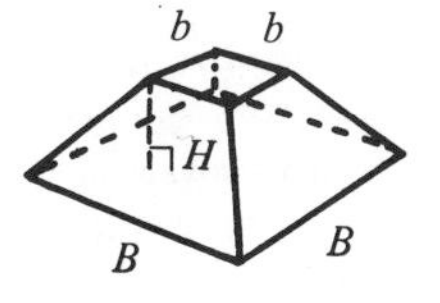

Unrelated Problems

7. Let A be any set of 20 distinct integers chosen from the arithmetic progression 1, 4, 7, . . . , 100. Prove that there must be two distinct intergers in A whose sum is 104.
8. Let a and b be given positive numbers. Find

$$\lim_{n\to\infty} (a^n + b^n)^{1/n}$$

9. In the diagram A, B, and C are the midpoints of $\overline{CF}$, $\overline{AD}$, and $\overline{BE}$, respectively. The area of ΔABC is three units. What is the area of ΔDEF?

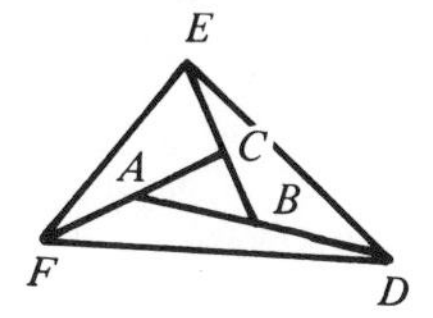

Figure 7.4

had been solved in class by exploiting the solution to its two-dimensional analogue. For that reason pretest and posttest Problems 3, which can each be solved by a similar technique, are called closely related to the instruction. Had the class discussed analogy in general but never solved geometric problems through the consideration of lower-dimensional analogues, the problem would have been called somewhat related. If nothing remotely similar had been discussed the problem would have been called unrelated. For

Measure 4: Posttest

Closely Related Problems

1. You are given a side A of a triangle T, the length of the altitude to A, and the length of the median to A. Can you construct T from the given infromation, using straightedge and compass? If so, how?

Side A

Altitude

Median

2. Show that the expression $n^5 - 5n^3 + 4n$ is divisible by 120 for all positive integers n.
3. Let P be a tetrahedron (pyramid) in space (to the right). You wish to circumscribe a sphere around the tetrahedron. What information do you need in order to do so, and how do you find it?

Somewhat Related Problems

4. Suppose C_1 and C_2 are two convex regions in the plane. Show there is a straight line which bisects the area of both regions.

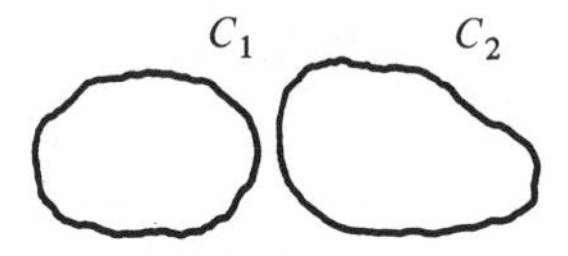

5. Let $P_1 + P_2$ be two consecutive prime numbers. If $P_1 + P_2 = 2Q$, show Q must be a composite number.
6. You are given the frustrum of a right circular cone, as in the figure. If the upper radius is r, the lower radius R, and the height H, find the surface area of the object.

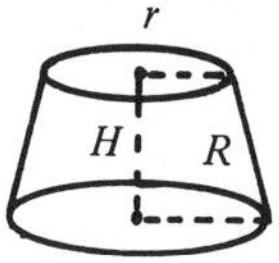

Unrelated Problems

7. A computer is used for 99 hours over a period of 12 days. Show that there are 2 consecutive days such that the combined computer usage on those 2 days exceeds 17 hours.
8. Let X_0 and X_1 be given numbers. We define $X_{n+1} = (\frac{1}{2})(X_{n-1} + X_n)$, so $X_2 = (\frac{1}{2})(X_0 + X_1)$, $X_3 = (\frac{1}{2})(X_1 + X_2)$, and so forth. Does $\lim_{n\to\infty} X_n$ exist? Find it.
9. Let P be the center of the square constructed on the hypotenuse AC of the right triangle ABC. Prove that BP bisects angle ABC.

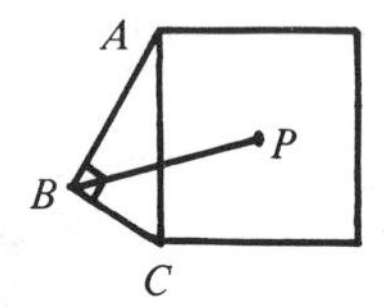

Figure 7.5

example, the class had never added auxiliary elements like those that help in either version of Question 9.

When Measure 3 was designed, the intent was only to tabulate the number of plausible approaches to each problem that came from each group. Under the assumption that there should be some transfer from the heuristic instruction, it was hypothesized that the number of plausible approaches gen-

Table 7.7
Scores on Measure 3[a]

				Degree of pursuit			
Problems	Treatment[b]	Evidence	Pursuit	Little	Some	Almost	Solved
1, 2, 3	Control	6 (6)	1 (0)	1 (0)	0 (0)	0 (0)	0 (0)
"Closely related"	Experimental	12 (30)	0 (23)	0 (5)	0 (4)	0 (5)	0 (9)
4, 5, 6	Control	7 (5)	1 (1)	1 (1)	0 (0)	0 (0)	0 (0)
"Somewhat related"	Experimental	15 (38)	3 (16)	3 (10)	0 (2)	0 (3)	0 (1)
7, 8, 9	Control	2 (2)	0 (0)	0 (0)	0 (0)	0 (0)	0 (0)
"Unrelated"	Experimental	13 (18)	6 (8)	6 (4)	0 (3)	0 (1)	0 (0)
Totals	Control	15 (13)	2 (1)	2 (2)	0 (0)	0 (0)	0 (0)
	Experimental	40 (86)	9 (47)	9 (19)	0 (9)	0 (9)	0 (10)

[a] Posttest scores appear in parentheses.
[b] For control treatment $n = 8$; for experimental treatment, $n = 11$.

erated by the students would correlate with the degree to which each problem was related to the instruction. The data turned out to be richer, and the analysis more subtle, than expected. Even though they were allotted only 6 or 7 minutes per problem, students in the experimental group outlined detailed solutions to some of the problems on the posttest and solved some others. Given these results, it was appropriate to use the scoring scheme developed for Measure 1. The summary data are presented in Table 7.7.

DISCUSSION OF RESULTS

The control group's performance was essentially constant from pretest and posttest, providing some indication that the two tests were roughly equivalent. Of all the tests used to examine the students' problem-solving performance, including the perception study described in the next chapter and the videotaping sessions, this is the only measure on which the treatment and control groups' pretest scores are discrepant. The "evidence" column indicates an average score of 1.9 for the 8 control students and an average of 3.6 for the 11 experimental students. I have no explanation for this.

Since the first three questions of Measure 3 were closely related to the instruction, one would expect (in parallel with Measure 1) to see the gains that were scored by the experimental group. As expected, the number of plausible approaches, the number of pursuits, and the degree of success with which the approaches were pursued all increased. The "somewhat related" problems provided more interesting results. Since these problems were further removed from the instruction than Questions 1, 2, and 3, it was expected that there would be fewer plausible approaches to Questions 4, 5, and 6. Instead there were more. I suspect that this is because, having found (and pursued) plausible approaches to the closely related problems, the students in the treatment group felt no need to generate additional approaches. Where they were less sure of their answers they generated more potentially useful approaches.

Examining the experimental group's performance on the "somewhat related" problems provides clear evidence of transfer. For example, the only problem discussed in class that was related in any direct way to posttest Question 4 called for "construct[ing] with straightedge and compass the common external tangent to two given circles." While the solution does use the "special case" or "easier, related problem" of the construction of the tangent from a point to a circle, extrapolating the solution method from that problem to posttest Question 4 is anything but straightforward.

Some, although less pronounced, improvement is seen on the "unrelated" problems as well. While the control group's performance was unchanged, the experimental group's plausible approaches increased by half. There was slightly more pursuit (which was not asked for), and slightly better success.

A Brief Discussion of Control Issues

There are numerous questions regarding the interpretation of the data presented here, and much more work along the lines explored in this chapter needs to be done. For example, the role of the control group in this study was limited. Since that course had a different instructor and covered different subject matter, any conclusions drawn on the sole basis of a comparison between the two groups are shaky at best. In particular, the control group's behavior at the "executive" level raises at least two fascinating issues.

First, the students in the control group were trained to be "organized and hierarchical in their approach to problem solving." Though there was no formal pretest-to-posttest comparison in the context of the programming class, the instructor indicated that there was a very strong change in the students' programming behavior. The students' subjective judgments on Measure 2 (which were accurate for the experimental group) were that they "plunged in" less and were better organized on the posttests than on the pretests, but the increase is fairly small—and may be illusory, given the lack of improvement in their problem-solving performance. Of course, the executive skills they learned in the computing course were fairly domain-specific, and it is not clear that they should transfer to any great degree. Exploring the conditions that promote such transfer will be an interesting and difficult task. Second, whatever improvement in executive behavior there might have been did not have a corresponding effect on performance; the pretest and posttest performance of the students in the control group was essentially identical on all of the problem-solving tests. In large part this is due to the fact that those students were not very good at generating plausible approaches to the problems. Simply put, an executive with few resources to manage will not be able to do very much. The relationship between resources and control rears its head once again.

Summary

This chapter introduced three paper-and-pencil measures of problem-solving processes. All three measures include matched pretests and posttests that focus on what the student does while trying to solve problems, rather than on the final solution he or she produces.

The two versions of Measure 1 contain five problems, each of which can be solved two or three, sometimes four or five different ways. Problems on the two examinations are matched not by form or by subject matter but by solution methods; the same problem-solving approaches yield solutions to corresponding pretest and posttest problems. The grading procedures developed for Measure 1 focus on (1) the degree to which students generate a range of relevant problem-solving approaches, (2) how far those approaches are pursued, and (3) with what success. The first grading procedure gives "multiple credit" for multiple solutions to a problem, in effect rewarding students for fluency in problem solving and providing an indication of the breadth of students' problem-solving knowledge. The second grading procedure rewards a student's best effort on any problem, and thus, while retaining its emphasis on solution processes, is similar to more customary grading scales.

Measure 2 asks for students' qualitative reactions to their work on the problems in Measure 1. It determines how familiar they perceived each problem to be, whether or not they planned their solution to the problem, whether their work on the problem was organized or disorganized, and how difficult they thought the problem was.

The two versions of Measure 3 each contain nine problems which, like those in Measure 1, were paired by solution type. Three subsets of three problems each were (1) closely related, (2) somewhat related, and (3) completely unrelated, to the problems studied in a problem-solving course. Thus Measure 3 provides a test of heuristic transfer; performance on these three classes of problems indicates the degree to which the students can apply what they learned in the course to problems increasingly distant from it.

All three measures were used to evaluate the "before and after" performance of students in a course in mathematical problem solving. Students with comparable backgrounds who were recruited from a concurrent course in structured programming served as a control for the experiment. The control group served primarily as a validity check on the difficulty of the problem-solving measures; the performance of the control group from pretest to posttest remained essentially constant, so any differences in the experimental group's performance could not be attributed to the suggestion that the posttests were easier than the pretests. There were differences in performance, and they were quite substantial.

Measure 1 explored performance on problems related to those studied in the course. On the posttest the experimental group was able to generate one and a half times as many relevant problem-solving strategies than on the pretest. They pursued these strategies, to at least some significant degree, more than three times as frequently. Almost 10 times as many complete problem solutions were obtained on the posttest as on the pretest; average scores on the "best approach" grading increased from 21 to 72%, while those of the control group increased from 14 to 24%. The experiment described in Chapter 6 indicates that mere problem-solving experience would not have contributed to such improvement in performance. Whether or not it would have, Measure 1 provides what may be the first clearly documented evidence that students in a problem-solving course can learn to employ a variety of heuristic strategies.

Measure 2 may be most interesting for what it says about students' planning and organization. On the posttest the experimental group had a good idea of how to start the problems twice as frequently as they did on the pretest, so one might expect them to jump into the solutions with increased frequency. Instead, they assessed themselves as having "plunged into" solutions only half as often. Similarly, the percentage of solution attempts they rated as "somewhat structured" or better increased from 36 to 70%.

Measure 3 provided clear evidence not only of heuristic mastery but also of transfer. The students were asked only to plan solutions to the nine problems on each examination, saying what they would do if they had a number of hours to work on each problem. Again, control group performance remained essentially constant from pretest to posttest. On the posttest the experimental group not only generated two and a half times as many relevant solution suggestions for the problems "closely related" to the course as they had on the pretest; despite time constraints and in excess of what the test instructions asked them to do, the experimental group actually solved a fair number of the posttest problems. On problems "somewhat related" to the course the number of relevant suggestions jumped from 15 to 38, with some (but not as vigorous) pursuit of the solutions. There was still improvement on problems unrelated to the course (relevant suggestions increased from 13 to 18) and a marginal difference in performance. This indicates a substantial degree of heuristic transfer, which, as one would expect, tails off as the problems become less and less familar. Measure 3, like Measure 1, provides clear evidence that students can master heuristic strategies—and use them in somewhat new situations.

8

Problem Perception, Knowledge Structure, and Problem-Solving Performance*

This chapter describes a study of problem perception conducted in parallel with the research described in Chapter 7. Its focus is on the resources that provide the support structure for problem-solving performance, particularly on the ways that individuals select solution methods to problems as a consequence of what they "see" in problem statements. Like the two chapters that precede it, this is a dual-purpose chapter.

The primary purpose of the chapter is to explore the relationship between problem perception and problem-solving performance. For the most part, studies in the literature have obtained indirect evidence regarding the two. For example, contrasting group studies such as expert–novice comparisons indicate that the two groups perceive problems differently. In such work, differences in perception have generally been attributed to differences in expertise. It should be noted, however, that such inferences are not rigorously justifiable. The comparison groups almost always differ on a number of attributes such as aptitude and maturity, and any such difference might serve to explain differences in perception. The study in this chapter exam-

* Chapter 8 is an expanded and revised version of Schoenfeld & Herrmann (1982b). Permission from the *Journal of Experimental Psychology: Learning, Memory, and Cognition* to reproduce parts of the article is gratefully acknowledged.

ines changes in students' problem perception as they become more proficient problem solvers. In this way it provides a direct exploration of the relationship between perception and proficiency. The secondary purpose of the chapter is to indicate the kinds of perceptual changes that can be induced by an intensive problem-solving course. It should be stressed that the course did not explicitly focus on problem structure or problem perception.

Students' perceptions of the structural relatedness of mathematical problems were examined before and after the students took an intensive course in mathematical problem solving. Those perceptions were measured with the use of a card sorting task. Students were asked to sort together problems that they thought would have similar solutions. The sort data were analyzed using cluster analysis. The experimental group's pre- and postinstruction sortings were compared with those obtained from two groups, a collection of experts (mathematics faculty) and a control group of students from a nonmathematical problem-solving course given at the same time.

The data obtained from the preinstruction sortings replicate and extend previously obtained results. Like experts in other fields, the mathematical experts appear to base their perceptions of problem relatedness upon principles of the discipline or an archetypal methods of solution, which have been called the problems' *deep structure.* Novices tend to classify problems by their *surface structure,* focusing on the words or subjects that are prominent in the problem statements.

The data obtained from postinstruction sortings indicate a strong shift in the students' perceptions of problem structure. The collections of problems sorted together by students after instruction are no longer homogeneous with respect to surface structure, and they appear to cluster according to deep structure characteristics. Various measures indicate that the students' postinstruction sortings much more closely approximate the experts' sortings than those same students' preinstruction sortings (or the sortings from the control group, which remained clustered by surface structure). These data permit the direct conclusion that problem perception shifts as students become more competent in a domain. Such "expert-like" perceptions can be seen as contributing to expert-like performance.

Background

The general context for the work discussed here was established in Chapter 2, where the question of how individuals routinely access relevant knowledge in familiar domains was addressed. A variety of different representational frameworks (scripts, schemata, frames, modeling via production systems, etc.) have been invoked to deal with the issue. Invariant across these frameworks, however, is the following underlying assumption: Individuals with extensive experience in any particular domain categorize their prior experiences in that domain and then use those categorizations both to interpret current situations and to access relevant methods for dealing with those situations. The categorizing process takes place at a variety of levels, from the microscopic (including, e.g., the perceptions that allow us to recognize words in our working vocabulary) to the macroscopic (as in the possession of "restaurant scripts" for understanding stories about meals away from home), some of which will be reviewed briefly below. Occupying a middle ground between these two extremes is problem perception. The reader may recall the following quotation from Hinsley, Hayes, and Simon:

> (1) *People can categorize problems into types. . . .*
> (2) *People can categorize problems without completely formulating them for solution.*
> (3) *People have a body of information about each problem type which is potentially useful in formulating problems of that type for solution . . .* directing attention to important problem elements, making relevance judgments, retrieving information concerning relevant equations, etc.
> (4) *People use category identifications to formulate problems in the course of actually solving them.* (Hinsley, Hayes, & Simon, 1977, p. 92)

One would expect individuals' perception of problem-solving tasks in a particular domain and their problem-solving performance in that domain to be closely tied. In particular, there is good reason to suspect that problem perception improves with the acquisition of expertise. Broadly speaking, problem perception is tied to problem representation. A variety of studies (Chi, Feltovich, & Glaser, 1981; Hayes & Simon, 1974; Heller & Greeno, 1979; Larkin, McDermott, Simon, & Simon, 1980; Simon & Simon, 1978) provide evidence that competent problem solvers' representations of phenomena in familiar domains differ from those of novices. There is an extensive literature on the issue of knowledge representation and problem-solving performance, much of which is reviewed in Greeno and Simon (to appear). As the authors noted, "Individuals with expert knowledge in a domain have been shown to have superior skill in recognizing complex patterns of information in their domains of expertise. Domains in which this phenomenon has been demonstrated include chess, Go, electronics, computer programming, and radiology" (Greeno & Simon, to appear, p. 60). It is hypothesized that the ability to perceive complex patterns allows experts to select efficient solution mechanisms for working their way forward to the solution of reasonably complicated problems. Without the benefits of such pattern recognition, the individual is constrained to use more cumbersome procedures.

A series of studies by Shavelson (1972, 1974; Shavelson and Stanton, 1975) provided early evidence regarding the broad relationship between perception and expertise. Those studies indicate that as students learn a discipline, their knowledge of the structural relationships among parts of the discipline becomes more like that of experts. That is, the students' perception of the organization of the domain becomes more expert-like. Working at a fairly global level, Shavelson did not directly address problem perception; his results do not specifically address the relationship between problem perception and performance. The most direct evidence regarding that relationship comes from studies of expert behavior (often for purposes of building expert systems) and from expert–novice comparisons. For example, expert chess players perceive board positions in terms of patterns or broad arrangements (in chunks such as "fianchettoed castled Black King's position") while novices do not (Chase & Simon, 1973; de Groot, 1965; Simon, 1980). Of particular relevance is the work by Chi, Feltovich, and Glaser (1981) on problem perception in physics.

Chi, Feltovich, and Glaser gave subjects (advanced graduate students in physics, called experts, and students who had completed a single course in mechanics, called novices) a set of 24 textbook physics problems to sort. The problems sorted together by the advanced students varied with regard to the superficial features in the problem statements but were consistent in that the principles of physics used to solve them (e.g., conservation of energy) were the same. The authors called this a sorting according to the problems' *deep structure*. In contrast to the experts' sortings, the novices' sortings tended to cluster problems together if the objects referred to in the problem statements (e.g., pulleys, gears, levers), or the terms of physics used in the statement (e.g., force, momentum, potential energy) were the same. The authors called these *surface structure* similarities. The implication of this research is that with experience one comes to recognize deep structure in a discipline and to categorize problems in the light of such perceptions.

Studies in elementary mathematics by Chartoff (1977) and Silver (1979) support the notion that perception becomes more expert-like with the acquisition of proficiency. Since there is a consensus regarding the underlying mathematical structure of simple work problems in algebra, neither Chartoff's nor Silver's study collected data from experts. In both, problem structures were assigned a priori by the experimenters. The results obtained in both cases indicate that students who have become proficient at solving word problems in algebra perceive the structures of those problems in much the same ways that experts do. That is, successful students exhibit a greater degree of agreement with the experts' perceptions of problem structures than do less proficient students.

Technically speaking, the evidence regarding the relationship between expertise and perception obtained from all of the studies discussed above is indirect, although it is fairly strong. The studies on algebra word problems are correlational, the rest expert–novice comparisons. These studies do show consistently that experts and novices differ in problem perception, but the design of these studies precludes unequivocal conclusions about the origins of those differences. The experts in problem-solving studies are often faculty or established professionals in a discipline, though they are occasionally advanced graduate students. They are usually older, more completely trained, and more experienced at problem solving in general than the novices (most often lower-division undergraduates) in those studies. They generally have more confidence than novices about their abilities and are likely to be more comfortable than novices in test-taking or experimental situations. In all likelihood they had, when they themselves were novices, better aptitude for the subject domain than the collection of novices with whom they are being compared. Presumably expert–novice differences in perception are rooted in differences in training and experience. However, it is reasonable to assume that such differences are influenced by differences in age, confidence, and the like. This spectrum of variables was not controlled for in the expert–novice comparisons described above. In particular, the contrasting group designs could have confounded expertise with aptitude. The ambiguous outcome of this kind of contrasting group design is not, of course, unique to studies of expertise and problem perception. The difficulties of the design are well known, and in some areas of psychology these difficulties are regarded as presenting insurmountable obstacles to inference (see, e.g., Schaie, 1977). The work described here avoids such methodological difficulties by examining the same students' perceptions of problem relatedness before and after problem-solving training.

As a final comment, it should be noted that the preceding discussion focused on the development of expert-like perceptions with the acquisition of proficiency. This relationship is not unidirectional, for the quality of one's problem perception is likely to have a strong effect on problem-solving performance. When things work well, an individual's correct perception of the essential aspects of a problem statement can cue access to relevant problem schemata. These may suggest straightforward methods of solution or may point the problem solver, more or less automatically, in directions that are fruitful to consider. The broad body of research on routine access to relevant solution procedures (see Chapter 2) supports this notion. The research on algebra provides direct evidence. When things do not work well, an individual's incorrect perception of the critical elements in a problem statement can have strong negative effects on performance. If inappropriate or irrelevant solution procedures are accessed and implemented, the individual's misperceptions may result in a solution attempt that starts off in the wrong directions. At best such misperceptions will result in deflections of time and energy that must be overcome by good control, and—even when the individual is ultimately successful—the resulting performance is less optimal than one would like. At worst such misperceptions can lead to wild goose chases from which there is no return. In sum, the perception-and-expertise pairing appears to be somewhat of a chicken-and-egg phenomenon. The research described here may help to clarify some aspects of the development of expert-like problem perception, but it leaves much unexplored.

Method

SUBJECTS

Three groups of individuals were involved in the study. The first consisted of nine faculty members from the mathematics departments of Colgate University and Hamilton College. These professional mathematicians, who participated without pay, will be referred to as experts. The second and third groups were the experimental and control groups described in the previous chapter. The experimental group consisted of 11 first- and second-year students at Hamilton College who had enrolled in the month-long, intensive course in mathematical problem solving given during the 1980–1981 winter term. Participation in this experiment was a condition of enrollment in the course. The control group consisted of eight students recruited from a course in structured programming, which was given concurrently and which made similar demands on the students. These students were paid $20 each for participating.

The two groups of students had similar backgrounds, as described in Chapter 7. All of the students had taken either one or three semesters of college mathematics, and their enrollment in winter term mathematics courses can be taken as a sign of positive attitude toward mathematics.

Some brief comments should be made regarding the use of the term *novice* in this study. In particular, it should be noted that all of the students in both the experimental and control groups had backgrounds that were more than adequate to deal with the problems used in this experiment. The problems are "nonstandard" in that problems like them rarely receive direct attention in the high school curriculum; the students were not likely to have encountered problems exactly like them during their academic careers. However, the solutions to the problems require neither tremendous mathematical sophistication nor any mathematical techniques with which the students were not familiar. At the time of the pretest, all of these novices knew more than enough mathematics to be able to solve all of the problems. What they lacked was extensive problem-solving experience.

MATERIALS

The study used 32 problems in the sorting task. The complete set of problems is given in the Appendix to this chapter. Each of the problems deals with objects familiar from the high school curriculum and is accessible to students with a high school background in mathematics. None of the problems requires calculus for its solution.

Before the problems were used in the sorting, each problem was assigned a surface structure and a deep structure characterization. The idea behind surface structure is to capture what a naive reader of a problem statement would find important. If a problem solver has little sense of the underlying structure of a mathematical problem, then that person is likely to focus on the objects in the problem statement (e.g., whole numbers, circles, polynomials) or the topic area it comes from (e.g., geometry, algebra) when asked what is important about it. In contrast, a sophisticated problem solver may regard those objects as incidental to the essential structure of the problem; what is important may be the form of the problem or the patterns of reasoning perceived as essential for solving it. The deep structure characterization is intended to capture what the sophisticated problem solver would deem important about the problem. As an example, let us consider Problem 1.

Problem 1 Show that the sum of consecutive odd numbers, starting with 1, is always a square. For example,

$$1 + 3 + 5 + 7 = 16 = 4^2.$$

This problem deals with whole numbers; more precisely, with a sum of odd numbers. A naive reader might focus on this aspect of the problem. A more sophisticated reader might see the implicit n in the problem statement and recast the problem as follows: "The sum of the first n odd numbers is a function of n (which you are told involves squares). What is that function?" Finding the function might involve trying values of n and looking for a pattern; the pattern, once found, would be verified by induction. As a result of these considerations, Problem 1 was assigned the surface characterization "sum of odd numbers," which is a subclass of "whole numbers problems"; it was assigned a deep structure characterization of "patterns; induction." Similar a priori judgments were made regarding all 32 problems used in the sorting.

The following examples serve to contrast surface and deep structures and to indicate the kinds of relatedness judgments that might be made by problem solvers of differing sophistication. Consider Problems 15, 17, and 19:

Problem 15 You are given the following assumptions.

i. Parallel lines do not intersect; nonparallel lines intersect.

ii. Any two points P and Q in the plane determine a unique line that passes between them.

Prove: Any two distinct nonparallel lines L_1 and L_2 must intersect at a unique point P.

Problem 17 Show that if a function has an inverse, it has only one.

Problem 19 How many straight lines can be drawn through 37 points in the plane, if no three of them lie on any straight line?

At the surface structure level Problems 15 and 19 deal with points and lines, while Problem 17 deals with functions and inverses. Thus a naive subject might perceive Problems 15 and 19 to be closely related, while not perceiving Problem 17 to be related to either of the other two. The situation is different at the deep structure level. Problems 15 and 17 both call for uniqueness arguments, namely showing that there is only one object (a point and a function, respectively) with specific properties. These problems share the same underlying structure, and a mathematician would find them to be closely related. Problem 19, which calls for a "counting argument," is altogether different. Hence different perceptions should result in different sortings.

The surface structure and deep structure characterizations for each problem are given in the Appendix to this chapter. Assigning the surface structure characterizations was a fairly straightforward task. For the most part one simply identified the most prominent characteristics of the problem statements in the same fashion as described above in the discussions of Problems 1, 15, 17, and 19. The deep structure characterizations of the problems were not nearly as straightforward to assign, and many of these were judgment calls. Some of the problems seemed to be best described by their form, for example, Problem 2 as a "linear Diophantine equation." For the majority of problems, the best characterization seemed to be the most appropriate method of solution, for example, the characterization of Problem 1 as "patterns; induction." The subjective nature of these judgments should be stressed. Along this dimension the differences between the domains of mathematics and physics are significant.

The categories representing deep structure in physics, namely, the principles of physics, are well agreed upon. More importantly, they are also the categories used in instruction. To put things another way, the answer to the questions (1) What is the underlying principle for the solution to this problem? and (2) What is the name of the topic under which this problem is likely to be taught? will most often be the same in physics. Calling a problem a "conservation of momentum problem," for example, means that the problem is solved by appealing to that principle and that similar problems, if not the problem itself, are likely to be found among the exercises for "conservation of momentum problems" in standard texts. One might say that physics is organized and taught according to deep structure. This consonance of principles and instructional organization does not hold in mathematics, where it sometimes appears that the underlying principles of solution and subject matter presentation are essentially independent of each other. "Finding patterns," "making uniqueness arguments," and "exploiting similar problems" may be universal techniques among competent mathematical problem solvers, but mathematics is not taught along those lines. A typical student will, for example, run across problems that call for finding patterns (or making uniqueness arguments, etc.) in virtually any advanced mathematics course. The problems will not be labeled as such, however. With time (and luck), the student may come to understand that such arguments are useful. But this will be an individual synthesis. No standard mathematics course has a unit on finding patterns, making uniqueness arguments, and so on; no standard text has a chapter on these topics. Indeed, if instruction were customarily organized in that fashion, Pólya's ideas would hardly have been considered revolutionary.

There was not necessarily reason to believe, when this experiment was being designed, that there would be a consensus among mathematicians regarding the deep structure characterizations of the problems used in the card sort. As noted above, many of these problems are nonstandard and do not, for the most part, come from easily recognizable parts of the curriculum. For that reason the "first thoughts" recorded by the mathematician JTA as he performed the card sort are of some significance. JTA was an accomplished problem solver who had not discussed any of the problems with me before performing the card sort. His classifications, which he recorded and used as the basis for his sorting, are compared with the deep structure characterizations in Table 8.1.

Of the 32 problems characterized in Table 8.1, JTA's "first thoughts" and the a priori deep structure classifications were discrepant on only five: Problems 4, 8, 16, 18, and 31. Moreover, the differences in the characterizations of two of those problems, Problems 4 and 31, are superficial. One can quibble about differences in the characterizations of the other three problems, but the differences are minor. Over all, JTA's "first thoughts" provide strong corroboration of the a priori classifications.

PROCEDURES

The experts performed the card sort once, at their convenience. The experimental and control groups performed the card sorts and took the mathematics tests twice, immediately preceding and immediately following

Table 8.1
Two Characterizations of the Problems

Problem	Deep structure characterization	JTA's "first thoughts"
1	Patterns, induction	Induction
2	Linear Diophantine equation	Diophantine equation
3	Patterns, induction	Induction
4	Analogy (fewer variables)	Note $x = y = z$; use $(x + 1/x \geq 2)$[a]
5	Diagram, analytic geometry	Analytic geometry
6	Unclear: Herron's formula or analytic geometry	Analytic geometry
7	Special cases	Factor out a[b]
8	Special Diophantine equation	Trial and error[c]
9	Contradiction	Contradiction
10	Contradiction	Contradiction
11	Linear Diophantine equation	Diophantine equation
12	Patterns; DeMorgan's law	Iterate (patterns)
13	Special cases (analogy)	Cases
14	Patterns; number repetitions	Modular arithmetic[d]
15	Uniqueness; contradiction	Contradiction
16	Special cases, diagram	Symmetry, analytic geometry
17	Uniqueness; contradiction	Contradiction
18	Auxiliary elements	Analytic geometry
19	Patterns, combinatorics	Combinatorics
20	Number representations	Modular arithmetic[d]
21	DeMorgan's law	Combinatorics
22	Easier problem, patterns	Iterate (patterns)
23	Diagram	1, 2, 3 collapses[e]; dumb problem
24	Special cases	Factor x^5
25	Linear Diophantine equation	Diophantine equation
26	Patterns, induction, number repetitions	Iterate (patterns)
27	Patterns, induction	Combinatorics, induction
28	Diagram, analytic geometry	Analytic geometry
29	Diagram	Dumb (do it graphically)
30	Analogy	Do two-dimensional case (i.e., analogy)
31	Analogy	Contradiction[f]
32	Number representations	Modular arithmetic[d]

[a] Here JTA describes how to use the fewer variables strategy.
[b] JTA's solution reduces the problem to the special case.
[c] Technically, this does differ from the other Diophantine equations (JTA is a number theorist.)
[d] JTA and I are saying the same thing in different words.
[e] Drawing the diagram shows that the triangle collapses.
[f] The two-dimensional analogue is solved by contradiction.

the intensive 1980–1981 winter term at Hamilton College. The sorting procedures were as follows.

Each of the 32 problems was typed on a 3 × 5 card. Each subject was given the problems in randomly shuffled order and instructed to decide which problems, if any, were "similar mathematically in that they would be solved the same way." Similar problems were to be returned to the experimenter paper-clipped together as a group. If a problem was not considered to be similar to any of the others, it was to be placed in a "group" contaning one card. The participants were told that they might return anywhere from 1 to 32 groups of cards to the experimenter. All of the subjects finished the task in approximately 20 minutes.

The experimental and control treatments were described in Chapter 7, and there is no need to repeat the details here. As noted above, all of the problems were within the students' reach at the time of the pretest. Needless to say, no mention of problem perception and no direct discussions of problem structure took place during the problem-solving course. It should be noted, however, that many of the problem-solving techniques discussed during the course are likely to contribute to improved abilities to perceive what is important in a problem statement. Students were encouraged to make certain that they had a full understanding of a problem statement before proceeding with a solution. They were told to examine the conditions of a problem carefully, to look at examples to get a "feel" for the problem, to check for consistency of given data and plausibility of the results, and so forth. This emphasis on attention to mathematical detail may have fostered improved problem perception. For the sake of balance, it should be noted that there was a comparable emphasis on "understanding" and perhaps greater emphasis on precision in the control course on structured programming.

Results of the Sortings

THE EXPERTS

We begin with the experts' card sort. Figure 8.1 presents a clustering analysis, using Johnson's (1967) method, of the sorting data from the nine professional mathematicians. The interpretation of clustering diagrams is fairly straightforward. One gets a rough sense of how closely related the subjects perceive two items (in this case two problems) to be by tracing the paths from the two items, from left to right, until they meet. The further to the left they meet (the higher the proximity level), the more frequently those problems were grouped together by the experimental subjects. Johnson's

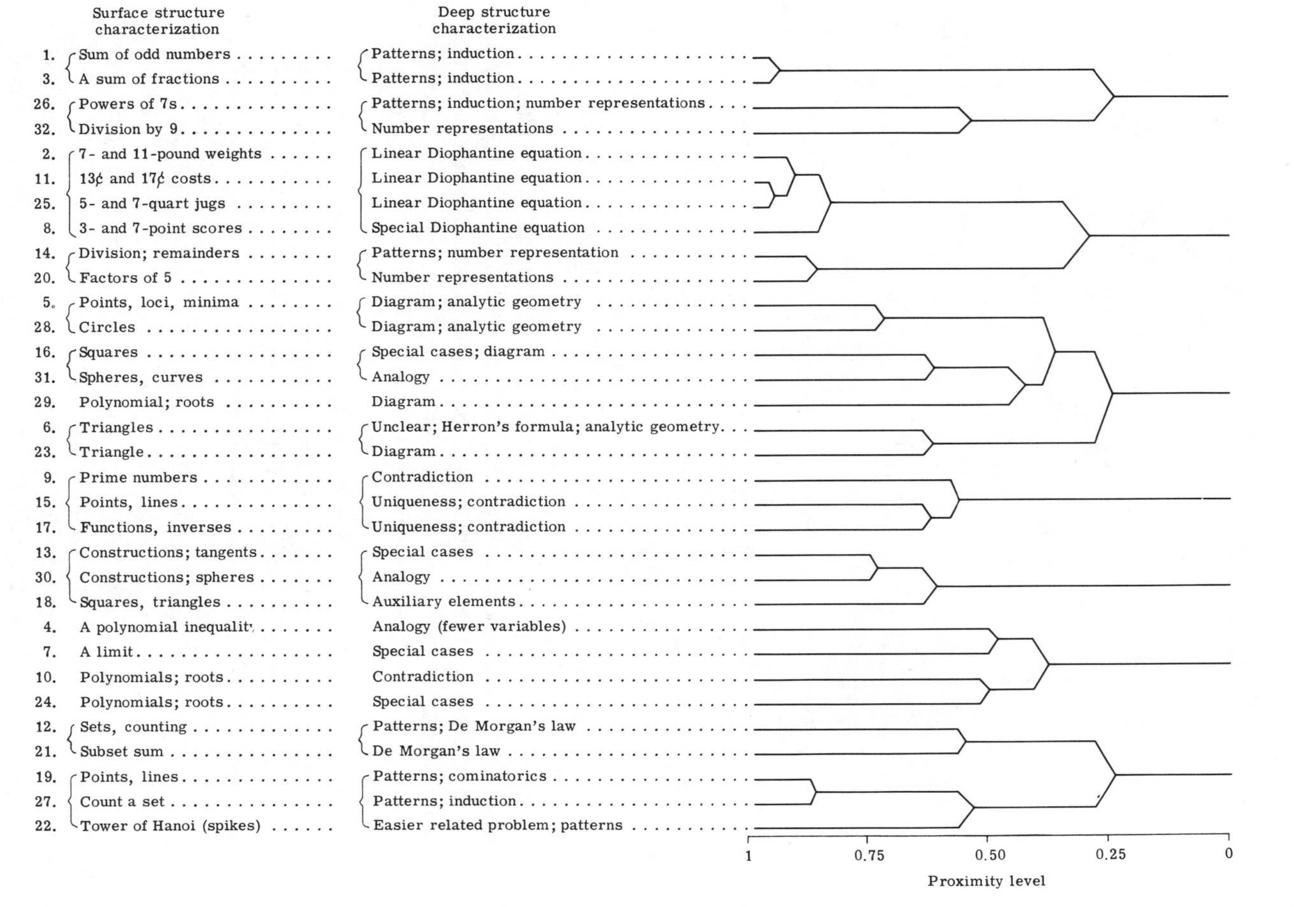

Figure 8.1 Cluster analysis of experts' card sort.

Table 8.2
Proportion of Strongly Clustered Pairs in Which Both Problems Share the Same Representation

	Measure 1 (all pairs)				Measure 2 (noncoinciding pairs)			
	Surface structure		Deep structure		Surface structure		Deep structure	
Test group	%	*n*	%	*n*	%	*n*	%	*n*
Experts	59	22	82	22	25	12	67	12
Experimental, pretest	81	26	58	26	58	12	8	12
Control, pretest	91	23	57	23	82	11	9	11
Combined, pretest	76	21	62	21	67	9	11	9
Experimental, posttest	58	24	79	24	09	11	55	11
Control, posttest	83	24	58	24	64	11	9	11

procedure starts with tightly clustered pairs of items and then expands these to groups of items that were frequently sorted together. Brackets on the left-hand side of Figure 8.1 indicate those sets of problems that had a proximity level exceeding 0.5, a minimum of 16 out of 32 possible clusters using Johnson's metric. Such bracketing can be taken as an indication of consensus on the part of the subjects that the problems are strongly related. Problems bracketed together are called a *strong cluster.*

A brief inspection of Figure 8.1 indicates that the experts' strong clusters are consistently homogeneous with regard to deep structure characterizations. In 8 of the 11 strong clusters, all of the elements share a common deep structure characterization. In contrast, only 4 of the 11 bracketed clusters are homogeneous with regard to surface structure—and 3 of those are also homogeneous with regard to deep structure.

Two measures of the degree of structural homogeneity in the cluster diagrams are given in Table 8.2.* Measure 1 provides, for surface and deep structure, respectively, the proportion of bracketed pairs that have the same structural representation. As noted above, some clusters are homogeneous with regard to both types of structure. To indicate perceptual preference when the two types of structures conflict, those pairs for which both surface and deep characterization coincide were deleted from the sample for Measure 2. The first row of Table 8.2 gives the data for the experts' sorting. Of the 22 pairs of problems bracketed together in Figure 8.1, 13 (59%) share the same surface structure and 18 (82%) the same deep structure. The surface and deep structures coincide in 10 of those problem pairs, however. When

* Jim Greeno deserves thanks for suggesting these measures.

only noncoinciding pairs are considered, only 3 of 12 pairings (25%) are homogeneous with regard to surface structure while 8 of 12 pairings (67%) are homogeneous with regard to deep structure.

THE NOVICES PRIOR TO INSTRUCTION

Figure 8.2 presents the cluster diagram of the sorting performed by the combined group of novices ($n = 19$) prior to instruction. In the interests of saving space, the cluster diagrams for the separate experimental ($n = 11$) and control ($n = 8$) groups are not given. All three of these diagrams were quite similar. [Johnson's procedure derives its cluster diagrams from the symmetric $n \times n$ matrices where the entry $a_{i,j}$ is the number of times that item i was sorted together with item j. The matrix from which Figure 8.2 was derived was strongly correlated with both the experimental group's pretest matrix ($r = .918$, $df = 496$, $p < .001$) and with the control group's pretest matrix ($r = .889$, $df = 496$, $p < .001$).]

Inspection of Figure 8.2 indicates a strong reversal from the pattern seen in Figure 8.1 Here the primary criterion for sorting problems together appears to be surface structure; 8 of the 10 strong clusters are homogeneous with regard to surface structure, while 6 of 10 are homogeneous with regard to deep structure. Of these 6, 5 are also homogeneous with regard to surface structure. Factoring these out, the surface-to-deep ratio of noncoinciding clusters is 5 : 1. The data in Table 8.2 confirm these impressions. Here the data are given for the experimental, control, and combined novice groups. In all three cases the proportion of surface structure homogeneity far exceeds the proportion of deep structure homogeneity.

POSTINSTRUCTION DATA

The effects of the problem-solving instruction were discussed at length in the previous chapter. Since this seems to be the place for recording statistical analyses, let it be noted that significance tests resulted in the following. Recall from Table 7.4 that the mean scores on the "best approach" scoring of problem-solving Measure 1 for the experimental and control groups prior to instruction were 20.8 and 14.0, respectively. After instruction, the means were 72.2 and 24.0, respectively. Analysis of variance on those means showed that scores increased across the term [$F(1, 17) = 47.5$, $p < .001$], were greater for experimental rather than control subjects [$F(1, 17) = 130.6$, $p < .001$) and that the increase across the term was significant only in the scores from the experimental subjects [$F(1, 17) = 48.2$, $p < .001$]. Simple effects tests indicated that the term effect was significant for the experimental

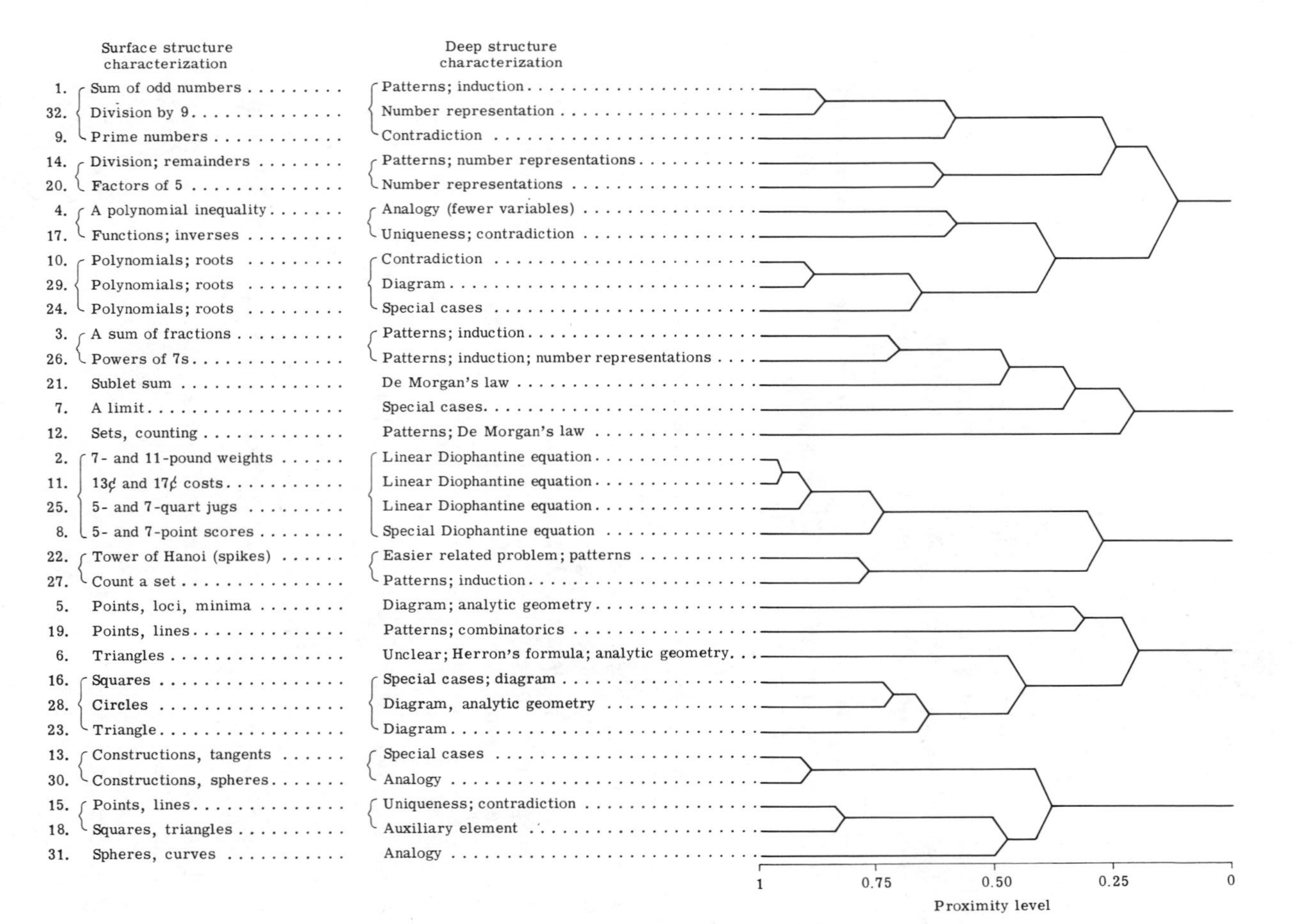

Figure 8.2 Cluster analysis of combined novices' card sort.

subjects ($p < .01$) but not for the control subjects. Or to put it another way, the students in the experimental group appeared to become more proficient at mathematical problem solving, while the students in the control group did not.

Figure 8.3 presents the cluster analysis of the experimental group's sorting after instruction. Inspection of Figure 8.3 indicates the shift in the students' perceptions. Six of the eight strong clusters in the experimental group's postinstruction sorting are homogeneous with regard to deep structure, while only four are homogeneous with regard to surface structure. Moreover, surface and deep structures coincided in all four of those clusters.

In contrast, the control group's sorting after the winter term indicates little change from their preinstruction perceptions. Once again to conserve space, the cluster diagram from that sorting, which closely resembles Figure 8.2, is not given. Of the 10 strong clusters in that cluster diagram, 7 are homogeneous with regard to surface structure and 4 with regard to deep structure; all 4 of the latter share common surface structures as well. Tables 8.2 and 8.3 offer statistical data characterizing the control group's postinstruction sortings.

The last two rows of Table 8.2 present the homogeneity proportions for the experimental and control groups after instruction. These data substantiate the impressions one obtains from Figures 8.1, 8.2, and 8.3. Differences between the deep and surface proportions were compared across the various conditions with the t approximation to the binomial. With only one exception, noted below, each of the following comparisons was significant to at least the $p < .05$ level, both in direction and size of the differences.

The difference between deep and surface proportions from the experts' sorting differed, in direction and magnitude, with the difference between deep and surface proportions obtained from the experimental group before instruction [$t(18) = 2.33$ for Measure 1; $t(18) = 4.09$ for Measure 2], the control group before instruction [$t(15) = 2.61$; $t(15) = 4.98$], the combined novice group before instruction [$t(26) = 1.88, p < .1$; $t(26) = 4.81$], and the control group after instruction [$t(15) = 2.31$; $t(15) = 3.99$]. Similarly, the differences between deep and surface structure proportions obtained from the experimental group after they finished the problem-solving course differed at the $p < .05$ level from their own scores prior to instruction [$t(11) = 2.38$; $t(11) = 4.51$], from the control group prior to instruction [$t(17) = 2.65$; $t(17) = 5.48$], and from the combined novice scores prior to instruction [$t(28) = 2.39$; $t(28) = 5.41$]. The experimental group's postinstruction scores also differed from the control group's postinstruction scores [$t(17) = 2.34$; $t(17) = 4.37$].

This comparison of surface and deep structure proportions provides an indirect indication that the experimental group's problem perceptions be-

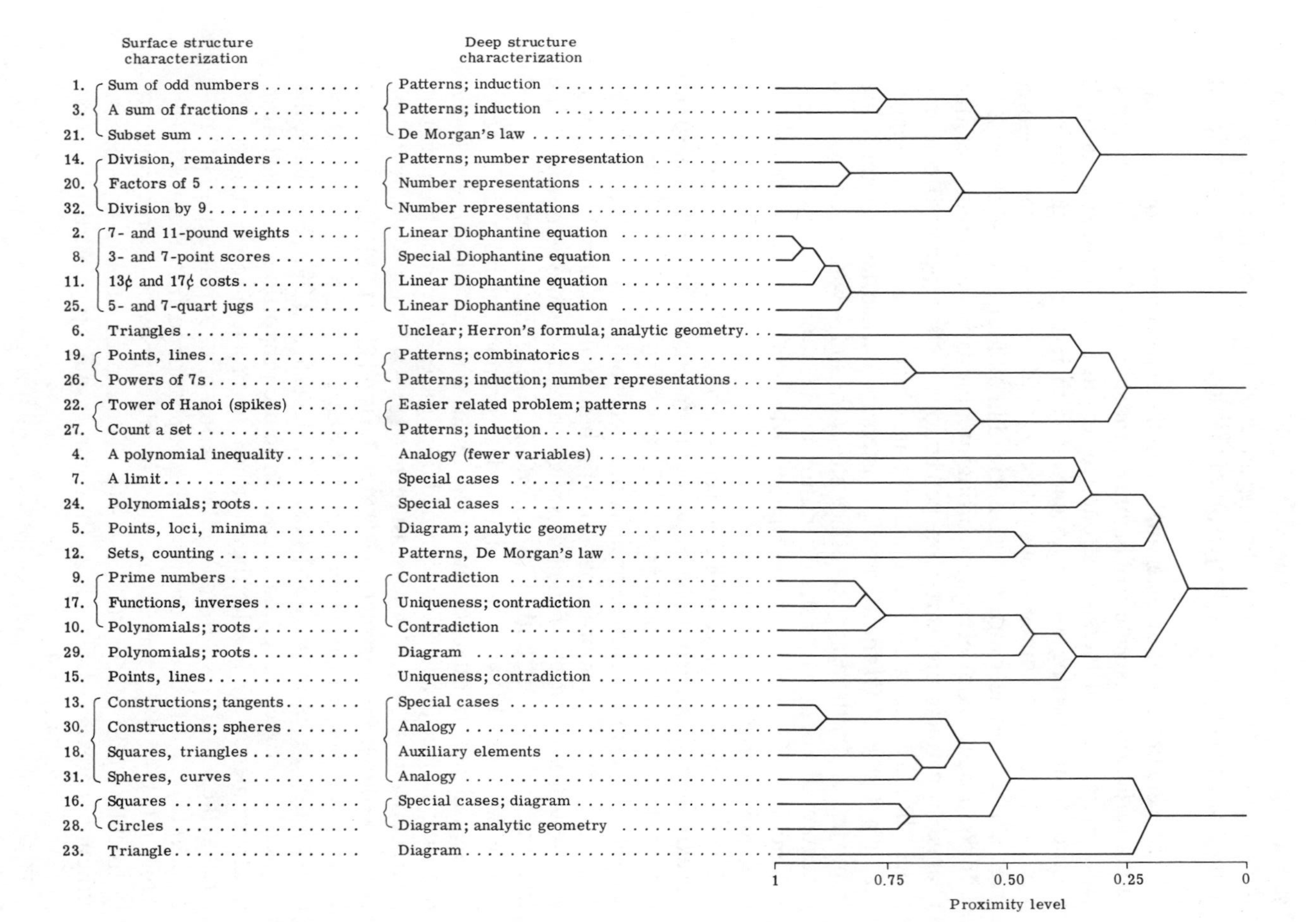

Figure 8.3 Cluster analysis of experimental group's card sort, after instruction.

Table 8.3
Correlations between Sorting Matrices of Novices with Expert Sort

Novice group	Correlation
Control, pretest	.551
Experimental, pretest	.540
Combined, pretest	.602
Control, posttest	.423
Experimental, posttest	.723

came more "expert-like" as they became better problem solvers. The control group's performance did not improve, and neither did their problem perception. A more direct measure of the degree to which the students' perceptions were expert-like was obtained by correlating the sorting matrices for each of the treatment groups, before and after instruction, with the sorting matrix obtained from the experts. The correlations are given in Table 8.3.

With $df = 496$, all of the correlations given in Table 8.3 are significant. The correlations between each of the novices' pretest matrices, and of the control group's posttest matrix, with the experts' matrix, are all significantly less ($p < .01$) than correlations between the experimental group's postinstruction matrix with the experts' matrix.

Discussion

To recall the technical point made above, the design of this study allows for the direct attribution of changes in the experimental group's problem perceptions to changes in their problem-solving proficiency. While there was little reason to doubt the evidence regarding perception and representation obtained from contrasting group studies, that evidence was not incontrovertible; in expert–novice studies the groups being compared usually differed along a number of dimensions that might have contributed to differences both in perception and performance. A clear understanding of how novices' performance improves in a discipline cannot be obtained by comparing them to a group of experts whose aptitude for the discipline is, in all likelihood, far beyond that of the novices. Similarly, an understanding of expert perception cannot be obtained by taking as the starting point of that development people whose performance alone makes it unlikely that they will ever be expert in that domain.

Before discussing the shift in the experimental group's perceptions, there needs to be a bit more discussion of those students' incoming knowledge. As the pretest scores indicate, the problems the students were asked to solve (and sort) were far from easy. The low pretest scores do not imply, however, that the students were ignorant of the relevant mathematics when they performed the sort.* The "novices" who participated in this study had taken 12 years or more of mathematics and had voluntarily enrolled in problem-solving courses. The problems used in the study, and their solutions, can be understood by students with high school backgrounds. In fact, the surface structure labels used in the sorting reflect this intermediate level of sophistication. For example, Problems 2, 8, 11, and 25 were all assigned the a priori surface structure label "combining integers" despite the fact that there are different objects in the statements of the four problems. Surely one would be surprised if college students could not see that integer combinations of weights, of costs, of liquid quantities, and of point scores all call for operations with whole numbers! (As the studies by Silver, 1979, and Chartoff, 1977, indicate, the same assumption cannot be made about school children.)

The data in Table 8.2 provide a clear indication that the experimental group's perceptions of problem relatedness shifted from a basis in surface structure to a basis in deep structure. As an illustration of the change in perception, let us consider the ways that the experimental group sorted Problems 9, 10, and 17, both before and after the problem-solving course.

Problem 9 Let n be a given whole number. Prove that if the number $(2^n - 1)$ is a prime, then n is also a prime number.

Problem 10 Prove that there are no real solutions to the equation

$$x^{10} + x^8 + x^6 + x^4 + x^2 + 1 = 0.$$

Problem 17 Show that if a function has an inverse, it has only one.

Before the problem-solving course, the students' sorting of each of these problems was consistent with the problems' surface structure characterizations. Problem 9 deals with primes, or, more generally, with whole numbers. It was sorted with two other whole numbers problems (Problems 1 and 32) in a homogeneous surface structure cluster. The deep structure characterizations of those three problems in that cluster all differed. Problem 10 deals with the roots of a complex polynomial. It was sorted together in a homogeneous surface structure cluster with Problems 24 and 29, which

* This point is important, for the preinstruction sortings would be of much less intrinsic interest if the students were unfamiliar with the techniques relevant for solving the problems. These students have had adequate experience with the appropriate mathematics. The question is whether they have codified it for themselves in the appropriate ways.

also ask about the roots of polynomials. These three problems in that cluster also have different deep structures characterizations. Problem 17 deals with (unspecified) functions and was (just barely) clustered with Problem 4, which confronts the reader with a complicated rational function for analysis.

Each of Problems 9, 10, and 17 is solved by the technique known as proof by contradiction. After instruction, all three problems were placed in the same cluster by the experimental group, despite the fact that they differed clearly with regard to surface structure. This clustering does not replicate the experts' contradiction cluster (Problems 9, 15, and 17, in Figure 8.1), but it comes close. It indicates a clear shift toward expert perception on the part of the experimental group. Substantiating evidence appears in Table 8.3, which shows that the correlation between the experts' sorting matrix and the experimental group's sorting matrices jumped from .540 (before instruction) to .723 (after instruction). This rather dramatic shift, after a short period of time, indicates that a problem-solving course that places a great emphasis on understanding can have a strong impact on both perceptions and performance.

Despite the strong change in the nature of their postexperimental sorting, the students' perceptions after instruction cannot be truly called expert-like. The experts' far more extensive knowledge and experience allow them to make distinctions inaccessible to the novices. Consider, for example, the bracketed clusters that include Problem 1 in each of the three diagrams: Problems {1, 3} in the expert cluster, Problems {1, 32, 9} in the novice cluster before instruction, and Problems {1, 3, 21} in the experimental students' cluster after the problem-solving course. After the course, the students drop Problems 32 and 9 from association with Problem 1. This is appropriate, in that they are similar to Problem 1 only in that they deal with whole numbers. Problem 3, which shares the same deep structure as Problem 1 (both can be solved by finding patterns and verifying them by induction), joins Problem 1. The mimicry of expert perceptions is not exact, however; Problem 21 joins the other two. The addition of Problem 21 provides an indication of the intermediate or "journeyman" status of the students at this point. As it happens, Problems 12 and 21 had been included in the card sort to see if the experts would cluster them together. Part of the mathematical structure of Problem 21 is the fact that multiples of 9 and multiples of 4 both include multiples of 36 (their intersection), and that when one subtracts multiples of 4 and 9, one must compensate for having subtracting the multiples of 36 *twice* by adding them in once again. At a formal level this observation is structurally similar to De Morgan's law,

$$N(A \cup B) = N(A) + N(B) - N(A \cap B),$$

upon which Problem 12 is based. This is a fairly subtle observation. The experts' experience with combinatorics problems might have made such an observation accessible, though this was far from a sure bet (indeed, Problems 12 and 21 just barely make the strong cluster cutoff in the experts' sort). After one course in problem solving, novices cannot be expected to see such subtle things. They did not. Lacking such knowledge, it was reasonable for them to think that "looking for patterns" will help to solve Problem 21—and to sort it with two other "patterns" problems as a result.

CONCLUSIONS, IMPLICATIONS

The work described in this chapter is consistent with and extends previous research on problem perception. Briefly stated, it supports these theses:

Thesis A Novices in a domain (but again, students with some experience) are likely to perceive problems in similar ways. Unfortunately this homogeneity in perception is not necessarily functional; students may focus on objects in the problem statements ("surface structure") that are peripheral to the essence of the problems and as a result may be led in wrong directions.

Thesis B Experts in a domain come to perceive the underlying structures ("deep structures") in problem situations. Their problem-solving performance is guided by those perceptions and is substantially facilitated as a result.

Thesis C As students become more proficient at problem solving, not only does their performance become more expert-like; so does their problem perception.

These broad assertions leave a host of serious issues wide open for (necessary) further investigation. In a sense, mathematics sits half-way between physics and chess, two disciplines in which a substantial amount of related research has been done. What that half-way position means is not clear to me. The research by Chi, Feltovich, and Glaser (1981) indicates that what physicists "see" in problem statements is their deep structure, the principles of physics. Fortunately for those who teach the subject, physics instruction is organized along those lines. Moreover, reasonable and implementable implications for teaching physics—for example, that the dangers of being misled by surface structure while solving problems should be explicitly demonstrated to students, and that the underlying principles (the deep structure) should be stressed in problem analysis—can be drawn from the research.

As noted above, reasoning patterns in mathematics and subject matter presentation in mathematics can be almost independent of each other. The implications drawn just above for physics are probably still reasonable in mathematics,* but how to implement them is not immediately obvious. Prescribing a reasonable balance between instruction according to mathematical topics and according to underlying mathematical structure is no easy task, and extant research provides little guidance. The research from chess (Chase and Simon, 1973; de Groot, 1965) and related fields makes things more confusing. While chess players talk broadly in terms of strategy ("protecting the center," etc.), much of their routine competence is based on "vocabularies" of some 50,000 chunks of extremely specific knowledge that is not at all principle-based (or based on deep structure, if you prefer). There are some parallels to knowledge organization in mathematics. I am not sure what the implications of those parallels might be. How much emphasis there should be on developing a vocabulary of specific bits of mathematical knowledge is not clear; nor is it clear how such a vocabulary might be efficiently acquired. The chess research casts doubts on the generality and power of broad strategies, but (since strategies are essentially ignored in conventional mathematics instruction) the right balance to strive for between *strategy* and *vocabulary* is very much an open question. A great deal more research on these issues needs to be done.

This having been said, it may be worth returning to the message that emerges from the details of this and the previous chapter. The "experimental treatment" in both chapters was a problem-solving course that focused on understanding—on making sure that problems are explored until one has a sense of their underlying structure, and of keeping one's explorations reasonable and well governed at the same time. These two chapters offer strong evidence that such attention to understanding pays off, with the students' problem perception and their problem-solving performance becoming more expert-like. Though there are many open questions in the research, there are also many useful implications for current practice.

Summary

This chapter explored the relationship between individuals' proficiency at mathematical problem solving and their perceptions of the structure of mathematical problems. Problem perception was assessed by a card sorting

* Note that one must be careful about drawing unreasonable implications from reasonable research. Work on problem schemata (e.g., Hinsley *et al.*, 1977) is quite solid, but it is dangerous to assume that experts' knowledge organization is a good thing to have naive students mimic. There have been some recommendations that the mathematics curriculum be organized around such schemata (e.g., at the secondary level, recommendations that we teach schemata for "mixture problems," "rate problems," and so on, as separate entities.) These notions are not only reductive and contrary to the spirit of real mathematical deep structure (the meaningful use of symbolism), they also do not work.

task using cluster analysis. Each of the 32 problems in the sorting task was assigned a "surface structure" characterization based on superficially prominent aspects of the problem statement, and a "deep structure" characterization based on the experimenter's assessment of relevant mathematical principles or relevant solution procedures. Deep structure characterizations were validated separately by a colleague. College mathematics faculty served as the experts for the study, with their sorting data serving as a baseline against which students' sorting data were compared. Two groups of students (novices) took part in the study. The experimental group consisted of students enrolled in an intensive problem-solving course, the control group of students with similar backgrounds from a mathematics course given concurrently. The novices performed the sortings before and after instruction.

Generally speaking, the experts sorted together problems that shared the same deep structure characterizations, independent of their surface structure characterizations. In the expert sorting, 8 of 11 strong clusters are homogeneous with regard to deep structure. Of those 11 clusters, 5 are homogeneous with regard to deep structure but not surface structure, while only 1 cluster is homogeneous with regard to surface structure but not deep structure. On a "conflicting pairs" measure that looked at strongly paired problems in which at least one of surface or deep structure differed, the experts sorted together 67% of the pairs with the same deep structure and 25% of the pairs with the same surface structure.

The novices' preinstruction behavior is very much at the other end of the spectrum, with an almost symmetric reversal of the experts' sort. In the novices' sorting, 8 of 10 strong clusters are homogeneous with regard to surface structure. Of those 10 clusters, 5 are homogeneous with regard to surface structure but not deep structure, while only 1 cluster is homogeneous with regard to deep structure but not surface structure. On the conflicting pairs measure, the novices sorted together 67% of the pairs with the same surface structure and only 11% of the pairs with the same deep structure.

The instructional treatment was a problem-solving course that focused on heuristic and control strategies. It emphasized *understanding,* in particular, analyzing problems carefully and monitoring and assessing one's work in the midst of problem solving. It did not explicitly address issues of problem perception, although many of the techniques studied in the course (examining "givens and goals," looking at examples, etc.) are likely to increase attention to problem structure and perhaps improve problem perception thereby. As reported in Chapter 7, the experimental students' problem-solving performance improved significantly.

The postinstruction sorting for the experimental group shows a dramatic shift toward expert-like problem perception. In the postinstruction sort, 6 of 8 strong clusters are homogeneous with regard to deep structure, 4 with regard to surface structure. Moreover, all 4 of those were homogeneous with regard to both. The conflicting pairs measure now has 55% of the pairs with the same deep structure and only 9% with the same surface structure. The correlation between the experimental group's sorting matrix with the experts' matrix jumped from .540 before instruction to .723 after instruction.

These data support earlier research indicating that experts' perceptions are based on deep structure while novices' are based on surface structure. They permit the direct conclusion that problem perception becomes more expert-like with the acquisition of problem-solving proficiency. They also indicate that a problem-solving course that focuses on understanding and analysis can have a significant impact both on perception and performance.

Appendix: Problems Used in the Card Sort

A problem's surface structure characterization is designated by SSC; that problem's deep structure characterization by DSC.

1. Show that the sum of consecutive odd numbers, starting with 1, is always a square. For example,

$$1 + 3 + 5 + 7 = 16 = 4^2.$$

SSC: Sum of odd numbers
DSC: Patterns; induction

2. You have an unlimited supply of 7-pound weights, 11-pound weights, and a potato that weighs 5 pounds. Can you weigh the potato on a balance scale? A 9-pound potato?
SSC: 7- and 11-pound weights
DSC: Linear Diophantine equation

3. Find and verify the sum,

$$\frac{1}{1 \times 2} + \frac{1}{1 \times 2 \times 3} + \frac{3}{1 \times 2 \times 3 \times 4} + \cdots + \frac{n}{1 \times 2 \times 3 \times \cdots \times (n+1)}.$$

SSC: A sum of fractions
DSC: Patterns; induction

4. Show that if x, y, and z are greater than 0,

$$\frac{(x^2 + 1)(y^2 + 1)(z^2 + 1)}{xyz} \geq 8.$$

SSC: A polynomial inequality
DSC: Analogy (fewer variables)

5. Find the smallest positive number m such that the intersection of the set of all points $\{(x, mx)\}$ in the plane, with the set of all points at distance 3 from (0, 6), is nonempty.
SSC: Points, loci, minima
DSC: Diagram; analytic geometry
6. The lengths of the sides of a triangle form an arithmetic progression with difference d. (That is, the sides are $a, a + d, a + 2d$.) The area of the triangle is t. Find the sides and angles of this triangle. In particular, solve this problem for the case $d = 1$ and $t = 6$.
SSC: Triangles
DSC: Unclear; Herron's formula; analytic geometry
7. Given positive numbers a and b, what is
$$\lim_{n\to\infty} (a^n + b^n)^{1/n}?$$
SSC: A limit
DSC: Special cases
8. In a game of "simplified football," a team can score 3 points for a field goal and 7 points for a touchdown. Notice a team can score 7 but not 8 points. What is the largest score a team cannot have?
SSC: 3- and 7-point scores
DSC: Special diophantine equation
9. Let n be a given whole number. Prove that if the number $(2^n - 1)$ is a prime, then n is also a prime number.
SSC: Prime numbers
DSC: Contradiction
10. Prove that there are no real solutions to the equation
$$x^{10} + x^8 + x^6 + x^4 + x^2 + 1 = 0$$
SSC: Polynomials; roots
DSC: Contradiction
11. If Czech currency consists of coins valued 13¢ and 17¢, can you buy a 20¢ newspaper and receive exact change?
SSC: 13¢ and 17¢ costs
DSC: Linear Diophantine equation
12. If $N(A)$ means "the number of elements in A," then $N(A \cup B) = N(A) + N(B) - N(A \cap B)$. Find a formula for $N(A \cup B \cap C)$.
SSC: Sets, counting
DSC: Patterns; De Morgan's law
13. Construct, using straightedge and compass, a line tangent to two given circles.

SSC: Constructions; tangents
DSC: Special cases

14. Take any odd number; square it; divide by 8. Can the remainder by 3? or 7?
SSC: Division; remainders
DSC: Patterns; number representations
15. You are given the following assumptions: (i) Parallel lines do not intersect; nonparallel lines intersect. (ii) Any two points P and Q in the plane determine an unique line which passes between them. Prove: Any two distinct nonparallel lines L_1 and L_2 must intersect at a unique point P.
SSC: Points, lines
DSC: Uniqueness; contradiction
16. Two squares s on a side overlap, with the corner of one on the center of the other. What is the maximum area of possible overlap?
SSC: Squares
DSC: Special cases; diagram
17. Show that if a function has an inverse, it has only one.
SSC: Functions, inverses
DSC: Uniqueness; contradiction
18. Let P be the center of the square constructed on the hypotenuse AC of the right triangle ABC. Prove that BP bisects angle ABC.

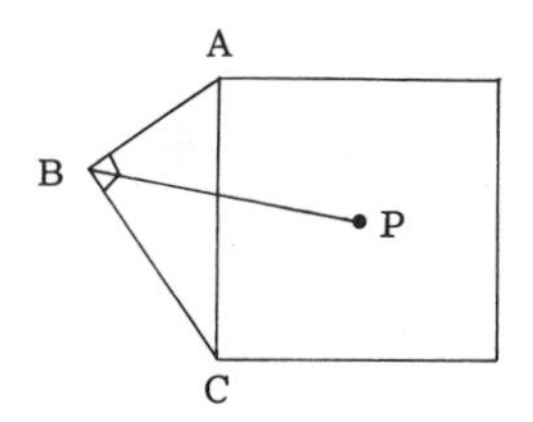

SSC: Squares, triangles
DSC: Auxiliary elements
19. How many straight lines can be drawn through 37 points in the plane, if no 3 of them lie on any one straight line?
SSC: Points, lines
DSC: Patterns; combinatorics
20. If you add any five consecutive whole numbers, must the result have a factor of 5?
SSC: Factors of 5
DSC: Number representations
21. What is the sum of all numbers from 1 to 200, which are not multiples

of 4 and 9? You may use the fact that

$$(1 + 2 + \cdots + n) = \tfrac{1}{2}(n)(n + 1).$$

SSC: Subset sum
DSC: De Morgan's law

22. Your goal is to convert the figure on the left to the one on the right. You may move only one disk at a time from one spike to another, and you may never put a larger disk on top of a smaller one. How to?

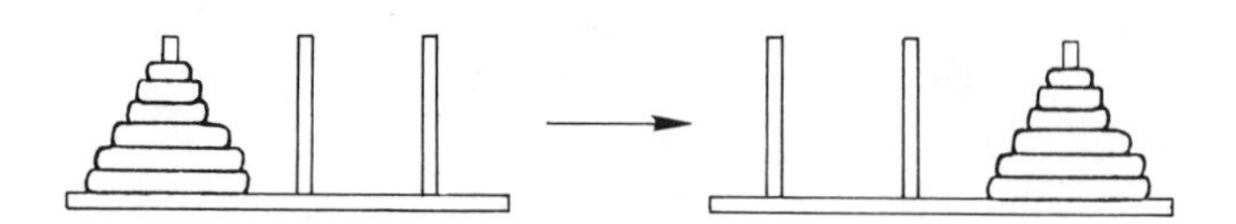

SSC: Tower of Hanoi (spikes)
DSC: Easier related problem; patterns

23. Determine the area of a triangle whose sides are given as 25, 50, and 75.
SSC: Triangle
DSC: Diagram

24. If $P(x)$ and $Q(x)$ have "reversed" coefficients, for example,

$$P(x) = x^5 + 3x^4 + 9x^3 + 11x^2 + 6x + 2,$$

$$Q(x) = 2x^5 + 6x^4 + 11x^3 + 9x^2 + 3x + 1,$$

what can you say about the roots of $P(x)$ and $Q(x)$?
SSC: Polynomials; roots
DSC: Special cases

25. You have two unmarked jugs, one whose capacity you know to be 5 quarts, the other 7 quarts. You walk down to the river and hope to come back with precisely 1 quart of water. Can you do it?
SSC: 5- and 7-quart jugs
DSC: Linear Diophantine equation

26. What is the last digit of

$$(\cdots((7^7)^7)^7\cdots)^7,$$

where the seventh power is taken 1000 times?
SSC: Powers of 7s
DSC: Patterns; induction; number representations

27. Consider the following magical configuration. In how many ways can you read the word "ABRACADABRA?"

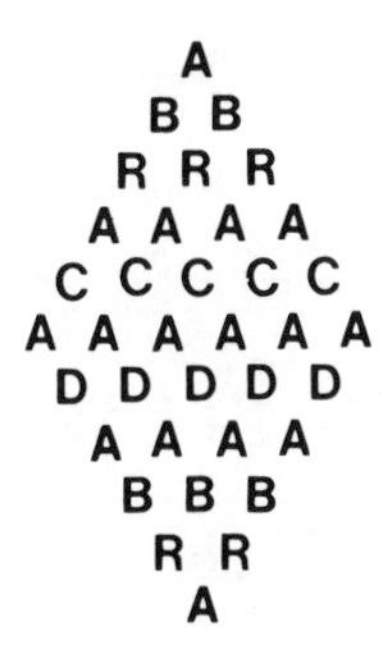

SSC: Count a set
DSC: Patterns; induction

28. A circular table rests in a corner, touching both walls of a room. A point on the rim of the table is 8 inches from one wall, 9 from the other. Find the diameter of the table.
    SSC: Circles
    DSC: Diagram, analytic geometry
29. Let a and b be given real numbers. Suppose that for all positive values of c, the roots of the question

$$ax^2 + bx + c = 0$$

    are both real, positive numbers. Present an argument to show that a must equal 0.
    SSC: Polynomials; roots
    DSC: Diagram
30. Describe how to construct a sphere that circumscribes a tetrahedron (the four corners of the pyramid touch the sphere.)
    SSC: Constructions; spheres
    DSC: Analogy
31. Let S be a sphere of radius 1, A an arc of length less that 2 whose endpoints are on the boundary of S. (The interior of A can be in the interior of S.) Show there is a hemisphere H that does not intersect A.
    SSC: Spheres, curves
    DSC: Analogy
32. Show that a number is divisible by 9 if and only if the sum of its digits is divisible by 9. For example, consider 12345678: $1 + 2 + 3 + 4 + 5 + 6 + 7 + 8 = 36 = 4 \times 9$, so 12345678 is divisible by 9.
    SSC: Division by 9
    DSC: Number representations

9

Verbal Data, Protocol Analysis, and the Issue of Control*

Overview

Many of the discussions in this book, in particular the discussions characterizing various aspects of the analytical framework advanced in Part I, have been based on summary descriptions of "out loud" problem-solving sessions. For the most part those descriptions consisted of brief synopses of problem sessions supplemented by a few lines of dialogue — dialogue clearly chosen to make a point. Such evidence might reasonably be considered anecdotal, serving perhaps to illustrate a perspective but hardly to document it in rigorous fashion. The analyses of such descriptions appear to be highly subjective. In the interests of accuracy, one would need to know much more about the analytical methods used before one would be willing to accept either the descriptions of the problem sessions, or the interpretations of what took place in them, as being accurate. Perhaps more important, one has good reason to question whether "verbal data" such as the transcripts of out loud problem-solving sessions can even be considered as reliable evidence in the first place. Such issues are the focus of this chapter.

Verbal methods, as they are called, are controversial. Serious questions have been raised as to whether transcripts of out loud problem-solving ses-

* Chapter 9 is a significantly expanded and revised version of Schoenfeld (1983b). Permission from Academic Press to reproduce parts of the chapter is gratefully acknowledged.

sions provide accurate reflections of the processes used by individuals as they solve problems, or of the processes that they would have used had they not been asked to work out loud. Such questions are the focus of the second and third sections of this chapter. The second section provides a brief review of background material on verbal methods and of trends in the literature that led to current controversies. The third section examines a particular verbal study in some detail. The examples described there indicate that verbal data are not always what they seem and that it is dangerous to take such data at face value—even when they are gathered in experimentally clean settings where appropriate methodological care has been taken. A summary discussion suggests that each methodology for gathering verbal data can be thought of as lens that may bring some aspects of behavior sharply into focus but that may also blur or distort other aspects of behavior. It suggests that great care must be taken in interpreting the data obtained from such methods. Within this context, the methodology used for generating the protocols in this book is then considered.

The bulk of this chapter is devoted to the discussion of an analytical framework for the macroscopic analysis of problem-solving protocols, with an emphasis on executive or control behavior. As noted above, the discussions of control in Chapters 1 and 4 can be considered anecdotal in nature. Even if one grants that attempts were made to be objective in summaries of problem sessions and to be balanced in discussions of transcript excerpts, there are issues of reliability and replicability that must be considered when dealing with such analyses. Would others see the same things in the protocols (or videotapes); would they ascribe to them the same importance? Ideally one would like to have an objective framework for such analyses. Independent analysts should identify the same "critical points" in a protocol and attribute the same importance to them.

The fourth section of this chapter reviews prior attempts at developing frameworks for protocol analysis. The fifth section returns to two problem sessions briefly introduced in Chapter 1, the attempts by two pairs of students (KW and AM, and DK and BM) to solve Problem 1.3. The transcripts of those sessions are given in full. An analysis of the students' work indicates the limitations of extant protocol analysis schemes. More importantly, it points to some essential aspects of problem-solving behavior that need to be captured by analytical frameworks.

A first attempt at the development of a comprehensive framework for protocol analysis at the control level is described in the sixth section. The basic idea underlying the analysis is as follows. Protocols are parsed into large chunks called *episodes.* These episodes are periods of time during which the problem solver (or problem-solving group) is engaged in a single set of actions of the same type or character, such as planning or exploration.

When a protocol is parsed in this way, the potential locations of strategic decisions become apparent: Individuals' decisions as to what to pursue, or what to abandon, are generally made at the junctures between episodes. Once these executive decisions have been identified, their effects are relatively easy to trace in subsequent work. In consequence, this procedure for the analysis of protocols at a macroscopic level provides a way of identifying the large-scale consequences of individual decisions made during problem-solving sessions, and of focusing on those key decisions that may in themselves determine success or failure. Equally important, it points to those places where executive decisions should have been considered or made. In that way the consequences of the absence of executive decision-making—in particular, of failures to curtail wild goose chases or to exploit potentially useful information—can also be traced.

The complete analysis of a problem-solving protocol is given in the seventh section. This detailed analysis supplements the discussions of the two sessions described earlier and demonstrates the precise workings of the analytical scheme. The eighth section provides capsule descriptions, derived from the analytical framework, of a number of problem-solving sessions. Although detail is sacrificed in the capsule descriptions, using a larger data base allows one to gain a better sense of the typicality of the executive behaviors that appear in those protocols. The problem-solving sessions that are summarized took place immediately before and immediately after the intensive problem-solving course that served as the experimental treatment in Chapters 7 and 8. In consequence, they provide one more "before and after" measure of the effects of that instruction.

The chapter concludes with a discussion of issues regarding the objectivity–subjectivity, the accuracy, the reliability, and the limitations of the analytical framework. There are many questions that need to be answered, and the framework only points to some of the answers. It is, however, worth mentioning one indication of the framework's robustness at this point. The parsings of the protocols that appear throughout the chapter were produced by a team of three undergraduates who were trained in the use of the framework but who were then asked to analyze the protocols independently. The analyses that they produced were good enough to use without change.

Background, Part 1: Verbal Methods

The antecedents of current verbal methods have had a long and often controversial history. This brief description can hardly do it justice, because it paints with a broad brush issues that could be painted in minute detail.

Introductions to the relevant literature, from an information-processing point of view, can be found in recent work by Ericsson and Simon (1980, 1981, 1984).

One of the most common methods of uncovering "rules of productive thinking" has been introspection, or reflecting on one's own thought processes. Descartes' *Discourse on Method* and *Rules for the Direction of the Mind* were the products of introspection, as were most of the philosophical–psychological treatises on thinking through the late nineteenth century. For centuries the use of introspection and the analysis of verbal reports were not only considered methodologically sound but were considered a primary source of information regarding complex human cognitions—especially in mathematics. This perspective can be seen in the Gestaltists' reliance on introspection and on retrospective analysis; see, for example, the works by Hadamard (1954), Poincaré (1913), and Wertheimer (1959).

By the middle of the twentieth century, Gestaltism and the methodologies associated with it had lost much of their luster. This happened in part because the Gestaltists, with their emphasis on inspiration as the product of the workings of the subconscious, had almost defined the major issues out of existence; when one begins by assuming that the most important cognitive phenomena are inaccessible, there really is not too much left to talk about.* But more importantly, it happened because introspection and retrospection themselves were shown to be unreliable as methods of inquiry into cognition.

With its focus on largely inaccessible mental procedures, Gestaltism was at one end of the psychological spectrum. At the other end was behaviorism, which waxed through mid-century and beyond as Gestaltism waned. From the behaviorist perspective, or at least that of some of its more notable expositors (see, e.g., Skinner, 1974), problem solving and other complex intellectual acts could be explained solely in terms of behavior. Ultimately, it was maintained, complete explanations of complex intellectual perform-

* One of the classic Gestalt manuscripts, Duncker's (1945) *On Problem Solving,* provides an example. Duncker discusses the "thirteen problem":

> Why are all six digit numbers
> of the form abc,abc divisible by 13?

As Duncker described it, the difficulty with the problem disappears when the common factor 1001 emerges from the subconscious. Of course it does; when one sees that abc,abc = (abc)(1001) = (abc)(7)(11)(13), then the thirteen problem is no longer a problem. That observation, however, is almost tautological. Defining the issue in those terms finesses the real problems: Where does (or might) the common factor 1001 come from? And how might one explore numbers of the form abc,abc so that the figure 1001 *will* emerge? Gestaltism, with its emphasis on inspiration as emerging by itself from the workings of subconscious mechanisms, could not get a firm grasp on such issues.

ance could be found in extended series of stimulus–response chains. From Skinner's perspective, "mind" and "mentalism" were superfluous theoretical constructs and investigating them was a misguided waste of time. What counted was tangible, observable behavior. Hypothetical mental mechanisms led to absurdities such as homunculi and were best avoided altogether.

The behaviorist perspective as just described may have been a reductio ad absurdum of scientism. It was not an aberration, however, but merely an extreme case of a general trend in psychology and the social sciences. In all of those fields, mathematics education among them, great emphasis was placed on the "science" in "social science." The idea was to be as rigorous and methodologically pure as possible, and the use of precise experimental and statistical methods was de rigueur. To be scientific, hypotheses had to be falsifiable; methods had to be reliable, and their results had to be replicable. To put things bluntly, introspection and retrospection did not meet such standards. The Gestaltists' methodologies stood in contradiction to the spirit of the times and were easy targets for attack.* Verbal methods were déclassé through the 1960s and much of the 1970s.

Perspectives on verbal methods have changed over the past decade. Perhaps the major cause of the change is the legitimation of protocol analysis as a consequence of its role as a major research tool in AI. Such research demonstrated that one can design successful problem-solving programs for computers based on principles abstracted from the analysis of human problem-solving protocols (see, e.g., Newell & Simon, 1972). Those programs' successes demonstrated the importance of certain kinds of problem-solving strategies, at least so far as machine implementation is concerned. Of equal importance, it gave credibility to the methodologies that uncovered them.

Another major cause was the impact of Piaget's work. Research in developmental psychology made it clear that careful clinical investigations could give rise to replicable results, to falsifiable hypotheses, and to predictions that could be tested experimentally. In short, it was found that clinical investigations could indeed lay the foundations for good science. A third cause, particularly influential for mathematics educators, was the impact of the Soviet psychologists' studies on mathematical thinking (*Soviet Studies in the Psychology of Learning and Teaching Mathematics,* Kilpatrick & Wirszup, 1969–1975; Krutetskii, 1976). Soviet psychology was anything but "science" in the unbiased, objective sense that we take the term. While it lacked rigor, however, it did not lack for ideas—among them the rather

* For example, a section in Hadamard's (1954) *Essay on the Psychology of Invention in the Mathematical Field* is almost embarrassing in its naiveté. Discussing the question of whether one can sense that an inspiration is in the offing, Hadamard essentially said the following: "It's an interesting idea. I'll have to ask my friends if they've ever had that experience." This kind of evidence is hardly the stuff of which science is made.

subversive notion (championed by Piaget as well) that one might learn something about people's problem-solving processes by observing them in the process of solving problems.

Indeed, one reason for the resurgence of verbal methods is the increased sophistication of the research community, which, after a few decades' experience, has developed a more balanced perspective on the utility of the "scientific" methodologies that supplanted verbal methods. As more and more ambiguous or contradictory studies dealing with the same issues piled up in the empirical literature, the limitations of those empirical methodologies began to emerge. It became clear, for example, that there are often many more variables in "treatment X versus treatment Y" comparisons than supposedly "tight" experimental designs control for. It became clear as well that the difficulties in extrapolating results from well-designed laboratory studies to more complex cognitive phenomena, and to more complex environments, had been seriously underestimated. Exploratory methodologies, including protocol analysis and clinical interviews, have been turned to with increasing frequency.

As Ginsburg, Kossan, Schwartz, and Swanson (1983, p. 7) observe, "While increasingly popular, protocol methods have not yet received thorough methodological analysis. Little is known concerning their fundamental natures, the rationales underlying their use, and their reliability." Opinions and evidence regarding these issues are divided. For example, *Psychological Review* has published two analyses of the effects of speaking aloud methods. Nisbett and Wilson's (1977) title, "Telling More Than We Know: Verbal Reports on Mental Processes," suggests its conclusions. Ericsson and Simon's (1980) "Verbal Reports as Data" concludes that certain kinds of talking aloud instructions—those that ask for verbalization as one solves a problem, without calling for explanations (elaborations or retrospections) of what one is doing—do not seem to affect people's performance while solving problems. The issue is still very much unresolved. It will be explored to a small degree in the following section, which focuses on one example of the subtle influences that can shape the generation of verbal data. The example suggests that Ericsson and Simon's conclusion needs to be strongly qualified and points to the great delicacy required in the analysis of verbal data. A bit of cross-cultural anthropology provides the context for the discussion.

Through a Glass Darkly: A Close Look at Verbal Data

Ulric Neisser's 1976 article "General, Academic, and Artificial Intelligence" begins with a discussion of Cole, Gay, Glick, and Sharp's (1971) study of cognition in a Liberian people called the Kpelle. Unschooled

Kpellans, while articulate and intelligent, often do poorly on tests and problems that seem easy to people with some formal education. A segment of dialogue suggests why:

> EXPERIMENTER: Flumo and Yakpalo always drink cane juice (rum) together. Flumo is drinking cane juice. Is Yakpalo drinking cane juice?
>
> SUBJECT: Flumo and Yakpalo drink cane juice together, but the time Flumo was drinking the first one Yakpalo was not there on that day.
>
> EXPERIMENTER: But I told you that Flumo and Yakpalo always drink cane juice together. One day Flumo was drinking cane juice. Was Yakpalo drinking cane juice that day?
>
> SUBJECT: The day Flumo was drinking cane juice Yakpalo was not there on that day.
>
> EXPERIMENTER: What is the reason?
>
> SUBJECT: The reason is that Yakpalo went to his farm on that day and Flumo remained in town that day. (Cole *et al.*, 1971, pp. 187–188)

Neisser stresses that the subject's answers are intelligent. The subject was asked about Yakpalo's behavior and his answers reflect sincere attempts, in response to the experimenter's questions, to provide information about that behavior. This is perfectly reasonable; in social circumstances, one expects such behavior. Cole *et al.* note that when subjects in similar circumstances are pressed by the experimenter and perceive that their answers are not considered satisfactory, they will often suggest that the experimenter clarify the issue by speaking directly to Yakpalo. As Neisser notes:

> Such answers are by no means stupid. The difficulty is that they are not answers *to the question.* The respondents do not accept a ground rule that is virtually automatic with us: "base your answer on the terms defined by the questioner." People who go to school (in Kpelleland or elsewhere) learn to work within the fixed limitations of this ground rule, because of the particular nature of school experience. (Neisser, 1976, p. 136)

A major point of Neisser's discussion is that intellectual behavior must be evaluated in context; behavior that may be regarded as "unintelligent" in some contexts may be considered quite natural and intelligent in others.* Consider, for example, the Kpellan's acting as if the particular details of the hypothetical situation described in a formal problem were indeed the details of a "real" situation requiring solution.

* These points are not new, of course. The context-dependent nature of "intelligence" is axiomatic among anthropologists and has been for some time; see for example, Bartlett's work in the 1920s, or for more recent examples, Cole *et al.* (1971) or Lave (1980). In recognition of such issues, some cognitive scientists (e.g., Norman, 1980) have called for extending studies in cognitive science beyond "clean" domains and beyond the "purely cognitive" domain.

> This may seem to be a poor strategy, from the problemsetter's point of view. In general, however, it is an extremely sensible course of action. In the affairs of daily life it matters whom we are talking to, what we are measuring, and where we are. The environment is endlessly complex and surprising. We do well to be alert to its opportunities and dangers, both the continuing ones and the ones that reveal themselves unexpectedly.
>
> Intelligent behavior in real settings often involves actions that satisfy a variety of motives at once—practical and interpersonal ones, for example—because opportunities to satisfy them appear simultaneously. It is often accompanied by emotions and feelings, which is appropriate in situations that involve other people. . . . "Intelligent performance in natural situations" . . . might be defined as "responding appropriately in terms of one's long-range and short-range goals, given the actual facts of the situation as one discovers them." Sometimes the most intelligent thing to do is to refuse a task altogether. (Neisser, 1976, pp. 136–137)

The dialogue with the Kpellan subject serves to make a methodological point as well. When experimenter and subject enter into dialogue (or any experimental interaction) with different expectations, or with different understandings regarding the rules of discourse that govern their interaction, there is a tremendous potential for misunderstandings and misinterpretations on both parts. The answers that a subject provides in an interview, or the behavior that a subject produces an experimental setting, must be interpreted in the light of that person's perception or understanding of the ground rules for the interaction.

One could imagine, for example, that the segment of dialogue quoted above were part of an IQ test. If the experimenter declared the subject to be unintelligent because the answers he provided were "wrong," we would argue that the experimenter had missed the point. The subject's understanding of the purpose of conversations—his sense of the ground rules for the exchange with the experimenter—was that one is supposed to respond to serious requests for information by providing the most accurate information possible. (This is quite natural in settings other than academic ones.) His answers were an attempt, in good faith, to provide information about Yakpalo's behavior. Given the psychological context within which those answers were generated, it would be inappropriate to evaluate their correctness as though they were examples of "purely cognitive" behavior.

I wish to suggest here that the potential for similar conflicts in ground rules is present in many of our methodologically clean laboratory studies, and that much of what we take to be "pure cognition" may be anything but pure. Subjects' behaviors in experimental settings can be shaped by a wide variety of subtle but extremely powerful factors. In studies that generate verbal data, factors affecting verbal behavior may include the subjects' responses to the pressure of being recorded (resulting in a need to produce *something* for the microphone), their beliefs about the nature of the experimental setting (certain methods are considered "legitimate" for solving problems in a formal setting, others not), and their beliefs about the nature of the discipline itself. Despite all the methodological safeguards one may employ, there are no guarantees that the participants in laboratory experiments will be acting in accord with the ground rules that were established for them. In consequence, appropriate caution needs to be taken in interpreting the verbal data that they produce. The following example offers a case in point.

In pilot studies in 1978 a series of recordings were made of students solving the following problem out loud.

> **Cells Problem** Estimate, as accurately as you can, how many cells might be in an average-sized adult human body. What is a reasonable upper estimate? A reasonable lower estimate? How much faith do you have in your figures?

The cells problem is a particularly interesting task, an excellent vehicle for examining the ways that people access and use the information that they have stored in long-term memory. Despite its apparent inaccessibility at first reading, the problem can be solved without any special technical information. One straightforward solution starts with the assumption that reasonable estimates for "average human body volume" and "average cell volume" should exist; the problem is to make those estimates. Since there will be a huge amount of guesswork on cell volume, the estimate of body volume can be quite rough: A box with dimensions 6 feet × 6 inches × 18 inches is probably within a factor of two of the actual average, and that should suffice. (A more accurate figure can be obtained by taking an estimate of average body weight and converting it to volume. There is, however, no need to be so precise. That degree of specificity is an indulgence.) Approximating an average cell's volume calls for more sophisticated guesswork. One can see the markings of a ruler down to 1/32 inch, so perhaps 1/50 inch is a lower limit to what people can see clearly with the naked eye. Cells were discovered with early microscopes, which (considering that magnifying glasses probably have about 5 power) were most likely greater than 10 power and less than 100 power. Thus a "canonical cell"—which one might as well assume to be a cube since the computations are easier that way—must be between 1/500 inch and 1/5000 inch on a side. The rest is arithmetic.

The first set of subjects recorded working on this problem were junior and senior college mathematics majors. The students knew me reasonably well and were familiar with my work. Some had done protocol recording themselves as parts of senior projects. All of the appropriate precautions were taken to set the students at ease for the recording sessions. The students were recorded working on the problem one at a time.

Protocol 9.1 (Appendix 9.1) was produced by AB, a senior mathematics major who graduated with honors. Student AB's work is typical of the attempts produced by the better students recorded working the problem. His solution, like most, starts with the outline of a plan and the decision to use volume (rather than weight) for comparison of human and cell size:

My first possible approach might be to . . . try to figure out the volume of each part of the body. And then make a rough estimation of what I thought the volume of a cell was and then try to figure out how many cells fit in there.

The order of computation, body volume before cell volume, is also typical—as are AB's remarkably detailed and time-consuming computations. An "average" body, most often the student's own, was approximated by a series of geometric solids. The more sophisticated students, like AB, used spheres (with volume computed as $\frac{4}{3}\pi r^3$) to approximate the shape of their heads, and cones (even truncated cones!) for their legs:

And now a leg . . . a cone might be more appropriate. And the base of my leg is approximately 6 or 7 inches in diameter so you would have $(3\frac{1}{2})^2 \times \pi$ and the height would be—what is my inseam size, about 32 or 34—so you've got to have a 34, and it's a cone, so you've got to multiply by $\frac{1}{3}$.

Other, less sophisticated students used cylinders for their legs and, when they could not recall the formula for the volume of a sphere, used cubes for their heads. But in all cases the volumes of the geometric figures were carefully calculated.

In sharp contrast to their meticulous calculations of body volumes, the students' estimates of cell size were extremely crude and brief. In Protocol 9.1, for example, AB chose his estimate of the diameter of a (spherical) cell as a "compromise" value somewhere between an undesignated upper limit and his lower limit of an Angstrom unit, the diameter of a hydrogen atom. Although AB did acknowledge that this was "a very rough estimate," he made no attempt to refine it. Another honors student, after spending 10 minutes on body volume, said, "All right, I know I can see 1/16 of an inch on a ruler, so say a cell is 1/100 of an inch on a side." In general the time that students spent on cell volume was a small fraction of the time spent on body volume. These patterns of behavior, though puzzling, were remarkably consistent.

Later that same year we began taping pairs of students solving problems together. Recordings were made of nearly two dozen pairs of students, who worked on the cells problem after receiving nearly identical out loud instructions. (The only modification in the instructions is that the pair of students was asked to work together as a team. The reason for this approach to generating protocols is discussed below.) Not once did a pair of students demonstrate the kind of behavior described above.

With hindsight it is apparent that the behavior in the single-student protocols was not at all a reflection of the students' "typical" cognitions. The behavior in those protocols was absolutely pathological, and the pathology was induced by the experimental setting itself. The cells problem was outside the range of the students' experience, and they found it disturbing. They read the problem statement and were stuck. They had not worked problems like this in a formal setting (if ever), and they had no clear idea of how to approach it. Knowing that one or more mathematics professors would be listening to the tapes of their work, the students felt great pressure to produce something substantial for the record. It goes without saying that the "something" they produced should be mathematical. So the students working alone responded to the pressure by doing the only formal mathematics related to the problem statement they could come up with under the circumstances: computing the volumes of geometric solids. Of course, the students in two-person protocols also felt the same kind of pressure. But when two students worked on the problem, it was typical for one student to turn to the other and say something like, "I have no idea of what to do. Do you?" This acknowledgment of mutual ignorance served as an escape valve, which lightened the burden of the environment pressure. As a result, the pressure became bearable, and extreme manifestations of pathology like those seen in Protocol 9.1 were avoided.

This example has been discussed at length because it indicates the kinds of subtle difficulties inherent in analyses of problem-solving processes. When I became aware of the environmental factors that caused the pathological behaviors to appear in the single-person protocols of the cells problem, I was on the verge of writing a paper describing (1) the students' surprising lack of ability to make "order of magnitude" calculations and (2) the remarkably poor executive judgment that the students demonstrated in their allocation of strategic resources. In hindsight this "purely cognitive" explanation of their verbal data makes no more sense than it would make sense to assign a low IQ score to the Kpellan native quoted earlier, in an ostensibly objective evaluation of his responses to the experimenter's questions. We need not travel to Liberia to observe clashes in the rules of discourse between experimenter and subject. They occur right here in our own laboratories, with powerful effects.

With such methodological issues as background, we turn to a brief discussion of the methods used for generating the protocols used in this book. In most of the student protocols, the students were recorded in pairs, and other than occasional prompts encouraging verbalization, there were no interventions during the problem-solving process.

WHY PAIRS?

As Protocol 9.1 (Appendix 9.1) indicates, strange things can happen in single-person protocols. One major reason for recording students in pairs is that doing so helps to alleviate the kinds of environmental pressures that weight so heavily on students as they solve problems individually. A second reason is that, odd as it may seem, two-person protocols will often provide better information about individual students' decision-making processes than do single-person protocols.

It is generally acknowledged that having subjects explain the reasons for their actions as they solve problems will disrupt the subjects' problem-solving processes (Ericsson and Simon, 1980). As a result, the subjects who generate protocols (mine included) are generally told to report, out loud, what they are thinking "on line"—without editorializing or providing additional explanations. When decisions are made rapidly in such single-person protocols, the result is often simply a statement of the form "I'm going to do X"; see, for example, AB's declaration of the outline of his solution plan in the segment of Protocol 9.1 quoted above. In such protocols there is no indication of whether options other than the one taken were considered, and what the grounds for selecting that particular option (or rejecting the others) might have been. (For example, did AB consider whether he might use weight instead of volume for estimates of human and cell size? Many students did. If he did consider both, why did he choose volume?)

When two students work together there is far more verbalization than in single-person protocols. In a single-person protocol an individual might just *do* X; in a two-person protocol that person is more likely to say "I think we should do X," and perhaps argue why. Also, suggestions of the form "let's do X" are occasionally met with a responses like "Why should we?", in which case the rationales for the decisions are made overt.

Of course, there is a price to pay for these advantages. First, there is the possibility that the question from one's partner may alter a subject's intended solution path—and the experimenter will never know what might have happened if the question were not asked. Second and more important, interpersonal dynamics may shape in any of a number of ways what takes place in the solution attempt. A strong personality can dominate the work of a pair, and not necessarily for the better; the students can confuse each other; the need to "cooperate" may prove distracting; and so on. These social dynamics make it difficult to tease out the more "purely cognitive" aspects of the students' behavior.

The reader may have noticed that the expert protocols discussed in this book were generated by individuals. My colleagues, for the most part, were much more comfortable with the idea of talking "on line" than were my students (after all, they do it for a living). While they certainly felt the pressures of the taping environment, they were not likely to be as severely affected by those pressures as were the students in single-person protocols. In this case, the potential disadvantages of social interactions seemed to outweigh the potential advantages, and it was decided (after some brief pilot work) to record the experts working alone.

WHY NO INTERVENTIONS?

Much of the work described in this book was exploratory. To document the importance of certain kinds of phenomena, it was essential to let the problem sessions run their course. The only way to see whether students would recover from potential wild goose chases, or whether such ill-chosen approaches would have devastating consequences (e.g., see KW and AM, Appendix 9.2) was to give the students free rein and to observe what happened. Similarly, the only way—at first—to see the degree to which students depended on straightedge and compass in geometric construction problems was to leave the students alone and to see how much of their time was spent with the "tools of the trade" in hand. Any interventions while the students were in the process of solving the problems ran the risk of altering their solution paths, a risk one could not afford until the students' behavior patterns had been adequately documented. A second reason for the noninterventionist stance was the potential for a training effect. If, for example, students are asked about the reasons that they pursued particular directions in one problem, they might begin to think about such issues (in preparation for the experimenter's questions) while working the next problem. This might well change their behavior on that problem. Similarly, asking "How do you know your construction will work?" might, for the first time, force a student to think about the issue. In consequence, the student's behavior on the next construction problem might be substantially different.

The noninterventionist methodology does have limitations that should be noted. While a hands-off approach may serve to document the frequency or importance of certain phenomena, it rarely serves to elucidate their workings. Filling in the details may call for any of a number of other methods. For example, we now have an elaborate protocol for interviews with students regarding their work on geometry problems. The interview takes place in four phases: (1) The students generate noninterventionist protocols. (2) The students provide retrospective reports of their work, with minimal interventions from the experimenter. (3) The students and the experimenter watch the videotapes of their work together, and the students are asked detailed questions about what happened during the problem session. (4) There is a clinical interview that explores any issues of interest that surfaced in the first three phases.*

The point of this section, in brief, is that any particular approach to studying intellectual behavior is likely to illuminate some aspects of that behavior, to obscure other aspects of it, and to distort some beyond recognition. Of necessity the same phenomenon must be investigated with a variety of methodologies, and from a variety of perspectives. Only then is there a chance that the artifactual behavior resulting from the use of particular methodologies can be separated from the behavior that is inherent to the phenomena being investigated. Needless to say these comments are not meant as a blanket a posteriori challenge to the accuracy of studies that have relied upon the interpretation of verbal data. It may well be that the points of concern raised here are moot in a number of contexts. One such context, for example, is the analysis of experts' verbal protocols for purposes of constructing AI programs. Experimenters tend to find their expert subjects among their colleagues, who are generally familiar with and sympathetic to the methodologies being used for protocol collection. It is unlikely that the kinds of pathologies that distorted AB's problem-solving work in Protocol 9.1 would take place in such circumstances. The point here is that verbal data cannot always be taken at face value. Verbal data serve as evidence, just as one person's testimony (given in good faith) serves as evidence in a trial. To get the whole picture, such evidence should be compared and contrasted with the evidence obtained from as many other sources as possible.

We now turn to the specific issue of control behavior.

Background, Part 2: Other Protocol Coding Schemes and Issues of Control

By definition, protocol coding schemes are concerned with producing objective traces of the sequence of overt actions taken by individuals in the process of solving problems. Such schemes have been used extensively in AI and in mathematics education, although they have been put to different uses in the two domains. In AI, protocol analysis is a tool for discovering regularities in problem-solving behavior. In much AI work the goal is to write a

* It should be noted that such interviews are only one way of exploring students' geometrical behavior. Other methods and techniques (classroom observations, questionnaires, etc.) serve to examine the same phenomenon from other perspectives.

program that simulates the behavior in a given protocol or the idealized behavior culled from a variety of protocols. In mathematics education research, coded protocols are generally subjected to qualitative analyses. In attempts to identify useful strategies or to document their effectiveness, correlations are often sought between the frequency of use of those strategies (e.g., the presence of goal-oriented heuristics) and problem-solving success. In both domains, however, the level of analysis is microscopic. This brief discussion of extant schemes begins with AI.

PROTOCOL ANALYSIS IN ARTIFICIAL INTELLIGENCE

As noted in Chapter 4, issues of control and of executive decision-making are of central concern in AI. The analysis of out loud problem-solving sessions has played a major role in AI work on problem solving, where protocols serve as a major source of data. Records of human problem solvers are examined for regularities in behavior, and these regularities are abstracted for use in problem-solving programs.

Early protocol analysis schemes in AI were extraordinarily fine-grained. A comprehensive and detailed description of such work is given in Newell and Simon's (1972) *Human Problem Solving,* which discusses the kinds of strategic analyses undertaken in symbolic logic, cryptarithmetic, and chess during the development of the general problem solver (GPS). In such analyses, control has a somewhat constrained technical meaning, which usually refers to conducting an efficient search through a problem space. Problem solving takes place within a carefully delineated search space. The initial state and the goal state for a problem are specified, and the act of solving the problem consists of identifying a path through the search space that leads from initial state to goal state. In this context, choosing the right problem-solving strategy means selecting an efficient method for navigating through the search space—or colloquially, deciding what to do next. Allen Newell (1966, p. 152, Figure 5) characterized those decisions, the "considerations at a position in problem space," as follows:

Select new operator:
Has it been used before?
Is it desirable: Will it lead to progress?
Is it feasible: Will it work in the present situation if applied

These questions are quite general, and they are relevant at both macroscopic and microscopic levels of analysis. When the questions are almost always very fine-grained and microscopic. For example, the following cryptarithmetic problem is typical of those studied during the development of GPS:

Cryptarithmetic Problem Each of the letters D, E, M, N, O, R, S, Y, in the expression

$$\begin{array}{r} \text{S E N D} \\ + \underline{\text{M O R E}} \\ = \text{M O N E Y} \end{array}$$

can be assigned the value of a different digit (one of the numbers 0, 1, 2, 3, 4, 5, 6, 7, 8, 9) so that the resulting numerical expression respresents a correct arithmetic sum. Find such an assignment.

Consider the sequence of steps one might go through in trying to solve this problem. One may begin by observing that the M in the bottom row must be 1. After assigning M the value of 1 in the second row, one may observe that, since S + M = S + 1 generates a carry, then S must be 8 or 9. One then selects either 8 or 9 as a tentative value assigned to S, and continues making other assignments. If a contradiction is reached, then the other value is the correct assignment for S; if no contradiction is reached, the the first value ultimately turns out to be the correct assignment. And so on, through the assignment of values for all of the letters in the problem statement.

In the formalization of the cryptarithmetic problem a location in the problem space represents a current knowledge state, and the act of "selecting a new operator" means "deciding which digit assignment to try next." In the terms used earler in this book, this kind of decision-making is *tactical* rather than *strategic.* Decision-making, and analyses of decision-making, are local and fine-grained. Protocol analysis in *Human Problem Solving* was done on a line by line basis. Each action taken by the problem solver (e.g., the choice of digits tentatively tried and the follow-up testing procedures) was coded as an individual entity for analysis.

As AI methodologies evolved and researchers turned their attention to more semantically rich domains, the level of analysis became more macroscopic. Issues of debate often centered around "strategy." Larkin provides a description of one study at the center of such discussions:

> Simon and Simon (1978), studying a novice solving kinematics problems, found that means-ends analysis of the known and desired equations explained quite well the order in which this solver applied various physics principles. She apparently assessed the difference between the equations she had and the equation she knew would be needed to solve the problem. She then used this information about the difference between her current means and the desired end to decide what new equations to generate so as to reduce this difference. (Larkin, 1980, p. 116)

In such studies, which examined problem-solving behavior in fairly complex problem-solving domains, the level of protocol analysis was to examine each equation selected for use by the problem solver. In Newell's terms, the "considerations at each position in the problem space" deal with selecting, at a particular point in a problem solution, which equations the problem solver should try to work with. Such analyses are still rather fine-grained.

Work in AI since the mid-1970s, specifically the work on planning as described in Chapter 4, has dealt with issues of control with increasing sophistication. Models such as the Hayes-Roths' (1979) opportunistic planning model clearly identify different levels of planning and deal with the interactions among them. To my knowledge, however, there have not been formally defined schemes for analyzing protocol data at the macroscopic (or stategic) level.

PROTOCOL ANALYSIS IN MATHEMATICS EDUCATION

Partly as a result of the success of early work in AI based on protocol analyses, protocol analysis has been adopted as a tool in mathematics education research on problem-solving processes. Perhaps the first rigorous coding scheme in mathematics education was the one developed by Kilpatrick in his (1967) dissertation. Unlike most AI schemes, Kilpatrick's analyses coded specifically for the kinds of heuristic behaviors that were hypothesized to be of importance in mathematical problem solving. Like such schemes, it analyzed problem-solving behavior at a rather fine level of detail. The result of the analysis was a long string of symbols representing the processes that were used during a solution attempt (see below). The coded sequence, once obtained, was used as a source of data for statistical analysis. The statistical tools of empirical research were used to explore correlations between problem-solving success and the frequency of occurrences of certain problem-solving processes.

A number of protocol analysis schemes for use in mathematics education have been based on Kilpatrick's (1967) scheme. For the most part, the researchers' intentions in coding protocols were to be as detailed and comprehensive as possible. For example, Lucas describes his (1972) modification of Kilpatrick's scheme as follows:

> A system of behavioral analysis was designed to record and evaluate many actions which could occur during a problem solution. The kind of notation used, the number of diagrams drawn, whether or not the diagrams accurately represented problem conditions, the number and kinds of diagram modifications, and whether or not the subject recalled a related problem or applied its method or results were examples of some of these activities. The frequency of checking was measured by a process-sequence code; the kind of checking was classified by seven checklist categories. Simi-

larly, two process codes were used to distinguish and sort errors of structure and execution; the checklist further distinguished four categories of executive errors. Instances in which errors were noted were also counted. Strategies by which a solution is produced (e.g. analysis, synthesis, trial and error, reasoning by analogy) had corresponding process-sequence codes. The by-products of the solution (equations, relations, and algorithmic processes) were also recorded. [There were codes for] separating or summarizing data, . . . looking back, . . . trying a different mode of attack. . . . When the composite picture reconstructed from a tape-recorded vocalized protocol was examined very carefully, little observable behavior was likely to escape scrutiny. . . .

To illustrate its appearance, the coding string

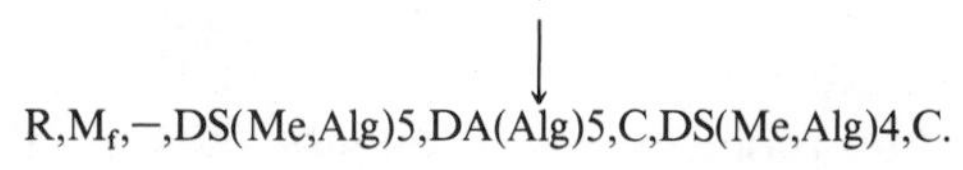

would be translated as follows: The subject read the problem (R), drew a figure (M_f), hesitated at least 30 seconds (—), started putting information together (DS) to yield an equation (Me) which was solved by a standard technique (Alg) to yield an intermediate result (5). Next, the subject looked at the goal and asked what was needed to obtain that (DA), followed by a brief calculation (Alg) in which a mechanical error (↓) was made. Upon checking back (C), the error was discovered and corrected (*), and the subject proceeded in a forward manner (DS) to derive another equation (Me) which was solved (Alg) to produce a correct final solution (4). This solution was verified (C) by checking against the conditions of the problem. (Lucas, 1980, pp. 72–74)

The quest for comprehensiveness and reliability persisted through much of the 1970s. The end product, in the case of Lucas' scheme, was a "process code dictionary" more than two pages long that was accompanied by coding procedures of comparable complexity (Lucas, Branca, Goldberg, Kantowski, Kellogg, & Smith, 1979). In part because of the cumbersome nature of such systems and the wealth of symbols that must be dealt with once the processes are coded, the end of the decade saw other researchers choosing to focus on more restricted subsets of behaviors. For example, Kantowski's (1977) work indicates some narrowing of focus. Her project used a "coding scheme for heuristic processes of interest" focusing on five heuristic processes related to planning, four related to memory for similar problems, and seven related to looking back. Kulm and his colleagues developed a revised and more condensed process code dictionary (Kulm, Campbell, Frank, & Talsma, 1981). Even so, the coding was still microscopic and analyses of the data were usually statistical. Correlations were sought between problem-solving success and the frequency of certain types of behavior as reflected in the protocol codings.

Like the schemes in AI, protocol analysis methods in mathematics education have generally coded behavior at the tactical level. Implications at the strategic level can, of course, be drawn from such studies. For example,

Kantowski's (1977) statistical analysis revealed positive correlations between students' use of certain goal-directed heuristic processes and their performance on measures of problem-solving success. To my knowledge, however, no system for protocol analysis has focused directly on strategic decisions and their impact on problem-solving performance.

One final point may appear almost trivial, but it has serious implications. Almost by definition, protocol coding schemes (and subsequent analyses of them) focus on the overt actions that take place during problem-solving attempts. In the words of Lucas *et al.* (1980, p. 359), "All behavior is required to be explicit; otherwise it is not coded." While this appears quite reasonable, it may result in serious omissions. As briefly described in Chapter 1 and as is discussed at greater length in the next section, the most important events in a problem session may be the ones that do not take place—for example, when a student does not assess the current status of a solution or the potential utility of a proposed approach, and as a result goes off on a wild goose chase that guarantees that the problem-solving attempt will fail. To my knowledge, no extant analytical framework has focused on determining where such decisions should have taken place during problem-solving attempts.

The Major Issues for Analysis: A Brief Discussion of Two Protocols

Protocols 9.2 and 9.3 (Appendixes 9.2 and 9.3) provide the complete transcripts of the attempts by KW and AM, and DK and BM, respectively, to solve Problem 1.3. The problem called for determining which of the triangles inscribed in a fixed circle has the largest area. The general outlines of these attempts were discussed in Chapter 1, and the reader may wish to review the discussion. The purpose of this discussion is to focus on the most significant aspects of executive behavior in those protocols, highlighting the kinds of issues that a framework for the macroscopic analysis of protocols should be able to deal with.

The following three comments deal with issues of control in KW and AM's attempt (Protocol 9.2).

Comment 1 The single most important event in the 20-minute problem-solving session, upon which the success or failure of the entire endeavor rested, was conspicuous by its absence: The students neglected to assess the potential utility of calculating the area of the equilateral triangle. (See Items 5 and 6 in the protocol.) The students simply embarked on the calculation. In consequence, the entire session was spent on a wild goose chase.

Comment 2 During the problem-solving process, KW and AM gave inadequate consideration to the utility of a number of potential alternatives that arose, including the related problem of maximizing a rectangle in a circle (Item 28), the potential application of the calculus for what can indeed be considered a max-min problem (Item 52), or making a variational argument on the shape of the triangle (Item 68). Any of these alternatives might have led to progress, if pursued. Instead, each of these alternatives simply faded out of the picture. (See, e.g., Items 27–31.)

Comment 3 Progress was not monitored or assessed during the solution, so there was no reliable means of terminating wild goose chases once they began. (This contrasts strongly with an expert protocol, in which the problem solver interrupted the implementation of an outlined solution with "This is too complicated. I know the problem shouldn't be this hard.")

How does one code such a protocol? At the risk of flogging a dead horse, I wish to stress that in this particular problem-solving session, matters of detail (such as whether or not the students will accurately remember the formula for the area of an equilateral triangle, Items 73–75) are virtually irrelevant. The military analogy is appropriate here: If one makes major strategic mistakes (e.g., opening a second front in World War II), then matters of tactics (e.g., the details of a battle on that front) are of little importance. A coding scheme should highlight major decisions.

A second and more crucial point is that not all major decisions are overtly made: KW and AM's problem-solving effort was a failure because of the *absence* of assessments and strategic decisions. Any framework that will make sense of that protocol must go beyond simply recording what did happen. It should suggest when strategic decisions ought to have been made and allow one to interpret the individuals' success or failure in the light of whether, and how well, such decisions were made.

Assessing the quality of the executive decisions in a problem attempt, and the ways those decisions shape a solution, is the central issue in the analysis of Protocol 9.3. The control processes in Protocol 9.3 were not muddled, as they were in Protocol 9.2; the decisions made by DK and BM were overt and clear. Those decisions also served to guarantee that the students would fail to solve the problem. The next paragraph provides a brief overview of what happened in the solution attempt, which appears in Appendix 9.3. The numbered comments refer to the commentary that follows.

SOLUTION SUMMARY

The students DK and BM quickly conjectured that the solution is the equilateral triangle and looked for ways to show it. Student DK, apparently wishing to exploit symmetry in some way, suggested that they examine triangles in a semicircle with one side as diameter. The students determined the largest triangle subject to these constraints, but then rejected it "by eye" as being smaller than the equilateral (Comment 4). Still focusing on symmetry, they decided to maximize the area of a right triangle in a semicircle, where the right angle lies on the diameter (Comment 5). This reduced the original problem to a one-variable calculus problem, which BM worked (see Comment 6 below). The attempt was unsuccessful, and it was abandoned 12 minutes later (Comment 7). At that point the solution attempt degenerated into an aimless series of explorations, most of which served to rehash the previous work (Comment 8).

Comment 4 Up to this point the students' actions in analyzing the problem were quite reasonable. Empirical investigations are entirely appropriate in helping one "get a feel" for a problem, as long as those investigations do not take up too much time. It was also reasonable for the students to reject the isosceles right triangle, in comparison with the equilateral. However, the nature of their rejection may have cost them a great deal. As we saw in Chapter 1, the variational argument that DK and BM used to find the isosceles right triangle (holding the base fixed and observing that the area is largest when the triangle is isosceles) is perfectly general and can be used to solve the original problem as stated. But the students simply turned away from their unsuccessful attempt without asking if they could learn from it. In doing so, they ignored a successful solution method.

Comment 5 This decision to work a different problem was made in a remarkably casual way (Items 24–27):

DK: [after one attempt at symmetry has failed] . . . you want to make it perfectly symmetrical—but we can, if we maximize this area, and just flip it over, if we can assume that it is going to be symmetrical.

BM: Yeah, it is symmetrical.

Although it turns out (serendipitously) to be correct, the students' assumption that the final answer will be symmetrical was not mathematically justifiable; in making that assumption the students had taken for granted part of what they were asked to discover and prove. The new problem of what they were asked to discover and prove. The new problem they choose to work on might or might not have been useful in solving the original problem. Yet the students proceeded, without apparent concern, to work on the altered version. Pursuing this new problem took more than 60% of their allotted time, all of which could have been completely wasted.

Comment 6 Student BM's tactical work here was quite decent, as was much of both students' tactical work throughout the solution process. The decision to scale down the problem to the unit circle (Item 37) is just one example of their proficiency. The students were aware of, and had access to, a variety of heuristics and algorithmic techniques during the solution. Unfortunately, an arithmetic mistake during the BM's calculation resulted in a physically impossible answer. He was aware of it, and local assessment worked well. However, global assessment (Comments 4 and 5) did not.

Comment 7 The decision to abandon the analytic approach (Items 74 and 75) was astonishingly casual:

DK: Well, let's leave the numbers for a while and see if we can do this geometrically.
BM: Yeah, you're probably right.

Given that more than 60% of the solution had been devoted to that approach (and that correcting a minor mistake would have salvaged the entire operation), this casual dismissal of their previous efforts had rather serious consequences.

Comment 8 There were a number of clever ideas in the earlier attempts made by the two students. Had there been an effort at a careful review of those attempts, something might have been salvaged. Instead, there was simply a "once over lightly" of the previous work that added nothing to what they had already done.

To address the kinds of issues raised in this discussion, the protocol coding scheme described in the next section is macroscopic in nature. The scheme is intended to identify the loci of important control decisions, both (1) those that did take place and (2) those that did not, but should have. It is intended to characterize the quality of those decisions, to describe the effects that the decisions (or their absence) had on the ways that solutions evolved. More broadly, it is intended to convey a sense of how well the problem solvers used the resources at their disposal.

A Framework for the Macroscopic Analysis of Problem-Solving Protocols

The method of protocol analysis described in this section focuses on decision-making at the executive or control level. The method provides a way of identifying three classes of potentially important decision points in a solution, and a way to examine the ways that individuals' behavior at the control level shapes the ways that solutions evolve.

Decisions at the control level are those that affect the allocation or utilization of a substantial amount of problem-solving resources (including time). It is thus appropriate to look for executive decision-making at points in problem solutions where there are major shifts in resource allocation. This observation suggests the method of protocol analysis described here and provides a means of identifying one generic class of potential decision points. Protocols are partitioned into macroscopic chunks of consistent behavior called episodes. An episode is a period of time during which an individual or a problem-solving group is engaged in one large task (e.g., KW and AM's calculation of the area of the equilateral triangle in Protocol 9.2, or DK and BM's use of the calculus in Protocol 9.3) or a closely related body of tasks in the service of the same goal (e.g., performing a number of empirical calculations to "get a feel" for a problem). Once a protocol has been parsed into episodes, one class of decision points becomes apparent. The junctures between episodes—the points at which the direction or nature of the problem solution changed significantly—are points at which, at minimum, action at the control level ought to have been considered.

The second generic type of situation where executive decision-making should be considered takes place when either new information or the possibility of taking a different approach comes to the attention of the problem solver(s). Three such instances occurred in Protocol 9.2. In Item 28, in the midst of calculating the area of the equilateral triangle, AM mentioned for the first time the related problem of maximizing the areas of rectangles inscribed in a circle. This observation, properly examined, might have led to a change of approach. Similarly, new information points arose in Items 51 and 68. As a technical issue, it should be noted that the arrival of new information does not necessarily signify that one episode has come to an end and another has begun. Often information that arises in the middle of an episode is ignored or dismissed (at least temporarily), and the problem solver continues working along previously established lines. The episode is terminated only when there is a clear break in direction.

The third category of executive decision points is far more subtle and difficult to handle. These are times in a solution where nothing has gone catastrophically wrong, but when a string of minor difficulties indicates that it is probably time to consider something else. When implementation bogs down, for example, or when the problem-solving process degenerates into more or less unstructured explorations, it may be time for an "executive review." It is difficult to say, however, when such reviews should be undertaken. As a case in point, it is difficult to say precisely where in their 20 minutes of calculating the area of the equilateral triangle in Protocol 9.2 the students KW and AM should have brought things to a halt — but it is certain that they should have reconsidered long before time ran out. As the examples in this chapter indicate, competent problem solvers consistently monitor and evaluate their solutions as they work. Novices do not, and the consequences can be severe. This issue is important, and one limitation of the framework described below is that it does not deal with the issue especially well.

Figures 9.1 and 9.2 present parsings of Protocols 9.2 and 9.3, respectively, into episodes. New information points within episodes are indicated. Figures 9.1A and 9.2A present the parsings in their original form, which focused on identifying the three categories of executive decision points described above. I am indebted to Don Woods (1983) for the time-line representations in Figures 9.1B and 9.2B. These provide a graphic illustration of some aspects of the same data.

In considering the parsings represented in Figures 9.1 and 9.2, it should be stressed that both protocols were parsed by examining the behavior in the problem sessions and dividing it into episodes — not by looking for points at which decisions were made. Yet both parsings point to the precise places in the protocols where decisions had — or could have had — a major impact on the solutions. In the case of Protocol 9.2, we see that KW and AM's tacit agreement to calculate the area of the equilateral triangle in Items 5 and 6 is followed by 20 minutes of exploration in the service of that goal. Moreover, opportunities arose at items 28, 51, and 68 to pursue alternative options, but they were not pursued. Similarly, the junctions between episodes in Figure 9.2 identify the places where the decisions that shaped DK and BM's attempt took place.

After a small amount of training, partitioning protocols into episodes and delineating new-information points are both rather straightforward processes. Reliability in parsing protocols is quite high. As an indication of that reliability, it is worth noting that the parsings of all of the protocols used in this chapter were derived, in consensus, by a team of three undergraduates who arrived at their characterizations of the protocols in my absence. No alterations in those parsings were necessary in order for them to be used here. (This does not, however, obviate the need for an appropriate formalism; see the following discussion.)

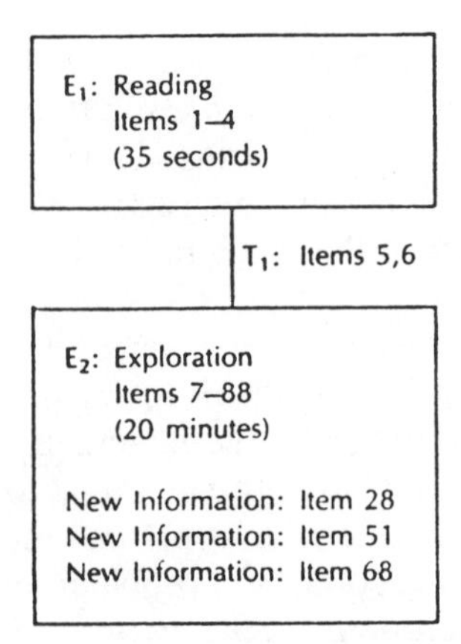

Figure 9.1A A parsing of Protocol 9.2. From the written protocol it might appear that Item 68 begins a new episode. In fact, the students had lost virtually all their energy by that point and were merely doodling; they returned (after the tape clicked off) to musing about the equilateral triangle. Thus Items 6–88 are considered to be one episode.

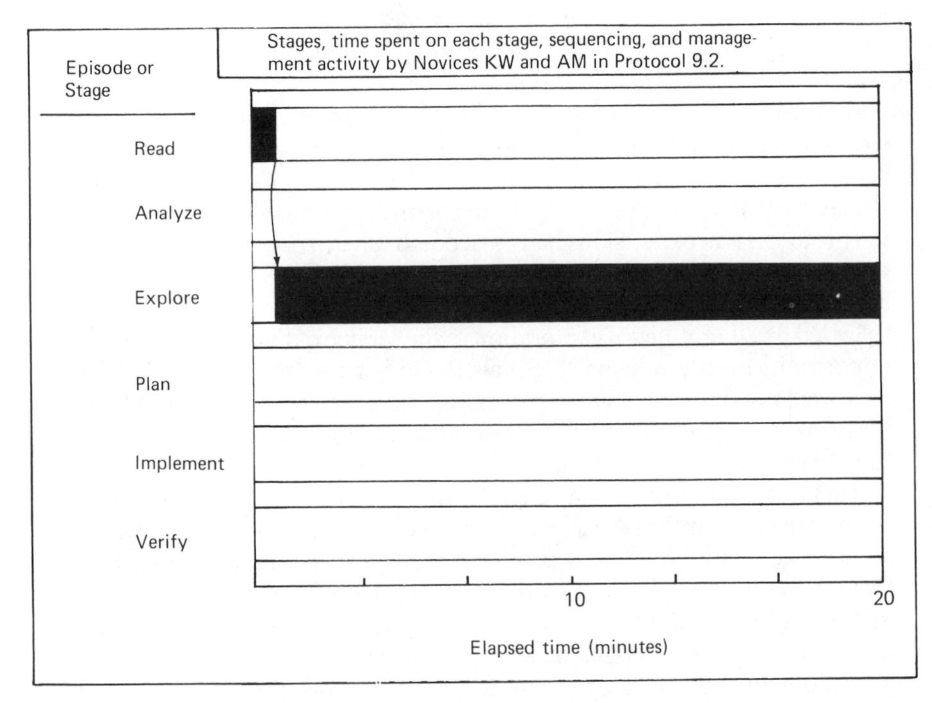

Figure 9.1B A time-line representation of Protocol 9.2.

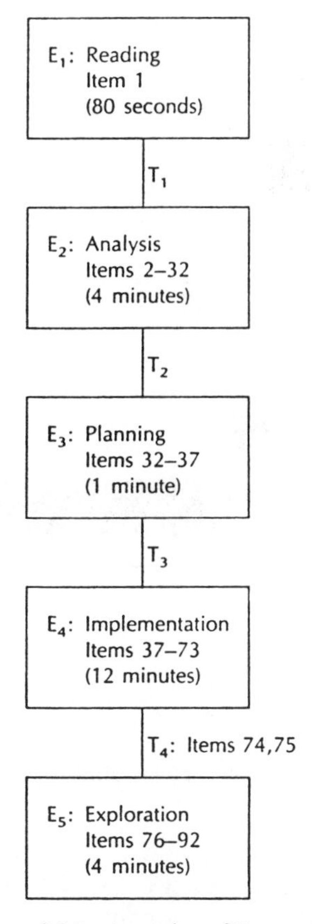

Figure 9.2A A parsing of Protocol 9.3.

Subjectivity lurks around the corner, however. It will be clearly present in the analyses of decision-making at transitions between episodes, as described below. In fact, it is already present in the way that the episodes were labeled (analysis, exploration, etc.) in Figures 9.1 and 9.2. This kind of labeling was essential, for without it the potential for combinatorial explosion in characterizing managerial behaviors is enormous. Managerial behaviors include selecting perspectives and frameworks for working a problem; deciding at branch points which direction a solution should take; deciding in the light of new information whether a path already embarked upon should be abandoned; deciding what (if anything) should be salvaged from attempts that are abandoned, or adopted from approaches that were considered but not taken; monitoring and assessing implementation "on line" and looking for signs that executive intervention might be appropriate; and much, much more. Our preliminary attempts at analyses of managerial behavior called for

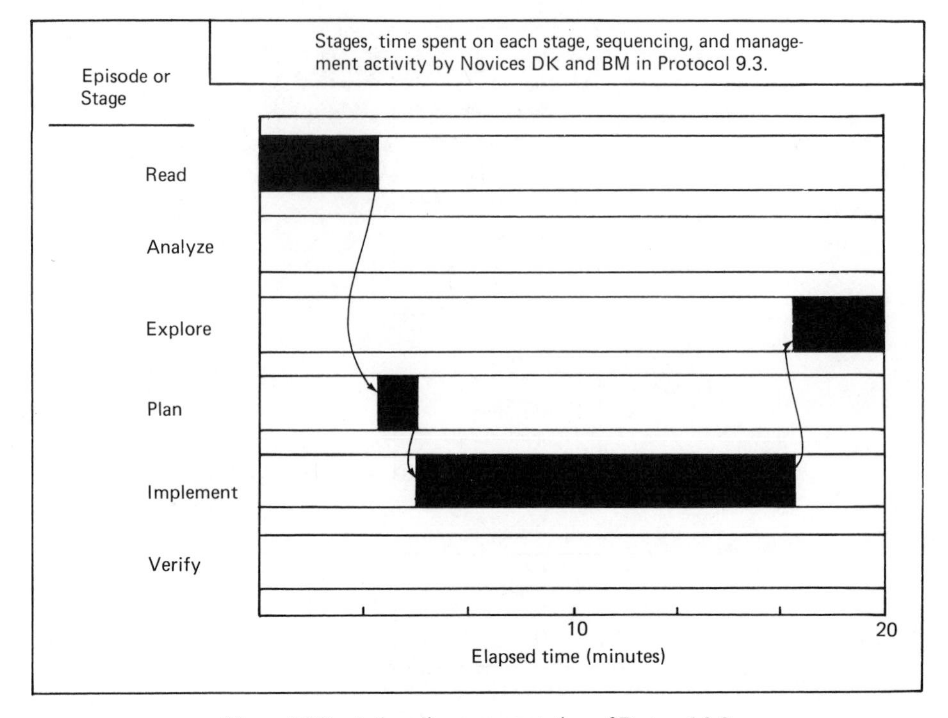

Figure 9.2B A time-line representation of Protocol 9.3.

examining protocols at all managerial decision points. At each decision point a series of questions exploring the issues mentioned just above was asked. This approach did prove reasonably comprehensive, but it was completely unwieldy. To give just one example, questions about assessing the state of a solution attempt when (1) one has just read a problem, (2) one is stuck, and (3) the solution has been obtained, are almost mutually exclusive. At any particular decision point, more than 90% of the questions about executive behavior that one might ask in general (i.e., control questions that are directly relevant in some context) are irrelevant. The preliminary attempts at comprehensiveness proved too difficult to implement, and the episode labeling described below was developed as a workable compromise.

Once a protocol has been parsed into episodes, each episode is characterized as one of the following: reading, analysis, planning, implementation (or planning–implementation if the two are linked), exploration, verification, or transition. Specific note is also taken of any new information or local assessments that appear during the solution. The analysis up to this point produces the (reasonably objective) characterizations like those given in Figures 9.1 and 9.2. At this point, each episode, and each transition between episodes, is analyzed separately.

The questions used in those analyses are of various types, and they are not nearly as uniform or as homogeneous as one would like. Some can be answered objectively at the point in the protocol at which they are asked, some in the light of later evidence; some call for inferences or judgments about problem-solving behavior. Further, some ask about the reasonableness of certain behavior. Asking questions in this way, of course, begs the significant question: What *is* a model of reasonable behavior? The creation of such models is the crucial long-term question, and there is no attempt to finesse it here. At present, however, the immediate goal is to deal with the notion subjectively in order to better understand managerial behaviors so that those models can be created. Although they are clearly subjective, the assessments of reasonableness can be reliably made; agreement between my ratings and the consensus scorings of my students was quite high. To quote Mr. Justice Stewart in a 1964 decision of the United States Supreme Court, "I shall not today attempt to further define the kind of materials I understand to be embraced within that shorthand definition; . . . But I know it when I see it."

What follows is the heart of the analytical framework. Each of the different kinds of episodes is briefly described, and the set of questions relevant for that type of episode is given. A full characterization of a protocol is obtained by parsing a protocol into episodes and providing answers to the associated questions.

Episodes and the Associated Questions

READING

The reading episode begins when a subject starts to read the problem statement aloud. It includes the time spent ingesting the problem conditions and continues through any silence that may follow the reading—silence that may indicate contemplation of the problem statement, the (non-vocal) rereading of the problem, or blank thoughts. It continues as well through vocal rereadings and verbalizations of parts of the problem statement. (Observe that in Protocol 1, reading included Items 1–4).

The reading questions are

R1. Have all the conditions of the problem been noted? Were they noted explicitly, or implicitly?

R2. Has the goal state been correctly noted? Was it noted explicitly, or implicitly?

R3. Is there an assessment of the current state of the problem solver's knowledge relative to the problem-solving task (see "Transition")?

ANALYSIS

If there is no apparent way to proceed after the problem has been read, the next (ideal) phase of a problem solution is analysis. In analysis an attempt is made to fully understand a problem, to select an appropriate perspective and reformulate the problem in those terms, and to introduce for consideration whatever principles or mechanisms might be appropriate. The problem may be simplified or reformulated. Often analysis leads directly into plan development, in which case it serves as a transition. Also, note that the analysis episode may be bypassed completely, as happens in schema-driven solutions where the individual already knows the relevant perspective and approach to take.

Analysis questions are

A1. What choice of perspective is made? Is the choice made explicitly, or by default?
A2. Are the actions driven by the conditions of the problem? (working forward)
A3. Are the actions driven by the goals of the problem? (working backward)
A4. Is a relationship sought between the conditions and goals of the problem?
A5. Is the episode, as a whole, coherent? In sum (considering Questions A1–A4), are the actions taken by the problem solver reasonable? Are there any further comments or observations that seem appropriate?

EXPLORATION

Both its structure and content serve to distinguish exploration from analysis. Analysis is generally well structured, sticking rather closely to the conditions or goals of the problem. Exploration, on the other hand, is less well structured and is further removed from the original problem. It is a broad tour through the problem space, a search for relevant information that can be incorporated into the analysis-plan-implementation sequence. If one comes across new information during exploration, for example, one may well return to analysis in the hope of using that information to better understand the problem.)

In the exploration phase of problem solving, one may find a variety of problem-solving heuristics—the examination of related problems, the use of analogies, and so forth. Ideally, exploration is not without structure; there is a loose metric on the problem space (the perceived distance of objects under consideration from the original problem) that should serve to select items for consideration (recall Figure 4.12). Precisely because exploration is weakly structured, both local and global assessments are critical here (see "Transition" as well). An unchecked wild goose chase can lead to disaster, but so can the dismissal of a promising alternative.

If new information arises during exploration, but is not used, or the examination of it is tentative, "fading in and fading out," the coding scheme calls for delineating new information within the episode. If, however, the problem solver decides to abandon one approach and start another, the coding scheme calls for closing the first episode, denoting (and examining) the transition, and opening another exploration episode.

Exploration questions are

E1. Is the episode condition driven? Goal driven?
E2. Is the action directed or focused? Is it purposeful?
E3. Is there any monitoring of progress? What are the consequences for the solution of the presence or absence of such monitoring?
E4. Is the episode, as a whole, coherent? In sum (considering Questions E1 – E3), are the actions taken by the problem solver reasonable? Are there any further observations or comments that seem appropriate?

NEW INFORMATION AND LOCAL ASSESSMENTS

New-information points include any items at which a previously unnoticed piece of information is obtained or recognized. They also include the mention of potentially valuable heuristics (new processes, new approaches). Local assessment is an evaluation of the current state of the solution at a microscopic level.

New information and local assessment questions are

N1. Does the problem solver assess the current state of his knowledge? (Is it appropriate to do so?)
N2. Does the problem solver assess the relevancy or utility of the new information? (Is it appropriate?)
N3. What are the consequences for the solution of these assessments, or the absence of them?

PLANNING – IMPLEMENTATION

Because the emphasis here is on questions at the control level, detailed issues regarding plan formation are not addressed. The primary questions of concern here deal with whether or not the plan is well structured, whether the implementation of the plan is orderly, and whether there is monitoring or assessment of the process on the part of the problem solver(s), with feedback to planning and assessment at local and/or global levels. Many of these judgments are subjective. For example, the absence of any overt planning acts does not necessarily indicate the absence of a plan. In fact, protocols of schema-driven solutions often proceed directly from the reading episode into the coherent and well-structured implementation of a nonverbalized plan. Thus the latitude of the questions below; the scheme should apply to a range of circumstances, from schema-driven solutions to those in which the subject develops an appropriate plan or even comes upon one by accident.

Planning–implementation questions are

PI1. Is there evidence of planning at all? Is the planning overt or must the presence of a plan be inferred from the purposefulness of the subject's behavior?
PI2. Is the plan relevant to the problem solution? Is it appropriate? Is it well structured?
PI3. Does the subject assess the quality of the plan as to relevance, appropriateness, or structure? If so, how do those assessments compare with the judgments in Question PI2?
PI4. Does implementation follow the plan in a structured way?
PI5. Is there an assessment of the implementation (especially if things go wrong) at the local or global level?
PI6. What are the consequences for the solution of assessments if they occur, or of their absence if they do not?

VERIFICATION

The nature of the episode itself is obvious. Verification questions are

V1. Does the problem solver review the solution?
V2. Is the solution tested in any way? If so, how?
V3. Is there any assessment of the solution, either an evaluation of the process, or an assessment of confidence in the result?

TRANSITION

The juncture between episodes is, in most cases, where managerial decisions (or their absence) will make or break a solution. Observe, however, that the presence or absence of assessment or other overt managerial behavior cannot necessarily be taken as being either good or bad for a solution. In an expert's solution of a routine problem, for example, the only actions one sees may be reading and implementation. This explains, in part, the contorted and subjective nature of what follows.

Transition questions are

T1. Is there an assessment of the current solution state? Since a solution path is being abandoned, is there an attempt to salvage or store things that might have been valuable in it?

T2. What are the local and global effects on the solution of the presence or absence of assessment as previous work is abandoned? Is the action (or lack of action) taken by the problem solver appropriate or necessary?

T3. Is there an assessment of the short- and/or long-term effects on the solution of taking the new direction, or does the subject simply jump into the new approach?

T4. What are the local and global effects on the solution of the presence or absence of assessment as a new path is embarked upon? Is the action there appropriate or necessary?

Since the preceding section suggested the answers to most of these questions with regard to Protocols 9.2 and 9.3, it would be redundant to write out the full analyses of them. The following section offers the full analysis of a third problem session.

The Full Analysis of a Protocol

Protocol 9.4 (Appendix 9.4) gives the entire transcript of KW and DR's attempt to solve the following problem:

Problem 9.1 Consider the set of all triangles whose perimeter is a fixed number *P*. Of these, which has the largest area? Justify your answer as best you can.

This protocol was taken at the end of the 1980–1981 problem-solving course, whereas Protocols 9.2 and 9.3 were taken as the course began. Subject KW is the same student who appeared in Protocol 9.2. On entering the problem-solving course he had completed three semesters of calculus. Student DR was a freshman with one semester of calculus behind him.

The parsing of Protocol 9.4 is given in Figure 9.3. The analysis given below follows that parsing.

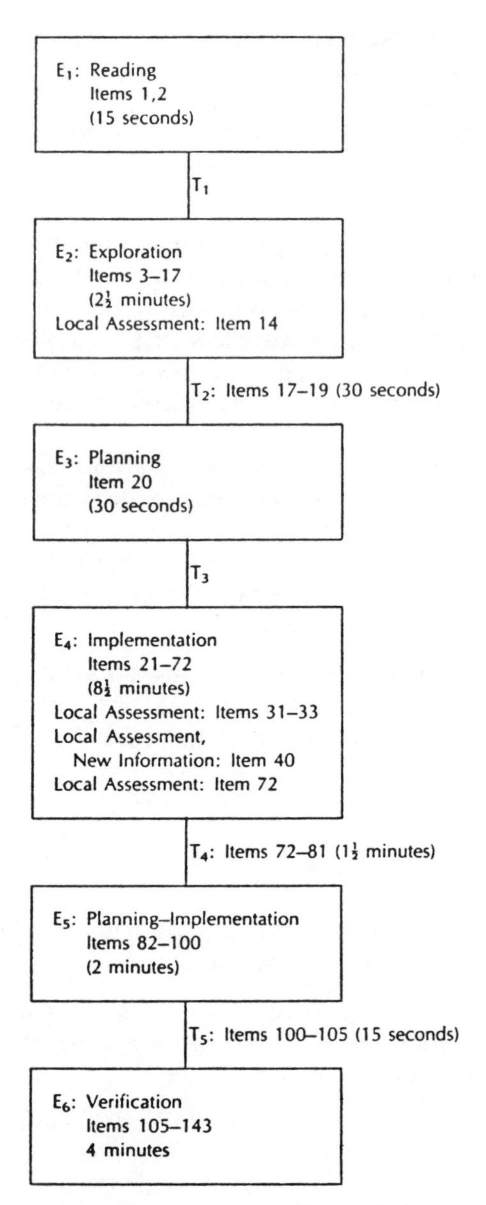

Figure 9.3A A parsing of Protocol 9.4.

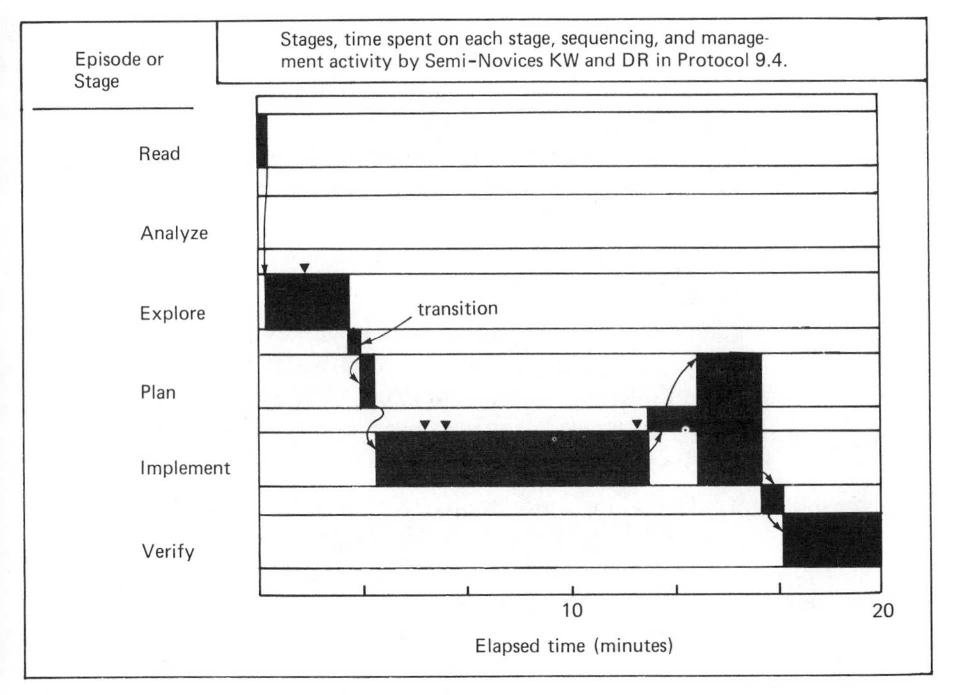

Figure 9.3B A time-line representation of Protocol 9.4. Overt signs of management activity are denoted by inverted triangles.

EPISODE 1: READING (ITEMS 1–2)

R1. KW and DR did make explicit note of the problem conditions.
R2. The goal state was noted, but somewhat carelessly (see Items 4–5, 9–11).
R3. There was no reflection upon, or assessment of, what they knew or where they were going. There was simply a jump into exploration.

TRANSITION 1 (NULL)

T1–T4. The students made no serious assessments of either their current knowledge or of directions to come. They should have done so, for this lack of assessment might have been costly. It turned out not to be, for assessments did come in Episode 2.

EPISODE 2: EXPLORATION (ITEMS 3–17)

E1. The explorations seemed vaguely goal driven.
E2. The actions seemed unfocused.
E3–E4. There was monitoring and assessment, at Items 14–17. This grounded the explorations, and led into Transition 2.

TRANSITION 2 (ITEMS 17–19)

T1–T4. Students KW and DR assessed both what they knew at that point and what they needed to do. One result was the establishment of a major direction, namely, the decision to try to prove that the equilateral triangle has the desired property. The second result is that they created a plan to do so (see Episode 3).

[Note: If I seem to be making a mountain out of a molehill in this discussion, contrast KW and DR's behavior with the first (null) transition in Protocol 9.2. The lack of assessment there, in virtually identical circumstances, sent the students on a 20-minute wild goose chase!]

EPISODE 3: PLAN (ITEM 20)

PI1. The plan was overt.
PI2. It was somewhat relevant and appropriate; it was well structured.
PI3. See the discussion of Transition 2.

TRANSITION 3 (NULL)

T1, T2. Little of value had taken place preceding the plan in Item 20; the questions T1 and T2 are moot.
T3. There was no assessment of the plan; they jumped immediately into implementation.
T4. Their plan was relevant, but it only dealt with half of the problem—showing that the largest isosceles triangle is equilateral. The other half of the problem is to show that the largest triangle must be isosceles, without which this part of the solution is worthless, a point they came to realize in Item 72, 8 minutes later. The result was a good deal of wasted effort. The entire solution was not sabotaged, however, because monitoring and feedback mechanisms caused the termination of the implementation episode (details are given in the sequel).

EPISODE 4: IMPLEMENTATION (ITEMS 21–82)

PI4. Implementation followed the lines set out in Episode 3, albeit in somewhat careless form. The conditions were somewhat muddled as the first differentiation was set up. The next two local assessments did correct for that (better late than never).

LOCAL ASSESSMENT (ITEMS 31–33)

T1–T4. The physically unrealistic answer caused them to take a closer look at the conditions but did not trigger a global reassessment (possibly not called for yet).

LOCAL ASSESSMENTS, NEW INFORMATION (ITEM 40)

T1–T4. The new information here was the realization that one of the problem conditions had been omitted from their implementation ("we don't set any conditions—we're leaving *P* out of that"). This sent them back to the original plan, without global assessment. The cost: Squandered energy until Item 72.

LOCAL–GLOBAL ASSESSMENT (ITEM 72)

This closes Episode 4. See Transition 4.

TRANSITION 4 (ITEMS 72–81)

T1, T2. The previous episode was abandoned, with reason. The goal of that episode, "show it's the equilateral," remained. This, too, was reasonable.

T3, T4. Students KW and DR ease into Episode 5, in Item 82. It is difficult to say how reasonable this transition is. Had they chosen something that did not work, it might have been considered meandering—but what they chose did work.

EPISODE 5: PLAN IMPLEMENTATION (ITEMS 82–100)

PI1, PI2. The plan is overt, and "set our base equal to something" is an obviously relevant heuristic.

PI3. They plunge ahead as usual. As noted above, this could have led to a wild goose chase.

PI4. The variational argument that they used seemed to evolve in a fairly natural way.

PI5. This time, local assessment (Item 95) worked well, making sure the students stayed on track. Local assessment led to a review of the argument in Item 96, at which point DR apparently "saw" the rest of the solution. Moreover, DR goes on to assess the quality of the solution and his confidence in the result (Item 100).

TRANSITION 5 (ITEMS 100–105)

T1–T4. What happens from this point on is most likely the result of a two-person dialectic. It appears that DR was (perhaps prematurely) content with his solution, although the clarity of the explanation he offers in Episode 6 suggests that he may have been justified in his confidence.

EPISODE 6: VERIFICATION (ITEMS 105–143)

V1–V3. Student DR was clearly confident, and the two students worked over the solution until its correctness was agreed upon by KW.

(Note: This was not a verification episode in the usual sense. Student KW's unwillingness to rest until he understood forced DR into a full rehearsal of the argument and a detailed explanation, the result being that they were both content with the (correct) solution. I suspect that DR, on his own, would have stopped at item 100. (Whether KW would have pressed as hard for an explanation before he took the course is also an open question.)

A Further Discussion of Control: More Data from Students, and the Analysis of an Expert Problem Solver's Protocol

There is always a trade-off between depth and breadth when one performs a small number of detailed analyses like the ones described in this chapter. From such analyses one begins to obtain a reasonably clear understanding of the problem-solving behavior of a few students working on a few problems. It is difficult, however, to get a sense of the typicality of that behavior. For that reason this section offers capsule descriptions of some additional problem-solving sessions. It also offers some data regarding the typicality of the behavior exhibited in Protocol 9.2, where KW and AM read the given problem and immediately jumped into an exploration episode that lasted a full 20 minutes.

As in previous chapters, the protocols discussed here serve more than one purpose. The problem-solving sessions that generated the protocols took place before and after the problem-solving course that served as the experimental treatment for the research described in Chapters 7 and 8. Thus the tools developed in this chapter, with their focus on control behavior, provide yet another method for examining the effects that a problem-solving course can have on students' problem-solving processes. In this context, two questions are of primary interest: (1) Was there a change in the students' behavior at the control level as a result of the course? (2) Did the students' control processes come to resemble those of experts?

In part to establish a standard of control behavior against which the students' work can be compared, an expert's problem-solving protocol is given in full and analyzed in some detail. That protocol stands on its own, however, as an object of intrinsic interest. The subject who recorded it, a professional mathematician, was relatively unfamiliar with the kind of problem he was asked to solve. He did not qualify as a domain-specific "expert" in the conventional sense, for he began the problem with little idea of what the appropriate method of solution might be. Rather, his problem-solving expertise was quite general; in a tour de force at the control level, he solved the problem by being remarkably efficient in using what he did know. Such behavior raises some interesting issues regarding the nature of "expertise," which are also explored in this section.

Protocols 9.2 and 9.3 were recorded at the beginning of a problem-solving course given during the 1980–1981 winter term at Hamilton College (Mathematics 161, described in Chapter 7). They are relatively typical of the 12 problem-solving sessions recorded at that time, which are discussed here.* Six pairs of students worked two problems each. The first problem that each pair worked is the same problem that was used in Protocols 9.2 and 9.3, to find the largest triangle that can be inscribed in a circle. The second problem tried by each pair is Problem 1.1, which was discussed extensively in Chapter 5. The problem solver is asked to show how to construct a circle tangent to two given lines, subject to the condition that a designated point on one of the lines must be one of the points of tangency.

Brief "snapshots" of a few representative preinstruction protocols are

* They are also typical of the more than a hundred tapes recorded before other versions of the problem-solving course, and for other purposes. As one indication of that fact, more than half of those tapes are of the read/explore type.

given below. These summaries are too condensed to be useful for purposes of detailed research, but they do serve to document the critical importance of decision-making at the control or strategic level; they also indicate the typicality of Protocols 9.2 and 9.3. As will be seen below, the students' work at the control level prior to instruction stands in partial contrast to their control behavior after instruction, and in stark contrast to an expert's work. For the sake of brevity in presenting the data, the diagrams that represent episode analyses are condensed into a sequential list of episode titles, with transitions deleted if there were none. Thus Figure 9.1 is rendered as Reading/Transition/Exploration, and so forth.

ET AND DR, FIRST PROBLEM: READING/TRANSITION/EXPLORATION

After a brief mention of max-min problems, and a brief caveat ("But will it apply for all cases? I don't know if we can check it afterward.") in transition, ET and DR set off to calculate the area of the equilateral triangle. So much for the next 15 minutes; in spite of some local assessments ("This isn't getting us anywhere.") they continued those explorations. Result: all wasted effort.

ET AND DR, SECOND PROBLEM: READING/EXPLORATION

In their initial explorations, ET and DR made a series of sketches that contained all the vital information they need to solve the problem. They made no attempt at review or assessment, and in consequence they overlooked the relevant information in their previous work. Their solution attempt was undirected and rambling. Possibly because they felt the need to do *something,* they tried their hand at an actual construction—one that was already shown to be incorrect by their sketches—and were stymied when it did not work. Overall their session demonstrated lost opportunities, unfocused work, and wasted effort. This was despite the fact that ET and DR were both bright. Both had just completed the first semester calculus course with A's.

DK AND BM, SECOND PROBLEM: READING/ANALYSIS/TRANSITION/EXPLORATION/ ANALYSIS (PROBLEM SOLUTION)/VERIFICATION

Analysis was extended and coherent, but it was followed by a poor transition into an inappropriate construction that deflected the students off track for $3\frac{1}{2}$ minutes. When that construction did not work, they returned to analysis and solved the problem. A detailed verification completed the session. Managerial decisions worked reasonably well here.

BC AND BP, SECOND PROBLEM:
READING/EXPLORATION/TRANSITION/EXPLORATION

A series of intuition-based conjectures led to a series of attempted constructions, the last of which happened to be correct—though neither student had any idea of why it might be, and they were content that it "looked right." This was a classic trial-and-error tape, and only because the trial space was small was there a chance that the right solution would be found. There was one weak assessment (rejecting a construction by eye) that constituted the transition recorded above, but the result was simply a continuation of trial-and-error search.

The kind of behavior indicated in these capsule descriptions is typical of the behavior demonstrated by the students at the beginning of the problem-solving course and in a variety of other recordings as well. For the most part solutions begin with the rapid choice of a particular direction, which is embarked upon without serious evaluation. Because students rarely if ever make global assessments of progress, little curtailment of inappropriately chosen solution paths takes place. Little is ever salvaged from an incorrect first attempt. In consequence, solutions are often doomed to fail by the actions taken in the first few minutes of exploration. The following statistic captures the essence of these students' behavior. Of the 12 sessions recorded at the beginning of the course, 7 were of the Reading/Exploration type, with null transition.

Protocol 9.4 was recorded by DR and KW after the problem-solving course. It is a representative although perhaps slightly better than average sample of postinstruction performance (another will be given in the next chapter). What makes this session better than virtually all of the pretest sessions is not that the students solved the problem, for they may simply have been lucky in coming upon the variational argument that solves it. What makes it a good solution is the fact that, by virtue of reasonable control decisions—by evaluating and curtailing a number of possible approaches while working on the problem, any of which could have occupied them for the full length of the solution attempt—DR and KW guaranteed that they would have the time to consider the variational approach. This kind of improved control behavior was typical of posttest performance. In general there was much more evaluation and curtailment of inappropriate solution paths after instruction than before it, and there was far less pursuit of wild goose chases. In some cases this allowed for a solution, in some not; but at least inefficient control behavior did not preclude the possibility of finding solutions. The episode analyses provide the clear contrast between pre- and postinstruction performance. *Although 7 of the 12 problem sessions reorded prior to instruction were of the Reading/Exploration type, only 2 of the 12 postinstruction protocols were of that type.* Not at all coincidentally, the students' performance improved on all of the measures described in Chapters 7 and 8. The answer to the first question asked at the beginning of this section is a clear "yes." Both the tapes, and the episode analyses, document substantially improved executive behavior after the course. (As with the other measures, there is similar documentation for the preceding and subsequent versions of the course.) The second question can also be answered in the affirmative, but an honest assessment is that the students still had a long way to go in the development of their executive skills. The overall quality of the students' monitoring, assessing, and executive decision-making on the postinstruction tapes was still relatively poor.

As an illustration of the way that executive skills can make a positive contributon to problem-solving performance, we examine the full text of the mathematician GP's solution to Problem 1.2. Segments of this problem session were discussed in Chapters 1 and 4. Subject GP is a number theorist who has a broad mathematical background but (as is apparent from the protocol) had not dealt with geometric problems for a number of years when he was asked to work Problem 1.2. When it is compared to JTA's schema-driven solution to the same problem (given in its entirely in Chapter 4), GP's solution appears clumsy and inelegant. But it is precisely because GP ran into difficulty in working Problem 1.2 that we have the opportunity to see the impact of his skills at the control level. Subject GP began working the problem with less domain-specific knowledge at his disposal than did many of the students I have asked to work it. Largely through efficient work at the control level, GP marshalled his cognitive resources and was able to solve it. None of the students have been able to do so.

The text of GP's solution is given in Protocol 9.5 (Appendix 9.5), with the corresponding episode analyses given in Figure 9.4. The contrast between Figure 9.4 and the three previous episode analyses speaks for itself. For obvious space limitations, the full analysis of Protocol 9.5 is condensed.

The critical point to observe in Protocol 9.5 is that a monitor-assessor-manager was always close at hand during the solution attempt. Rarely did more than a minute pass without there being some clear indication that the entire solution process was being watched and controlled, both at local and global levels. He began the solution with an attempt to make certain that the given problem was fully understood. By Item 3 GP was aware that some other information, or observation, would be necessary in order for a solution to be obtained. The actions in Items 4 and 5 were clearly goal driven and, in Item 6, yielded the necessary information. This was utilized immediately in

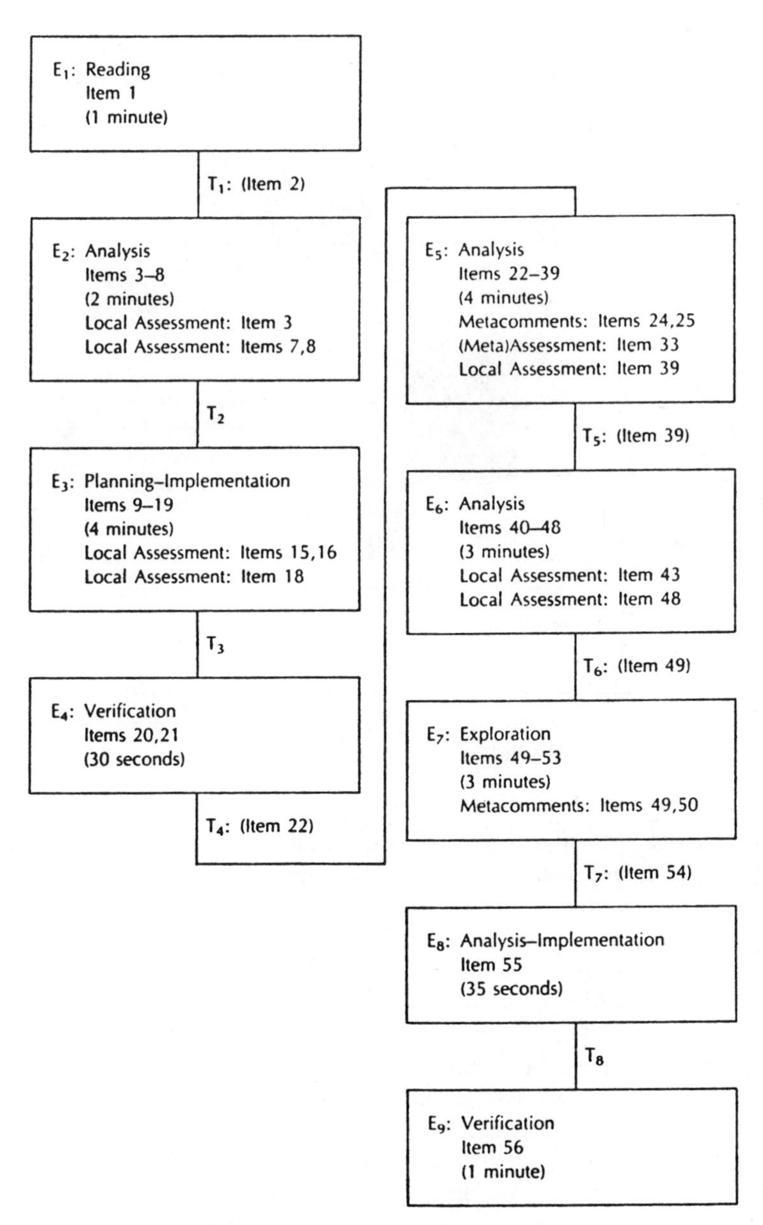

Figure 9.4A A parsing of Protocol 9.5.

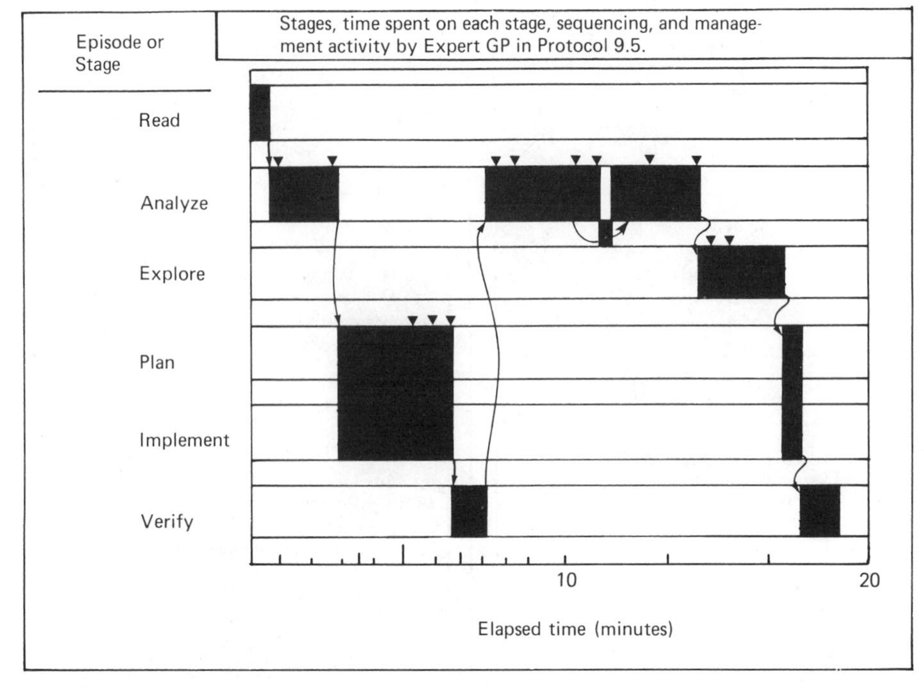

Figure 9.4B A time-line representation of Protocol 9.5. Overt signs of management activity are denoted by inverted triangles.

Items 7 and 8. Plan assessment is evident in the comment, made in Item 9, that solving the problem would depend on his ability to construct $\sqrt{2}$; since he did know how to do that construction, the plan could proceed. The full plan was stated later in Item 9. Implementation of that plan was interrupted twice with refinements (Items 15 and 16; Item 18). This is yet another indication that GP was on guard for clarifications and simplifications at almost all times. The first part of the solution concluded with a quick but adequate rehearsal of the argument.

The second part of the solution also began with a qualitative analysis. It is worth noting GP's comment, in Item 24, that "this is going to be interesting" (i.e., difficult). Such a preliminary assessment of difficulty is, I believe, an indication of an important element of experts' metacognitive behavior. It appears that experts assess how difficult a problem appears to be early in their solution attempts and that they later judge their progress on the problem against this template of expectations. These early estimates of the level of problem difficulty (which are, of course, subject to revision) may be major factors in the experts' decisions to pursue or curtail various lines of exploration during the problem-solving process.

Subject GP's solution of the second part of the problem continued, well structured, with a coherent attempt to narrow down the number of cases that must be considered. This case analysis was an implementation of "that kind of induction thought" from Item 29. It appeared to be a "forward" or "positive" derivation, verifying that all the cases can be done. Yet the phrase "no contradiction" in Item 33 reveals that even while working forward GP retained an open mind about whether the constructions could actually be implemented and that he was still probing for trouble spots. The potential for a reversal using argument by contradiction, if he should come to believe one of the constructions impossible, was very close to the surface. That he continues to entertain both possibilities, and that he maintains a somewhat impartial and distanced evaluative stance regarding the progress of his solution, are confirmed in Item 49.

Similarly, there were quite frequent assessments of the current state of the solution. The comment "if so this can be done in one shot" that appears in Item 40 indicates not only that solutions were planned ahead, but that the plans were assessed and reassessed during the solution attempt. Even the rather unusual excursion into quadratic extensions (Item 54) was preceded by a comment about "knocking this off with a sledgehammer" and was quickly curtailed.

The solution attempt can be summed up as follows. On the surface it appears to be clumsy and unstructured, containing what looks at times to be more or less undirected meandering through the solution space. In reality, however, there was an extraordinary degree of executive control at all times during this solution. There was constant monitoring of the solution process, both at the tactical and strategic levels. Plans and their implementation were continually assessed, and then acted upon in accordance with the assessments. As a result the problem solver could consider a wide range of possible options while working the problem, without squandering energy on wild goose chases; he could discover valuable information through his explorations and then utilize it. Tactical, subject-matter knowledge was important in this solution, of course — but metacognitive or control skills provided the key to success.

Subject GP's work in Protocol 9.5 provides a good opportunity to discuss the ways that the terms *expert* and *novice* have been used in the literatures of AI and IP psychology, and to discuss the potential misinterpretations of such usage.

Most definitions of expert and novice are domain-specific: An expert is someone who knows the domain inside out, whereas a novice is someone who does not and is usually new to it. The domain-specific definition of expertise is narrow and exclusive, requiring that experts be proficient in the domains in which they are studied. Because such experts have a domain at their fingertips, they have ready access to a range of problem-solving schemata and their performance is nearly automatic. An archetypal example of this kind of expertise was illustrated by the mathematician JTA's solution to the same problem (given in its entirety in Chapter 4). For JTA the task of partitioning the triangle was not a problem; it was an exercise. Access to the relevant solution mechanisms was nearly automatic.

When expert performance is defined in this way, GP does not count as an expert, and the kinds of skills he demonstrated in solving Problem 1.2 are not included in the inventory of skills that contribute to expertise. The implications of this perspective are disturbing. Although GP may not have been an expert geometrician, he was an expert problem solver; it was precisely the collection of metacognitive skills at his disposal that allowed him to succeed where others failed, although many of the others began working the problem with more domain-specific knowledge than he. In my opinion the ability to make great progress when faced with unfamiliar problems is the hallmark of talented problem solvers. Such abilities deserve as much attention as the routinized competencies of good problem solvers working in familiar domains, and it is unfortunate that they have not received it.

The standard definition of novice is broadly inclusive, meaning "generally new to this domain." In consequence, it allows people who are expert problem solvers to serve as novices in somewhat unfamiliar domains. According to this convention, a professional problem solver (e.g., a research psychologist who does computer simulations of problem solving) could serve as a "novice" in an experiment on kinematics—provided that he or she had not studied physics for years. This person might be a novice in the domain-specific sense, but as a problem solver with vast experience he or she would come to the physics problems armed with a broad collection of general problem-solving techniques and strategies. This inclusive definition of novice is a potential source of confusion about the capacities of problem-solving novices and about the skills that they use. For example, the (1978) Simon and Simon study has been frequently quoted to support the claim that means-ends analysis is a weak technique used by problem-solving novices. (See the quotation from Larkin in the section on Background, Part 2, this chapter.) The novice in that study was a problem-solving expert, and her use of means-ends analysis is still rather sophisticated by some standards. What a real novice might do in such situations is still open to question.

Brief Discussion: Limitations and Needed Work

The framework for protocol analysis presented in this chapter raises a host of questions. While its presentation may give it the appearance of a finished product, it is best thought of as a work in progress.

The idea behind the framework is to identify major turning points in a solution. This is done by parsing a protocol into macroscopic chunks called *episodes,* and then examining the junctures between them. But what, precisely, constitutes an episode? As noted in this chapter, a team of undergraduate coders can be trained to parse the protocols with accuracy and reliability. That the coding can be done, however, is only part of the story. There is still the need for a rigorous formalism for characterizing such episodes. Attempts have been made to adapt representational frameworks for story understanding, and for the analysis of episodes in memory (see, e.g., Bobrow and Collins, 1975), in ways that would enable them to deal with these kinds of macroscopic problem-solving episodes. Those attempts have been unsuccessful. The question of precise definition is an open issue, and a troubling one.

Equally important and far more thorny are questions regarding the objective characterization and evaluation of the monitoring, assessing, and decision-making processes that take place during problem solving. The role of such processes was quite clear in Protocol 9.5; it was constant overseeing at the executive level that assured that GP's resources were efficiently utilized and that his solution stayed on track. But what was the basis on which those executive decisions were made? The analysis of Protocol 9.5 suggested that early assessments of problem or subtask difficulty serve as one basis for later executive intervention. But the criteria for making such assessments, the nature of the monitoring processes, the question of how intervention is triggered, the question of what degree of tolerance there is in the system—and many other issues—all remain to be elaborated.

Each such issue leads to a host of others when it is closely examined. Consider, for example, the role of assessment, which made a clearly positive contribution to the evolution of Protocol 9.5. Assessment is not always desirable or appropriate. In a schema-driven solution, for example, one should simply implement the solution without assessment unless or until something untoward occurs. A simple-minded model that simply checks for assessments at a series of previously designated places (such as transition points between episodes and new information points) would miss the point entirely. The utility of assessment is context-dependent, and much work needs to be done in specifying the contexts in which it is appropriate.

In the long run there is a need for a detailed model of behavior at the executive level, a model that deals with these issues rigorously and reliably. Many of the questions asked in this framework, for example, Are the actions taken by the problem solver reasonable? in question A5, and Is the action (or lack of action) taken by the problem solver appropriate or necessary? in Question T2, are highly subjective and can only be answered at this point by making judgment calls. It is hoped that further work will elaborate the bases for such judgments and provide the empirical base from which the more detailed model can be constructed.

Summary

This chapter began with a discussion of verbal methods and of the subtle difficulties one may encounter when trying to interpret what takes place in out loud problem-solving sessions. As Protocol 9.1 indicates, it is dangerous to take verbal data at face value, even in apparently clean experimental settings. Clashes in the "rules of discourse" between subjects and experimenters, or environmental pressures that impinge on the subjects, can distort subjects' behavior in ways that are not apparent in the protocols themselves. Hence analyses of behavior based on such protocols can misrepresent what actually took place in the problem sessions. Generally speaking, any methodology (protocol analysis included) may highlight some aspects of behavior and may obscure or distort others. It is thus prudent to examine particular instances of problem-solving behavior from as many perspectives as possible, to help separate what resides in the behavior from what resides in the interactions between behavior and methodology.

Following the general discussion of verbal data, the chapter turned to the examination of executive or control behavior in problem solving. Transcripts of two problem-solving sessions illustrated the ways that students' problem solving can be hampered by poor control. Protocols 9.2 and 9.3 demonstrated in particular that the absence of monitoring and assessment at the control level can guarantee failure in problem solving. The students in Protocol 9.2 embarked on a series of computations without considering their utility and failed to curtail those explorations when (to the outside observer) it became clear they were on a wild goose chase. While pursuing wild geese they failed to examine and exploit potentially useful ideas that arose periodically during their solution attempt. These absences of executive behavior guaranteed that they would be unsuccessful. Similarly, the lack of plan assessment and the absence of review when discarding a potentially salvageable approach were major contributing factors to the students' failure in Protocol 9.3.

A framework for the macroscopic analysis of problem-solving protocols was introduced, and its use was demonstrated. Protocol analyses are carried out as follows. Problem-solving sessions are parsed into episodes, periods of time during which the problem solver or problem-solving group is engaged in either one large task or in a closely related body of tasks in the service of the same goal. (Episodes are classified as belonging to one of six categories: reading, analysis, exploration, planning, implementation, and verification.) Because of the way that episodes are defined, junctures between episodes delineate points at which there are major shifts in resource allocation or in the direction of the problem solution. The framework also calls for delineating points in a solution where new information or the possibility of an alternate approach occurs, or when sufficient time has elapsed in the pursuit of any one direction that it is time for an "executive review."

The framework allows for significant extensions of previous analyses. Prior analytical schemes, in coding only overt behaviors, could focus only on decisions that were made and on their impact. This framework explicitly identifies places in the problem solution where action at the executive level should be considered. By identifying places where control decisions would be of vital importance, the framework provides a mechanism for tracing the consequences when they are absent.

In addition to the detailed analyses of a number of protocols, this chapter provided summary descriptions of additional problem sessions. Those sessions were recorded before and after the course in mathematical problem solving that served as the experimental treatment for the research described in Chapters 7 and 8. The following statistic summarizes the results. Prior to instruction 7 of the 12 sessions recorded were of the Reading/Exploration type, meaning that the students went off on a 20-minute long wild goose chase immediately after reading the problem statement. After the instruction only 2 of 12 protocols were of that type. Similar results have been documented before and after other versions of the problem-solving course. These data indicate that courses in which there is a significant emphasis on metacognitive concerns (recall Table 7.1) can have a significant effect on students' behavior at the control level.

Appendix 9.1: A Single-Person Protocol of the Cells Problem

AB: [reads problem] Estimate, as accurately as you can, how many cells might be in an average-sized adult human body. What is a reasonable upper estimate? A reasonable lower estimate? How much faith do you have in your figures?

I'll think of some approaches I might take to it.

The first one might be just to go by parts of the body that are fairly distinct and try to figure out—

My first possible approach to the problem might be to look at them as approximations to geometric shapes and try to figure out the volume of each part of the body. And then make a rough estimation of what I thought the volume of a cell was and then try to figure out how many cells fit in there.

I would say that the arm from the wrist up to the shoulder and it's approximately a cylinder and it's, I don't know, about 3 or 4 inches in diameter. So you would have, it's about 2 or $1\frac{1}{2}$ inches in radius, squared, times π and the volume of my arm in square inches. So I've got two arms, so I've got two of those.

And now a leg. A leg—think this might be better—there's a little more variance, so I would say a cone might be more appropriate. And the base of my leg is approximately 6 or 7 inches in diameter so you would have $3\frac{1}{2}^2 \times \pi$ and the height would be—what is my inseam size, about 32 or 34—so you've got to have a 34, and it's a cone, so you've got to multiply by $\frac{1}{3}$.

And now the head is very, very roughtly a sphere. And so you've got a sphere of—I don't know how many—I don't know, maybe on the average 6 inches in diameter. That may be a little small, maybe 7 inches in diameter. And so quick recognition of the formula was $\frac{4}{3}\pi r^3$. So I've got $\frac{4}{3}$ of whatever my head is cubed, I've got $3\frac{1}{2}^3$, and what am I missing now?

Oh, torso—very important. Well, a torso is—you could say is approximately like a cylinder except with an oval base. So I could figure out what the area may be around is, and I won't calculate this explicitly. Say my waist is about 34 inches and I could approximate it across here. And if I worked on it I could figure out what the geometry of it of the volume of that ellipse.

AHS: Well, make a ballpark estimate. I would like to have a number just out of curiosity.

AB: So I've got an ellipse. This may take a while though because my geometry is bad. I've got an ellipse with a perimeter of about 34, and major axis is something along the lines of 18 inches, and the minor axis is maybe—I don't know—18 inches—And—Oh, geez—

Yeah, it's going to be very messy. So I will dispense with that and instead make another rough estimate, and rather assume myself to be—well, I'm not going to bother to do this since it's not very exact, anyway. But I could draw a circle, a little bit smaller than that maybe. Well, that circle has got—how much—something between 8 and 18, and looking at this I guess you have to stretch and elongate it in the width more than the height—closer to 18—and say 14 in diameter. So that would mean 7 inches in radius. So, I've got $\pi \times 49$. And that would be my guess for that, and the height would be—I don't know—about 15.

Now, I've covered the torso, the two legs, and the two arms.

Ok, for the hands. I'm going to have to make another rough estimate. If I put my hand into a fist, I get a little cylinder of maybe an inch and a half and a height of about 4. So I've got two hands with a height of 4, π and the radius of $\frac{3}{2}$.

And I have no idea what I'm going to do with my feet. Well, you could make these into little rectangular prisms. $4 \times 2 \times 10$. No, actually that looks about right.

Well, maybe the neck; if we're going to be precise about it is going to be 4 inches in diameter, so we've got a 2 inch radius neck. So that would be 4π in area, in volume of it. Yeah, $4\pi r^2$. And now I would add all these up. Do I have to add them up too?

AHS: We'll just call that number capital N, and then I'll get a calculator and we'll actually do it out of curiosity.

AB: OK, the number is N. OK, now that I've got the volume of a body, now I've got to figure out what the volume of a cell might be.

And it seems to me something along the lines [unclear]. The diameter of a hydrogen atom is like an angstrom unit, and that's something like 10^{-10} centimeters. And that's not going to be anything close to the size of a cell. So, if I had to go with the size of a cell—this is a very rough estimate, it might not even be in the right magnitude—it should be 10,000 to the inch or 10,000 cells to the centimeter. Maybe I'll make a compromise and say 100,000 cells to the inch is right. So that would give me 10^5. So each one is 10^5 in diameter, so we should call them spheres since that would make it simpler. I would have $(10^5/2)^2 \times \pi$. Is that right? $10^5/2$—you've got 10^5 to the inch so it would be 10^{-5} inches over 2 for the radius—so square that and multiply by π. So you take that and divide it by π.

And I'm going to say that that should give you the volume, but somehow I'm not convinced that that's the case. Well, maybe it would be right because you're going to have a 10^{-10} in the denominator there, and you multiply these things are going to come out to a good 1000 or so. So hopefully a couple thousand square inches or so when you multiply it.

AB was told that he had computed the area of a circle rather than the volume of a sphere. He made the correction, and then computed all the volumes with the help of a calculator (to four-place accuracy before rounding off).

Appendix 9.2: Protocol 9.2

1. KW: [reads problem] Three points are chosen on the circumference of a circle of radius R, and the triangle containing

them is drawn. What choice of points results in the triangle with the largest possible area? Justify your answer as best you can.

You can't have an area larger than the circle. So, you can start by saying that the area is less than $\frac{1}{2}\pi r^2$.

2\. AM: OK, so we have sort of circle—three points in front and r here, and we have let's see—points—

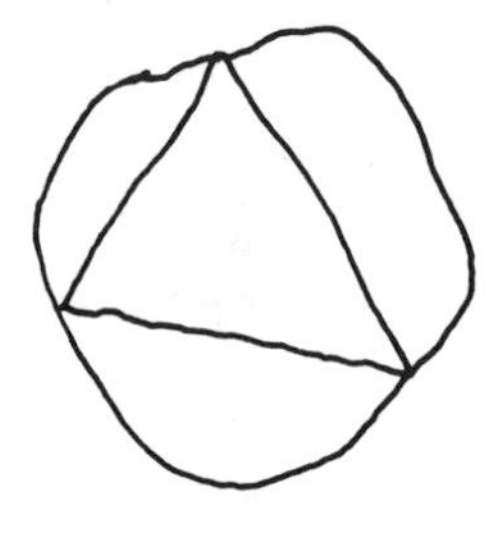

3\. KW: We want the largest one—

4\. KW: We want the largest one—

5\. AM: Right, I think the largest triangle should probably be equilateral. OK, and the area couldn't be larger than πR^2.

6\. KW: So we have to divide the circumference of the three equal arcs to get this length here. That's true. Right. So, 60–120 are degrees. OK. So, let's see, say that it equals r over s—this radius doesn't help.

7\. AM: Do we have to justify your answer as best as you can? Justify why this triangle—justify why you—OK. Right.

8\. KW: OK. Let's somehow take a right triangle and see what we get. We'll get a right angle.

9\. AM: Center of circle of right triangle. Let's just see what a right triangle—is this point in the center? Yep, OK. Yeah.

10\. KW: This must be the radius, and we'll figure out that'll be like that, right?

11\. AM: So the area of this—

12\. KW: —is R, is R—$\frac{1}{2}$ base times height, that's s and $2R$, height is R so it is $\frac{1}{2}R^2$. It's off by a factor of 2.

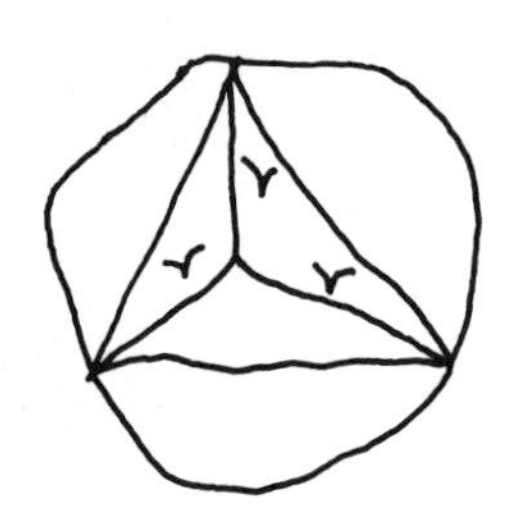

13\. AM: OK. But what we'll need is to say things like—OK. Let's go back to the angle—probably we can do something with the angle.

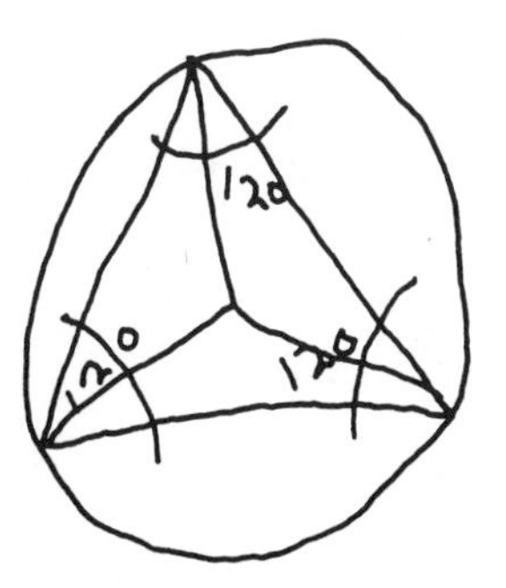

14. KW: Oh, I got it! Here, this is going to be 120–the angle of 120 up here—
15. AM: Right! Yes, this is 120 and this is 120.
16. KW: Right!
17. AM: So—
18. KW: We have to figure out—
19. AM: Why do we choose 120? Because it is the biggest area—we just give the between the biggest area—120.
20. KW: Umm. Well, the base and height will be equal at all times.
21. AM: Base and height—right—
22. KW: In other words, every right triangle will be the same.
23. AM: Ah, ah—we have to try to use *R*, too.
24. KW: Right.
25. AM: OK [seems to reread problem]. Justify your answer as best as you can. OK. [pause]
26. AM: So—there is the picture again, right? This is—both sides are equal—at this point—equal arc, equal angles—equal sides—this must be the center and this is the radius *R*—this is the radius *R*—
27. KW: So we have divided a triangle with three equal parts and—
28. AM: There used to be a problem—I don't know about something being square—the square being the biggest part of the area—do you remember anything about it?
29. KW: No. I agree with you—the largest area—of something in a circle, maybe a rectangle, something like that—
30. AM: Oh, well—so—
31. KW: Since this is *R*, and this is going to be 120, wouldn't these two be *R* also?
32. AM: Right.
33. KW: This is 120.
34. AM: Ah, ah.
35. KW: Like a similar triangle—120 and 120

are the same angle—so these two should be R.

36. AM: OK. Maybe they are.
37. KW: Why can't they be?
38. AM: [mumbles]
39. KW: See, look—this is the angle of 120. Right?
40. AM: Right.
41. KW: And this is an angle of 120. Right? This is like similar triangles—
42. AM: Wait a second—I think if you—this is true 120 but I don't think this one is—

43. KW: It is an equilateral triangle—that's—
44. AM: No. It should be a 60.
45. KW: That's right. It should be a 60. [mumbles] That's $\frac{1}{2}$ of it—that's right —$2R$.
46. AM: What are you trying to read from?
47. KW: What if we could get one of these sides, we could figure out the whole area.
48. AM: Ah, ah.
49. KW: Right?
50. AM: Presume this to be $\frac{1}{2}$ that side, we've got $\frac{1}{2}$ base times height. We'll get the area. All we have to show is the biggest one.
51. KW: When we take the formula πR^2 minus $\frac{1}{2}$ base times height and then maximize that, then take the derivative and set it equal to 0, we can get that function—then we can get this in the form of R.
52. AM: OK.
53. KW: Then we can try this as the largest area.
54. AM: Do you want to get this function, this as a function of R?
55. KW: Yeah.
56. AM: We can, I think. So you want this. Right?

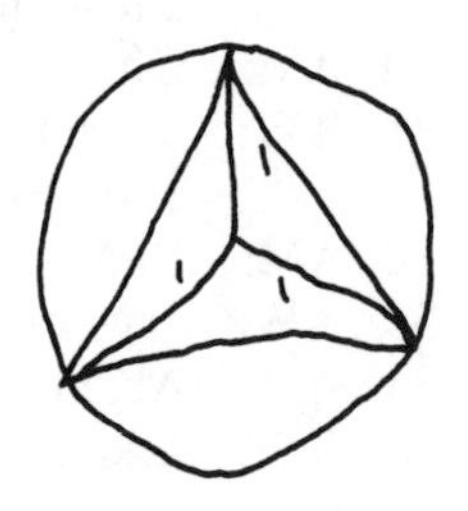

57. KW: Well, it is kind of obvious that with H and H you are still going to have an R in it. So you can subtract it.
58. AM: You have H in it. Well, we have this

one here. [mumbles; repeats the problem] Try this to be $2R$.

59. KW: No. It can't be. It has to be between R and $2R$.

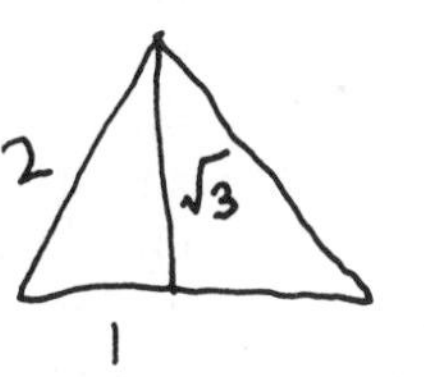

60. AM: Yeah.
61. KW: Helps us a lot! Set R equal to 1.
62. AM: $R = 1$?
63. KW: Right.
64. AM: OK.
65. KW: That's 1, that 1 that's 1—it'll equal S over R. The area of the triangle is equal with $R = 1$, it's 2.
66. AM: Well, height equals—
67. KW: That's for the sides of the triangle—that's obvious: $R = 1$.[1]
68. AM: OK. Divided into equal parts [lots of mumbling]. This form—well, you know—OK. If you see we probably try to fix one point and choose the other two—OK. We are going to go from something that looks like this all the way down—
69. KW: Right.
70. AM: Right. OK, and here the height is increasing where the base is decreasing.
71. KW: Right. [mumbles]
72. AM: When we reach—OK.
73. KW: What is the area, side squared over 4 radical 2 for an equilateral triangle? Is it like that?
74. AM: You want the area for an equilateral triangle.
75. KW: The area? I don't know. Something like side squared over radical 2, or something—
76. AM: If you can probably show—at a certain point where we have the equilateral triangle the base and the—well, you know—the product of the base since the base is decreasing and the height is increasing every time we move the line. If you can show a certain point,

this product is the maximum—so we have the area is a maximum at that point. So this one is decreasing, and at this point we have R, R, and R.

77. KW: Ah, ah.
78. AM: OK. This is the base—is $2R$—a right angle.
79. KW: It wouldn't be $2R^2$.
80. AM: [mumbles] One more—I mean—
81. KW: OK.
82. AM: It should be R^2. But base times height [mumbles] and this one, say this is $R + x$.
83. KW: The height equals $R + x$, so the base equals $R - x$.
84. AM: [mumbles] Those two things are equal to this—
85. KW: Right.
86. AM: All right.
87. KW: I don't know.
88. AM: We want this product of h as a maximum—as a maximum—and this one—I don't know.

Appendix 9.3: Protocol 9.3

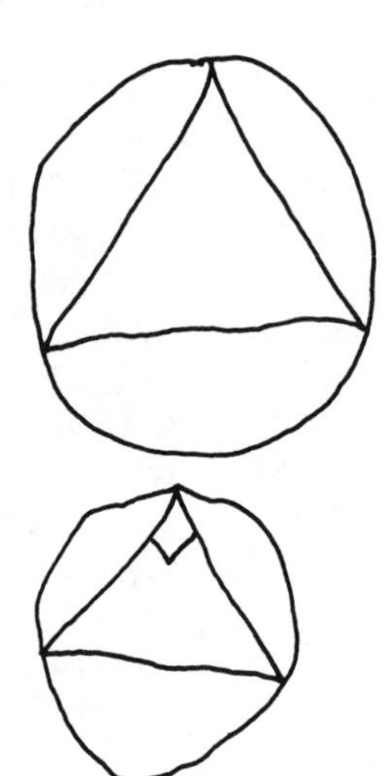

1. DK: [reads the question]
2. BM: Do we need calculus for this? So we can minimize, or rather maximize it.
3. DK: My guess would be more like [mumbling] my basic hunch would be that it would be—
4. BM: An equilateral—
5. DK: 60, 60, 60.
6. BM: Yeah.
7. DK: So what choice of points has to be where on the triangle—these points are gonna be.
8. BM: Try doing it with calculus—see if you can—just draw the circle—see what we'll do is figure out the right triangle—

9. DK: Yeah, or why don't we find—or why don't we know the—some way to break this problem down into—like, what would a triangle be for half the circle?

10. BM: 60 degrees here?
11. DK: Why don't we, why don't we say that—OK—why don't we find the largest triangle with base—one of the diameters, OK.
12. BM: Base as one of the diameters?
13. DK: Yeah.

14. BM: OK. That would be just a family of right triangles—that go like this.
15. DK: And they're all the same area?
16. BM: No, no they're not all the same area. The biggest area would be in one like that. See if we could figure out—make it into sort of like a—if we could do it with calculus and I know there is a way. I just don't remember how to do it.
17. DK: I have a feeling we wouldn't need the calculus. So this area then this is R and this would be—R^2—that would be the area of this—so then the distance here has got to be—45 degrees—

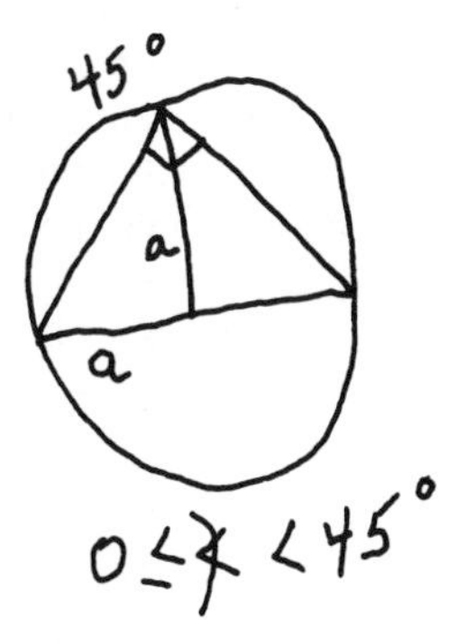

18. BM: Right. That's got to be 45 degrees because they are the same. That's A—$A/\sqrt{2}$. Right?
19. DK: Umma.
20. BM: If that's radius—A—and this is A, too, so that would be A^2, that would be R^2, wouldn't it?
21. DK: Right.
22. BM: But I think this would be bigger.
23. DK: Oh, of course it would be bigger—I was just wondering if [pause]

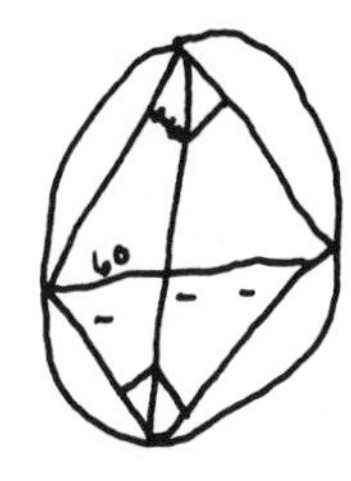

24. DK: Well, we can't build a diamond, so we can't build a diamond that would go like that—obviously you want to make it perfectly symmetrical—but we can,

if we maximize this area, and just flip it over, if we can assume that it is going to be symmetrical.

25. BM: Yeah, it is symmetrical.
26. DK: And if we can find the best area—
27. BM: You mean the best—cut it in half in a semicircle.

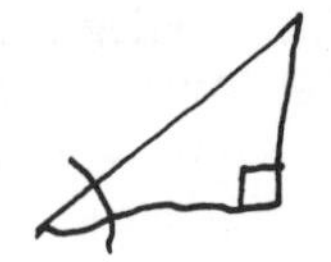

28. DK: Right. And if we can find the best area of—
29. DM: And triangle that fits in a semicircle—well, it shouldn't be a semi-—
30. DK: No, it's a semicircle.
31. BM: Largest triangle that fits in there?
32. DK: Yeah, but it would have to be—if it is going to be symmetrical, though, then you know this line has to be flat. It is going to have to form a right angle. So all we really have to do is form a right angle. So all we really have to do is find the largest area of a right triangle—inscribed in a semicircle.

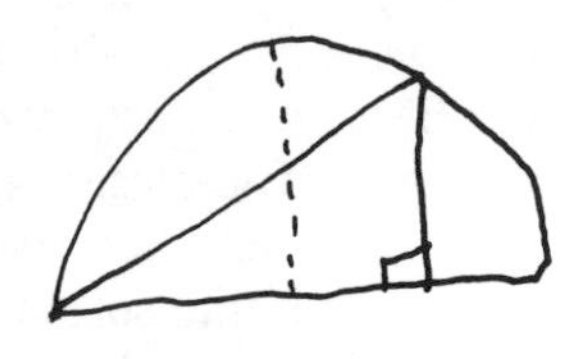

33. BM: Largest area of a right triangle. Yeah, but obviously it is this one which is wrong.

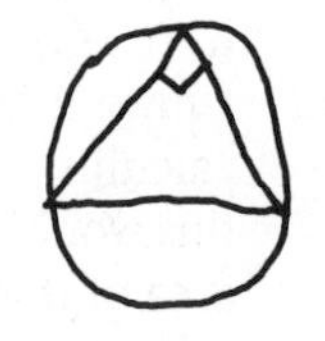

34. DK: No—No—
35. BM: One like this.
36. DK: Yeah, with that angle, right.
37. BM: OK. How we go about doing that? Hey, like, we can—use the unit circle, right?

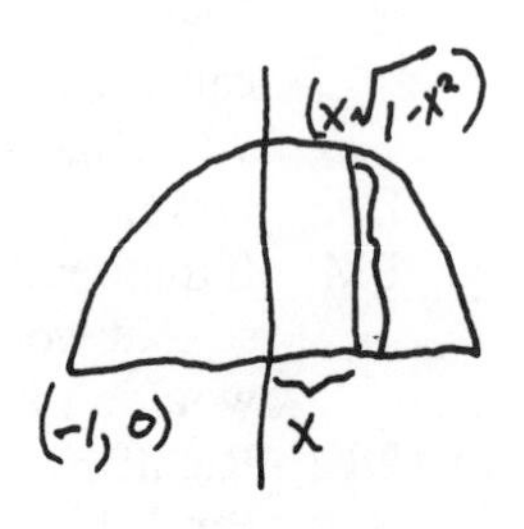

38. DK: Umma.
39. BM: So that means—this is $\sqrt{1-x^2}$—this point right here—will be $\sqrt{1-x^2}$. OK. This squared [mumbling] I'll just put some points down to see if—pick an arbitrary—
40. DK: Yeah, yeah, just to find this point—
41. BM: All right, this is 1. Now I've got to find that point, OK. What is the area of this? This is the distance right here times that distance, right? Product of those distance—area equals from this

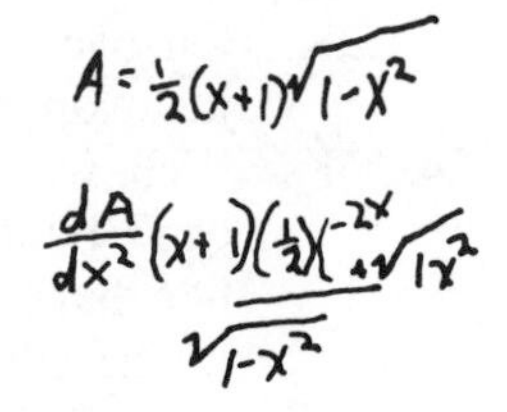

distance would be this, would be x value which would be $x - 1$ or $x + 1$? OK, it's $x + 1$, this distance right here times this distance right there, which would be the y coordinate, which is x^2. Want to take the derivative of that — to the x [mumbling].

$$\frac{1}{\sqrt{1-y^2}} = \frac{\sqrt{1-y^2}}{1}$$

42. DK: OK.

43. BM: Times $2 - x$. Did I have, oh, the 2 is crossed out so I just have an minus x— or, that was over $\sqrt{2-x}$, plus all this stuff. And set that equal to 0 and you get that—oh, this is just one, isn't it—this is just one—so one of that, plus that equals 0, right?

44. DK: I think we're getting a little lost here. I am not sure. Well, you go ahead with that—

45. BM: Well, I'll just think about it, as it is just mechanical. There is a minus in here, isn't there? [mumbling] OK, x equals the $\sqrt{2}$, and what was this distance, we said? That was x. So that means it would be $\sqrt{2} + 1$. That's impossible.

46. DK: Times R.

47. BM: If x equals plus or minus the $\sqrt{2}$—

$$-x^2 = -2$$

48. DK: Umma—

49. BM: This y thing would be $1 - x^2$, right?

$$x = \pm\sqrt{2}$$

50. DK: This is just the distance. Therefore this right here has to be $\sqrt{2}$. Guess your calculations are all right.

51. BM: Yeah, if I got x equals $\sqrt{2}$. We've got a semicircle here, right? OK. And I have the points—right, it's a unit circle and I said that $x^2 + y^2 = 1$, so $y = \sqrt{1 - x^2}$. OK? And [pause] the x can't equal the square of the 2 because it would be out there. I know this has to be right but—

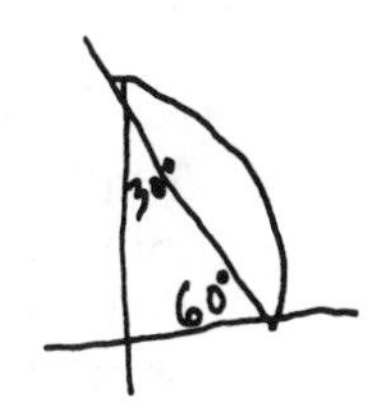

52. DK: But all kinds of—let's see—well, we know already, OK, that the triangle is

not 45, 45 because that would make it too small, OK?

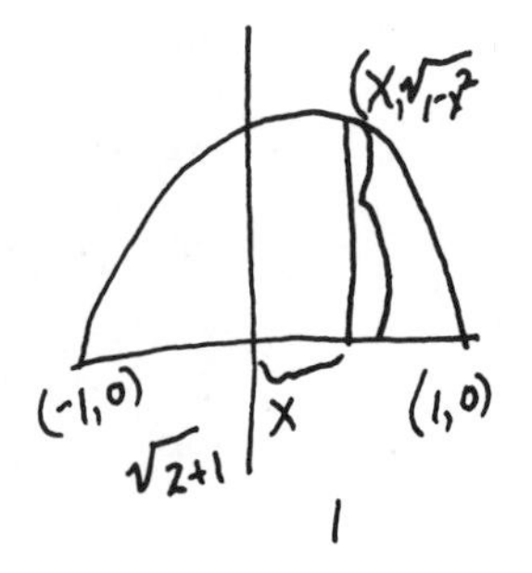

53\. BM: Um—

54\. DK: So we know this angle is greater than 0 and less than 90 degrees—

55\. BM: I just want to make sure I didn't—so this is $x + 1, x + 1n$—and cross multiply to set $1 - x^2 = 1$, which means $x = \sqrt{2}$.

56\. DK: No, it has to be a 60, 60, 60—right triangle—no, I am sorry, not a right triangle—has to be a 60, 60, 60 triangle. Because no matter where you move these vertices, it has to be a 60, 60, 60 triangle—because no matter where you move these vertices—

57\. BM: OK.

58\. DK: —you are going to add area to this—like the [mumbling] you are going to add area to this.

59\. BM: All right, OK. I understand, but I don't understand why it didn't work for this. I mean that—is there no solution for this equation?

60\. DK: I don't know. Are you sure what you are looking for in that one?

61\. BM: Yeah, I marked off these, and I just wanted to mark off these dimensions.

62\. DK: OK, what were you looking for? The length of this?

63\. BM: I was just looking for the maximum area of this—I said $A = (x + 1) \times \sqrt{\frac{1}{2} - x^2}$. That's this height, which is $(\sqrt{1 - x^2})$. This is the right circle. That's this distance right here—this minus the x value that I used—x value that is just x. OK—' cause it is all in terms of x—x minus the x value here, which is $x - 1$, which $x + 1$—so area — ah shoot—I should have put $\frac{1}{2}$ that is, well [mumbling], I'll get it. That should be $\frac{1}{2}$ there, but I don't think that makes any difference—so that's all in terms of 1.

64. DK: So—if—
65. BM: Oh, wait a minute, there's a difference — so 1 for 2 is $\frac{1}{2}$ the first part—
66. DK: So if you find the maximum area equal to—
67. BM: It doesn't make any difference. It is just a factor of $\frac{1}{2}$ here—because the area equals $\frac{1}{2}$ that.
68. DK: No. What's the next move?
69. BM: See, I get x—see, I get a value of x with a $\pm\sqrt{2}$, right?
70. DK: Umma.
71. BM: If I plug x back into this I get $\sqrt{2} + 1$, right? Then I plug x back into there and I get $1 - (\sqrt{2})^2$, which is -1, which doesn't work.
72. DK: Umma.
73. BM: Which doesn't seem right. Plus r^2 [mumbling]. Let me just check my derivative over again. Now I know my mistake—hold it. I added this x—it's supposed to be times so we've still got a chance. So let me go from there. It is just a derivative mistake. Let me see, it will be $1 - x^2$—no it will be—$-x + 1$. This might work. If it does, we solve that and cross out this minus 1. That means $x + 1 + x^2 - 1$, that makes $x^2 + x$—cross this out [mumbling]—all right? It still doesn't work.
74. DK: Well, let's leave the numbers for a while and see if we can do this geometrically.
75. BM: Yeah, you're probably right.
76. DK: Well, we know that these two are some kind of symmetry.
77. BM: Yeah.
78. DK: I still say we should try—yeah—what

we were doing before—just try to fix two of the points and let the third one wander around.

79. BM: Yeah, we were going to fix them. Yeah, I know what happens if you fix them on the diameter—than you have a family of right triangles.

80. DK: Those the maximums.

81. BM: Well, I don't see how—where are you going to fix the two points?

82. DK: Well, you just fix them on any diameter. You find the largest triangle.

83. BM: That would—obviously that would be the 45, 45 triangle if you fix them on the diameter. If you fix them on any chord.

84. DK: Yeah, why though? Well, we know that if we put two of the points too close together—OK—OK. No matter where we put the third point—

85. BM: Yeah.

86. DK: —it's going to be too small. OK. If we put them too far apart—OK. No matter where we put the third point, we are only using half a triangle.

87. BM: OK.

88. DK: So it's got to be—OK. So—two of the points, at least, well, matter of fact if you've got three points, each two of the points have to be between 0 and $\frac{1}{2}$ of the circle distance away from each other.

89. BM: OK.

90. DK: See how I got that? OK, so therefore each two of the *p* points has to be like that—so how can we construct a circle that's like that? OK, so we stick one point here—arbitrarily—so now the second point has to be somewhere, OK — within—OK, in other words, it can't be right here—it can't be right here—it can be anywhere else. We've

got to place it so that the third point is going to be within half—

91. BM: Half of what? I don't get you there.

92. DK: OK. Now wait a minute—let's see. You know when I said that [pause]. OK. In other words, the relationship between every pair of the three points.

[At this point the interviewer (I) terminated the session and asked the students to sum up what they had done. Student B focused on the algebraic computations he had done in trying to differentiate $(1 + x) \times \sqrt{1 - x^2}$. The following dialogue ensued.]

I: So what do you wind up doing when you do that? You wind up finding the area of the largest right triangle that can be inscribed in a semicircle.

DK: We determined that.

I: My question is, how does that relate to the original problem?

BM: Well—

Appendix 9.4: Protocol 9.4

1. KW: [reads problem] Consider the set of all triangles whose perimeter is a fixed number P. Of these, which has the largest area? Justify your assertion as best you can. All right now, what do we do?

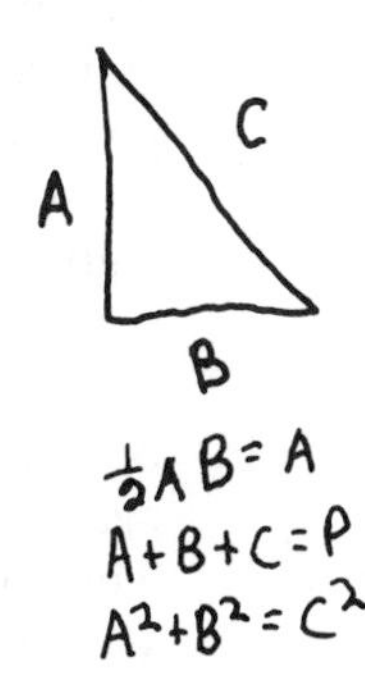

2. DR: We got a triangle. Well, we know we label sides A, B, and C.

3. KW: Right. I'll make it a right triangle—all right—A, B, C and the relationship such as that $\frac{1}{2}AB$ equals area and $A + B + C = P$ and $A^2 + B^2 = C^2$, and somehow you've got an area of one of these in the perimeter.

4. DR: Yeah, except for somehow—I mean, I don't really know—but I doubt that's the triangle of minimum area — well, OK, we'll try it.
5. KW: Largest area. Well, it is the only way we can figure out the area.
6. DR: All right.
7. KW: But for an isosceles we can do almost the same thing. This is $\frac{1}{2}A$ so that we know that the area is $A/2$ times $\sqrt{C^2 - (A/2)^2}$. The perimeter equals $A + B + C$ and the height equals $\sqrt{C^2 - (A/2)^2}$.
8. DR: All right.
9. KW: Now what do we do? We've got to figure out the largest area.
10. DR: Isn't it the minimum?
11. DR: The largest area.
12. DR: So actually if we can get A—we have to get everything in terms of one variable and take the derivative, right? Basically?
13. KW: Yeah, well—
14. DR: Well, I still don't know if we should do—I mean, we can find an area for this and can find an area for that, granted, but if we ever come to a problem like this—I mean, we don't know—we have no idea as of yet with a given perimeter what's going to be that.
15. KW: Right.
16. DR: So, there—I mean—you can do that again, but then what do you do?
17. KW: Then we're stuck, right? Usually, you know, you could probably take a guess as to what kind of triangle it would be—like you could say it is a right triangle or an isosceles. I think it is an equilateral, but I don't know how to prove it.
18. DR: Umma.

19. KW: So we have to figure out some way to try to prove that.
20. DR: All right, a good guess is that it is an equilateral, then why don't we try an isosceles, and if we can find that these two sides have to be equal to form the maximum area, then we can find that — then we should be able to prove that side also has to be equal.
21. KW: OK, so B will be equal to C, so the perimeter $P = A + 2B$, or $A + 2C = P$.
22. DR: All right.
23. KW: Ummmm.
24. DR: See what we've got.
25. KW: Fix A as a constant then we can do this, solve that for C.
26. DR: All right.
27. KW: For a maximum area we've got $\frac{1}{2}$, let's say $A = 1$; $\sqrt{C^2 - \frac{1}{4}}$. Right? Maximum area: $\frac{1}{2}\sqrt{C^2 - (\frac{1}{4})^{1/2}} = 0$.
28. DR: C^2 minus what?
29. KW: $\frac{1}{2}^2$, yeah, $\frac{1}{2}^2$. $A/2$, where $A = 1$. OK?
30. DR: Ah, ah.
31. KW: [mumbling] This is $\frac{1}{4}(C^2 - \frac{1}{4})^{-1/2}$. $2C$, so we know that $2C$ has to equal 0 and $C = 0$, and we are stuck!

$$1/2\sqrt{c^2 - 1/4}$$

$$1/2(c^2 - 1/4)^{1/2} = 0$$

$$1/8(c^2 - 1/4)^{-1/2}(2c) = 0$$

$$2c = 0$$

$$c = 0$$

32. DR: We should have taken a derivative in it and everything, you think?
33. KW: Yeah, that's the derivative of that. So does it help us? My calculus doesn't seem to work anymore.
34. DR: The thing is [pause] you are letting C be the variable, holding A constant. So what was your formula?—$\frac{1}{2}$ base times square root.
35. KW: The base A times the square root times the height, which is a right triangle to an isosceles which is—so it is $C^2 - (A/2)^2$, which would give you this height.
36. DR: A to the $\frac{2}{4}$, no, A to the $\frac{2}{2}$, no, $(A/2)^2$.

37\. KW: How about P equals no, $C = P - (A/2)^2$? Should we try that?

38\. DR: No, see part of the thing is, I think that for here we're just saying we have a triangle, an isosceles triangle, what is going to be the largest area? Largest area.

39\. KW: Largest area—set its derivative equal to 0.

40\. DR: All right. Well, the largest area or the smallest area—I mean—what's going to happen is you have a base, and it's going to go down like that—I mean—we don't set any conditions. We're leaving P out of that.

41\. KW: Ah, ah.

42\. DR: That's absolutely what we have to stick in.

43\. KW: We've got C and $P - A/2$.

44\. DR: $P - A/2$.

45\. KW: Formula—isosceles.

46\. DR: $A + 2B = P$. All right?

47\. KW: Shall we try that [mumbling]; $-A$ over 2—we've got to have a $-\frac{1}{4}PA$—

48\. DR: Well, then you can put A back in—then you can have everything in terms of A, right? Using this formula, we have the area and we have a—

49\. KW: All right—P—so that's

$$\frac{A}{2}\left(\frac{P^2 - 2A + A^2}{4} - \frac{A^2}{4}\right)^{1/2}$$

and that's

$$\frac{A}{2}\left(\frac{P^2 - 2A}{4}\right)^{1/2}$$

[mumbling and figuring].

50\. DR: Wait a minute. You just took the derivative of this right here?

51\. KW: This times the derivative of this plus this times the derivative of this.

52\. DR: Oh.

$$\frac{A}{2}\left(\left(\frac{P-a}{2}\right)^2 - \frac{a^2}{4}\right)^{1/2}$$

$$\frac{a}{2}\left(\frac{p^2-2a-a^2}{4} - \frac{a}{4}\right)^{1/2}$$

$$\frac{a}{2}\left(\frac{p^2-2a}{4}\right)^{-1/2}(2p-2)+$$

$$\left(\frac{p^2-2a}{4}\right)^{1/2}(1/2)=0$$

$$\frac{2ap-2a}{4}+\frac{p^2-2a}{8}=0$$

$$\frac{8}{p^2}$$

53. KW: [mumbling and figuring]

$$\frac{A}{4}\left(\frac{P^2-2A}{4}\right)^{1/2} 2(P-2)$$

$$+\left(\frac{P^2-2A}{4}\right)^{1/2}$$

Let's see . . . $\frac{2AP-2A}{4}$

$$+\frac{P^2-2A}{8}=0.$$

54. DR: So can we get A in terms of P?

55. KW: P^2—

56. DR: $8P^2 - 8P^2$ bring the P^2 on this side and multiply it by 8 and we'll have a quadratic in terms—no we won't—than we can just have A we can factor out in the equation—you see.

57. KW: OK. $P^2 = P^2$.

58. DR: $-8P^2$—oh, are we going to bring everything else to the other side?

59. KW: Yeah, $2A - 4A - 4AP \times B$—No—

60. DR: That's not right. Well, the—we can just multiply—

61. KW: P^2 equals all this.

62. DR: Right.

63. KW: $P^2 - 4AP$—this isn't getting us anywhere.

$p^2 = 2a + 4a - 4ap$

64. DR: P^2 equals—factor out the A—then we can get A in terms of P.

$p^2 = 2a(3+2p)$

65. KW: $P^2 = 2A$. So you've got $P^2/(6 + 4P)$.

66. DR: So if we have an isosceles triangle and A is equal to—

$\frac{p^2}{3+2p} = 2a$

67. KW: B is equal to that—

68. DR: And if A has to be equal to that and B and C are equal—

$a = \frac{p^2}{6+4p}$

69. KW: So. B equals [whistles]

70. DR: B equals P minus that.

71. KW: $2B = (P - A)/2$.

72. DR: No, we aren't getting anything here. We're just getting—thing is that we

assumed B to be equal to C so, of course, I mean, that doesn't—we want to find out if B is going to be equal to C, and we have a certain base. Let's start all over and forget about this. All right, another triangle. Certain altitude.

73. KW: Well, let's try to assume that it is an equilateral.
74. DR: All right.
75. KW: Sides [mumbling] perimeter equals $3S$, right?
76. DR: Yeah, but wait a minute—that's still not going to really help us. What are we going to do?—simply assume that it is an equilateral? We're just going to get that it is an equilateral if we assume that.
77. KW: True.
78. DR: We want to prove that it is an equilateral if we think it is. If we want to do anything, we can—
79. KW: Yeah, how do you prove it?
80. DR: Well, we can make up a perimeter. We don't need a perimeter P, do we? So—
81. KW: Where are you going to get area formula in the form of P?
82. DR: We want to maximize the area so that we can prove—OK, we have the given base—we'll set our base equal to something.
83. KW: Yeah [mumbling], P, or something —I don't know.
84. DR: Then the other two sides have to add up to P.
85. KW: We—how about we say—let's start with an equilateral just for the hell of it—see what happens. You get $\frac{1}{3}P$, $\frac{1}{3}P$, and $\frac{1}{3}P$. And this is $\frac{1}{9} - \frac{1}{36}$, which is the height—
86. DR: Now the thing we want to do is say—

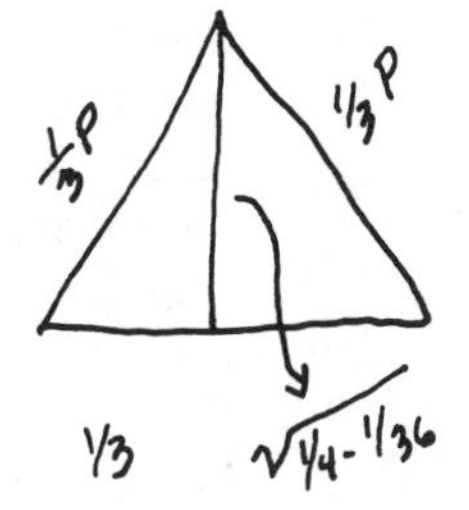

OK, if we shorten this side at all and then what's going to happen to the height—if we leave this the same?

87. KW: We can't shorten it.

88. DR: And we shorten this side—sure we can—

89. KW: Well—

90. DR: We can have a—this equal to $\frac{1}{3}$ and then a—this equal to—well, you're going to have—I mean—

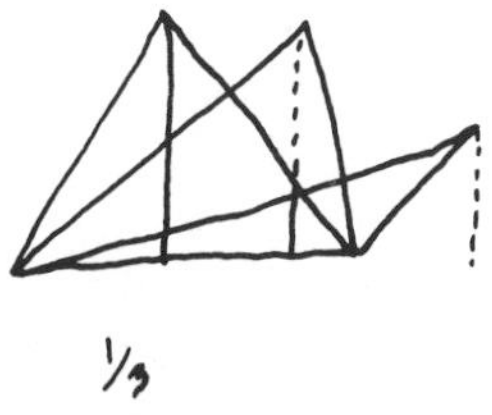

91. KW: Aha.

92. DR: This is going to get longer like that. Now we can see from this that all that is going to happen is that the base is going to get shorter, so we know from that as far as leaving the base constant goes if we move—if we shorten this side then it is going to—somehow the point's going to go down in either direction.

93. KW: Semicircle.

94. DR: Right. That proves that we have to have an equilateral.

95. KW: No, it proves an isosceles.

96. DR: No, isosceles, I mean. All right, from that if we set—we know that those two have to be equal so if we set his base equal to anything—it doesn't have equal to be $\frac{1}{3}P$—we can also show that if goes down—the area is going to get smaller, so the constant base then the height is going to get shorter and shorter and is getting smaller and smaller actually.

97. KW: OK.

98. DR: In this case if it goes down to his side, we're going to have again a smaller angle here, shorter base here—and [noise].

99. KW: So we get—so we know it is an equilateral—well, prove it.

100. DR: I don't know, that's not a rigorous

proof, but it is proof—good enough for me.

101. KW: Proves that an equilateral has the largest area.

102. DR: Oh, we're talking about the largest area.

103. KW: Yeah.

104. DR: Oh, we just did.

105. KW: We have to prove it has fixed number *P*—perimeter.

106. DR: Well, we already—we assumed that we have a fixed *P*, all right? I mean, this is a proof as far as I—

107. KW: Well, we've shown that an equilateral has the largest area. We haven't shown that if you have a certain set perimeter, let's say a right triangle, with a perimeter which is the same—we will not have a larger area.

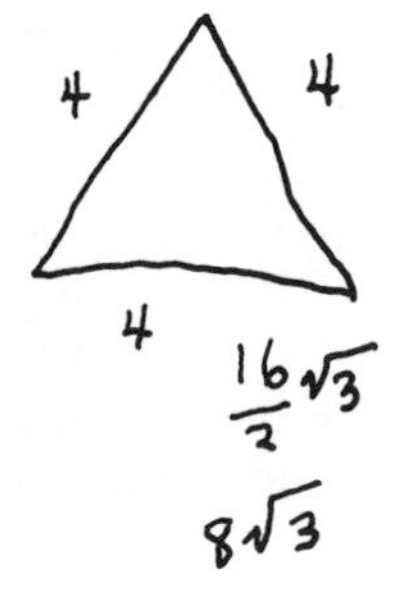

108. DR: No, but we have because we have shown with the set perimeter—OK, we know that—

109. KW: Well, what if we have 3, 4, 5 with an equilateral being 4, 4, 4?

110. DR: 3, 4, 5 is what? [mumbling]

111. KW: 12. So this area will be 6 and this area will be side squared 16. OK, that will have the largest area.

112. DR: What's that, 1.7?

113. KW: Yeah, 11 is still greater than 6 and that's greater than 1.

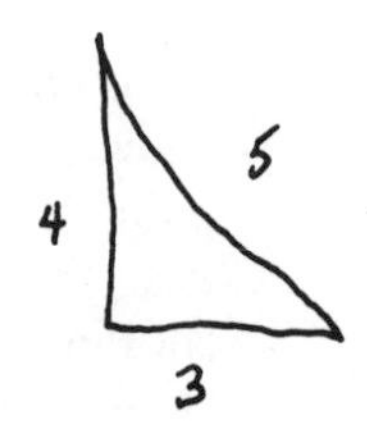

114. DR: Oh, yeah, that's right. Yeah, but the thing is if we have a fixed dimension, we already showed that, OK, what is going to happen is as this side gets longer—say we use 4 as a base here, so then what's going to happen—well, say we use 3 as a base, just so we won't have an equilateral when we are done—what's going to happen as 4 gets longer and 5 gets shorter—it's going to go upwards. The optimum

area—the maximum area is going to be right there. Because you've got—

115. KW: Right.

116. DR: —this angle and that height. If you make this angle any less—maybe let me draw a picture.

117. KW: I can understand that—this will give us largest area—but how can we prove this bottom is $\frac{1}{4}$—$\frac{1}{3}$ the area of the perimeter?

118. DR: Well, remember all the problems we've done where we say—OK, let me just start from here once more—so that we have 3, 4, 5—is that what you have?—because that's going to be 5. Wasn't a very good 3, 4, 5, anyway. So you start out with 3, 4, 5—all right, we pick the 3 has the base, right?

119. KW: Aha.

120. DR: All right, it's 5 [mumbling] if we have 3 as the base—as this is a little bit off an isosceles, but if we draw an isosceles as 3 as the base—OK, we've got a right angle—that's got to be the maximum [mumbling]—height? because if it goes any—

121. KW: Right.

122. DR: Over this way, it is going to go down.

123. KW: OK.

124. DR: All right, so remember the argument we've used—well, if we—

125. KW: Yeah, I can show that, but what you're not showing is—what you're not proving is that—

126. DR: —that it has to be an equilateral?

127. DR: Right. But you're not showing that this side is $\frac{1}{3}$ the perimeter.

128. DR: Right. I'm showing—first of all, it has to be an isosceles. Right.

129. KW: Right.

130. DR: It has to be an isosceles. That means

that we've got these three sides and those two are equal, right?

131. KW: Umma.

132. DR: Right. So now I pick this side as my base—I already picked—if that side is my base then the maximum area would have to have an isosceles—so I turn around—this side is my—

133. KW: That I understand as proof, but you're not showing me that this is $\frac{1}{3}$ the perimeter [mumbling].

134. DR: If we have an isosceles triangle—if we have an equilateral triangle—then each side has to be $\frac{1}{3}$ the perimeter. That's the whole thing about an equilateral triangle.

135. KW: I know. OK.

136. DR: First, we know it must be an isosceles, right?

137. KW: Umma.

138. DR: OK.

139. KW: I understand this.

140. DR: If it is an isosceles, it must be an equilateral, right?

141. KW: All right.

142. DR: And if it must be an equilateral, all three sides must be equal, and if the perimeter is P, all three sides must be $\frac{1}{3}P$.

143. KW: OK. I've got it.

Appendix 9.5: Protocol 9.5

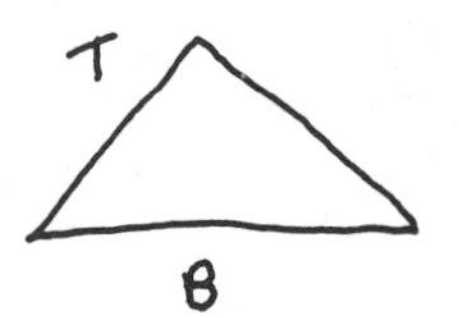

1. [reads problem] You are given a fixed triangle T with base B. Show that it is always possible to construct, with ruler and compass, a straight line parallel to B such that that line divides T into two parts of equal area. Can you similarly divide T into five parts of equal area?

2. Hmm. I don't know exactly where to start.
3. Well, I know that the—there's a line in there

somewhere. Let me see how I'm going to do it. It's just a fixed triangle. Got to be some information missing here. *T* with base *B*. Got to do a parallel line. Hmmm.

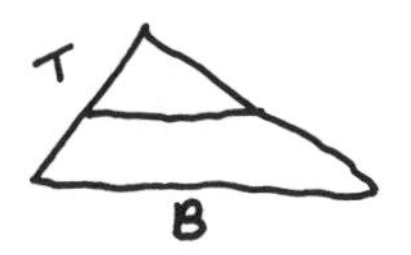

4. It said the line divides *T* into two parts of equal area. Hmmm. Well, I guess I have to get a handle on area measurement here. So, what I want to do—is to construct a line—such that I know the relationship of the base—of the little triangle to the big one.
5. Now let's see. Let's assume I just draw a parallel line that looks about right, and it will have base little *b*.

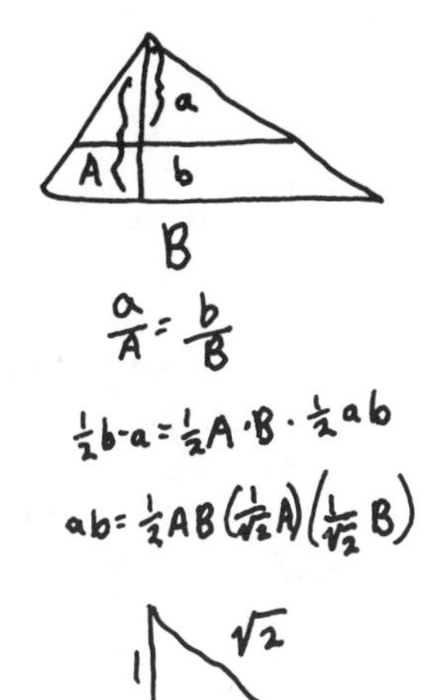

6. Now, those triangles are *similar.*
7. Yeah, all right, then I have an altitude for the big triangle and an altitude for the little triangle so I have little *a* is to big *A* as little *b* is to big *B*. So what I want to have happen is $\frac{1}{2}ba = \frac{1}{2}AB - \frac{1}{2}ba$. Isn't that what I want?
8. Right! In other words I want $ab = \frac{1}{2}AB$. Which is $\frac{1}{4}$ of *A* times [mumbles; confused] $(1/\sqrt{2}) \times A \times (1/\sqrt{2}) \times B$.
9. So if I can construct the $\sqrt{2}$, which I can! then I should be able to draw this line—through a point which intersects an altitude dropped from the vertex. That's little $a = A/\sqrt{2}$, or $A = a \times \sqrt{2}$, either way.
10. And I think I can do things like that because if I remember I take these 45-degree angle things, and I go 1, 1, $\sqrt{2}$.

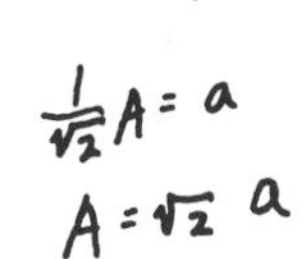

11. And if I want to have $a \times \sqrt{2}$—then I do that — mmm. Wait a minute—I can try and figure out how to construct $1/\sqrt{2}$.
12. OK. So I just got to remember how to make this construction. So I want to draw this line through this point and I want this animal to be—$(1/\sqrt{2}) \times A$. I know what *A* is, that's given. So all I got to do is figure out how to multiply $1/\sqrt{2}$ times it.

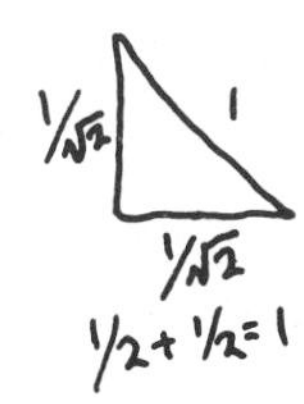

13. Let me think of it. Ah huh! Ah huh! Ah huh! $1/\sqrt{2}$—let me see here—ummm. That's $\frac{1}{2}$ plus $\frac{1}{2}$ is 1.

14. So of course if I have a hypotenuse of 1—
15. Wait a minute: $(1/\sqrt{2}) \times (\sqrt{2}/\sqrt{2} = \sqrt{2}/2$—that's dumb!

16. Yeah, so I construct $\sqrt{2}$ from a 45, 45, 90. OK, so that's an easier way. Right?
17. I bisect it. That gives me $\sqrt{2}/2$. I multiply it by A—now how did I used to do that?
18. Oh heavens! How did we used to multiply times A. That—the best way to do that is to construct A—A—then we get $\sqrt{2}$ times A, and then we just bisect that and we get A times $\sqrt{2}/2$. OK.

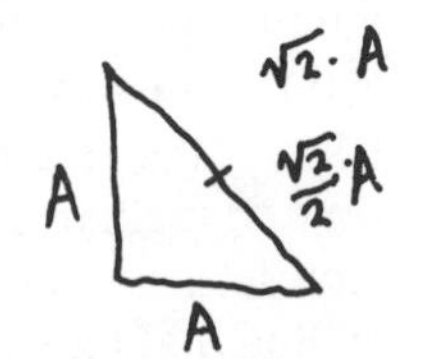

19. That will be—what!—mmm—that will be the length. Now I drop a perpendicular from here to here. OK, and that will be—ta, ta—little a.
20. So that I will mark off little a as being $A \times \sqrt{2}/2$. OK, and automatically when I draw a line through that point—I'd better get $\sqrt{2}/2$ times big B. OK.

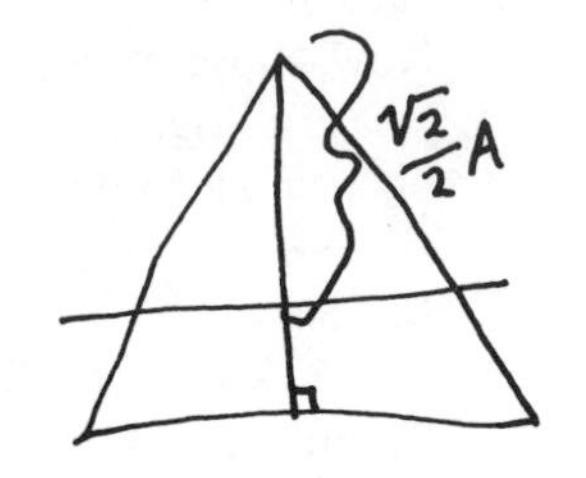

21. And when I multiply those guys together I get $\frac{2}{4}AB$. So I get half the area—what?—yeah—times $\frac{1}{2}$—so I get exactly $\frac{1}{2}$ the area in the top triangle, so I better have half the area left in the bottom one. OK.
22. OK, now can I do it with 5 parts?
23. Assuming 4 lines.
24. Now this is going to be interesting since these lines are going to have to be graduated—that—
25. I think, I think, that rather than get a whole lot of triangles here, I think the idea, the essential question is can I slice off—$\frac{1}{5}$ of the area—mmm—
26. Now wait a minute! This is interesting. Let's get a—how about four lines instead of?—
27. I want these to be—all equal areas. Right? A_1, A_2, A_3, A_4, A_5, right?
28. Sneak! I can—I can do it for a power of 2. That's easy because I can just do what I did at the beginning and keep slicing it in half all the time.
29. Now can I use that kind of induction thought?

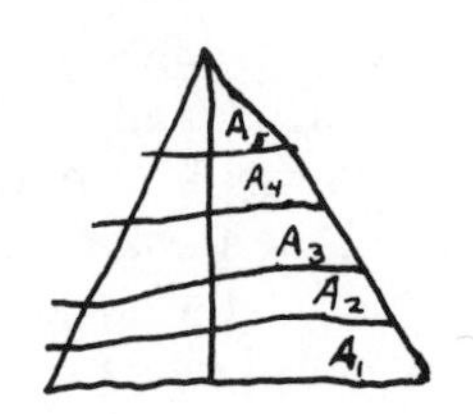

30. I want that to be $\frac{2}{5}$. And that to be $\frac{3}{5}$.
31. So let's make a little simpler one here.
32. If you could do that then you can construct $\sqrt{5}$. But I can construct $\sqrt{5}$ to 1—square root of 5, right?
33. So I can construct—OK. So that certainly isn't going to do it. No contradiction—

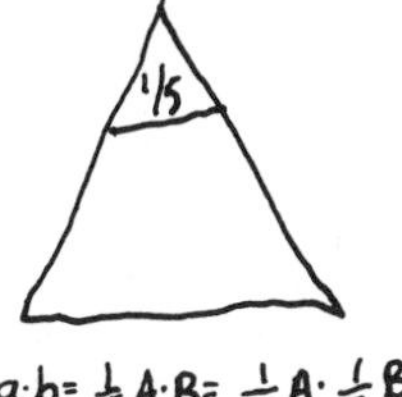

34. Now, I do want to see, therefore, what I have here.
35. I'm essentially saying is it possible for me to construct it in such a way that that is 1, 2, 3, 4, 5, $\frac{1}{5}$ the area—OK.

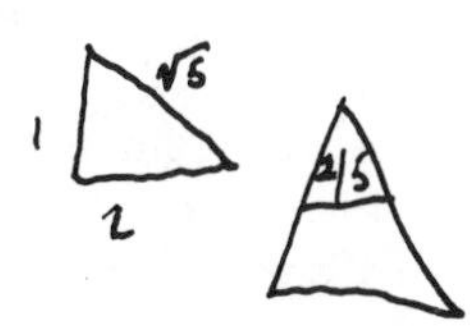

36. So little a times little b has got to equal $\frac{1}{5}AB$. So I can certainly chop the top piece off and have it be $\frac{1}{5}$ of the area. Right? Right?
37. Now, from the first part of the problem, I know the ratio of the next base to draw—because it is going to be $\sqrt{2}$ times this base. So I can certainly chop off the top $\frac{2}{5}$.
38. Now, from the first part of the problem I know the ratio of the top—uh, OK, now this is $\frac{2}{5}$ here, so top $\frac{4}{5}$—OK.—all right. So all I got to be able to do is chop off the top $\frac{3}{5}$ and I'm done.

39. It would seem now that it seems more possible — let's see—
40. We want to make a base here such that little a times little b is equal to—the area of this thing is going to be $\frac{3}{5}$—$\frac{3}{5}AB$—in areas, right! and that means little a times little b is $[(\sqrt{3}/\sqrt{5})A][(\sqrt{3}/\sqrt{5})B]$. OK, then can I construct $\sqrt{3}/\sqrt{5}$. If so then this can be done in one shot.
41. Well let's see. Can I construct $\sqrt{3}/\sqrt{5}$? That's the question. $\sqrt{3}/\sqrt{5} \times \sqrt{5}/\sqrt{3} = \sqrt{15}/5$.
42. $\sqrt{15}$, $\sqrt{15}$. Wait a minute! $\sqrt{15}/5$. Is $\sqrt{15}$ constructable? $\sqrt{15}$ is—

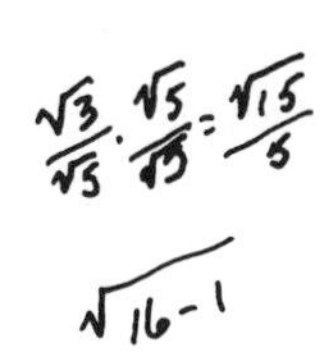

43. It is $\sqrt{16-1}$. But I don't like that. It doesn't seem the way to go.
44. $16^2 - 1^2$ equals [expletive deleted]
45. Somehow it rests on that.
46. [expletive] If I can do $\sqrt{15}$. Can I divide things and get this?
47. Yeah, there is a trick! What you do is you lay

off five things. One, two, three, four, five. And then you draw these parallel lines by dividing them into fifths. So I can divide things into fifths so that's not a problem.

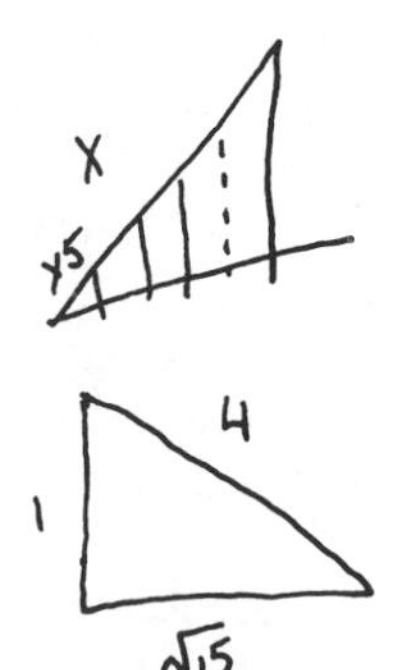

48. So it's just constructing $\sqrt{15}$, then I can answer the whole problem.
49. I got to think of a better way to construct $\sqrt{15}$ than what I'm thinking of—or I got to think of a way to convince myself that I can't—ummm—$x^2 - 15$.
50. Trying to remember my algebra to knock this off with a sledgehammer.
51. It's been so many years since I taught that course. It's 5 years. I can't remember it.
52. Wait a minute! Wait a minute!
53. I seem to have in my head somewhere a memory about quadratic extension.
54. Try it differently here. mmm—
55. So if I take a line of length one and a line of length—And I erect a perpendicular and swing a 16 [transcriber's note: for mathematical clarity he really means 4 instead of 16] here. Then I'll get $\sqrt{15}$ here, won't I?
56. I'll have to, so that I can construct $\sqrt{15}$ times anything because I'll just multiply this by A and this by A and this gets multiplied by A divided by 5 using that trick. Which means that I should be able to construct this length and if I can construct this length then I can mark it off on here and I can draw this line, and so I will answer the question as YES!!

10

The Roots of Belief*

This chapter focuses on the origins of students' beliefs about mathematics. It begins with the analyses of two problem-solving sessions recorded by the students LS and TH before and after the 1980–1981 problem-solving course. These sessions are examined in terms of all four categories of the analytical framework, with an emphasis on the ways that belief shaped the students' behavior. These analyses provide the opportunity to see how the full framework is implemented. The "before and after" comparisons of the students' work also provide one last view of the effects of the problem-solving instruction.

The discussion then turns to an exploration of the causes of the students' behavior. The protocols of the two problem sessions are as important for what they indicate about the nature of LS and TH's previous mathematical experiences as for what they reveal about the students' behavior at the time the recordings are made. The major lesson that LS and TH seem to have learned from their encounters with geometry is that geometric argumentation has nothing to do with learning or discovery. Although they were capable of making deductive arguments, the students avoided making them in situations where they could be used to advantage. As indicated by the research discussed in Chapter 5, this kind of behavior is much more the rule than the exception. The prevalence of such behavior raises some serious

* The two protocols (and the discussion of them) in the first section of Chapter 10 are taken and modified from Schoenfeld (1983a). Permission from *Cognitive Science* to reproduce parts of the article is gratefully acknowledged.

questions about what students are really learning in our mathematics classrooms. Selections from some ongoing classroom studies suggest that there is cause for concern. They indicate that the roots of students' empiricism can be traced directly to the ways that they are being taught mathematics.

A brief concluding section points to some work that lies ahead.

A Discussion of Two Geometry Protocols

CONTEXT: ENVIRONMENTAL ISSUES

When they recorded Protocol 10.1, LS and TH had just completed their first semester of calculus. They were good students who enjoyed doing mathematics and had both received A's in the calculus course. They were also good friends who worked well together. Protocol 10.1 (Appendix 10.1) was recorded the second day of the 1980–1981 problem-solving course. Protocol 10.2 (Appendix 10.2) was recorded the last day of the course.

In the light of the methodological discussions that began Chapter 9, one should check Protocols 10.1 and 10.2 for signs of environmentally induced difficulties before exploring the content of the protocols themselves. Might the verbal data in these protocols be significantly distorted, as it was in AB's work on the cells problem (Appendix 9.1)?

Segments of dialogue in both solution attempts make it clear that LS and TH felt the pressure of the testing environment. For example, in Protocol 10.1 the students' enthusiastic adoption of the perpendicular hypothesis H1 (Items 43–50), and their declaration that using the corner of a ruler to make a right angle is "legal" (Items 62–63), were clearly brought on by their awareness that they had been unsuccessful and that they were running out of time. There were similar kinds of pressures in Protocol 10.2. For example, TH's comment "We're running out of time. Draw faster, draw faster" in Item 74 indicates that the students were very conscious of the clock when they worked. As far as can be determined, however, neither of the two recording sessions was plagued to any significant degree by the kinds of environmental difficulties that plagued AB in Protocol 9.1. LS and TH wanted to solve the problems in both problem sessions and tried hard to do so. They concentrated on their work for the full length of the problem sessions. Moreover, they remained in the laboratory for more than half an hour after completing the recording session that generated Protocol 10.1, in order to discuss the solutions to the problems they had worked. Had they found the testing environment overly oppressive, they would have tried to escape it as rapidly as possible (as many students did). Perhaps, however, the clearest evidence regarding the veridicality of the tapes comes from a segment of the protocol itself. Protocol 10.1 was recorded the second day of the problem-solving course. The first homework assignment had been given the previous day. The assignment included a problem that asked the students to inscribe a circle in a given triangle. In Items 84–94 of Protocol 10.1, LS and TH discuss their work on that problem:

LS: All right, what are we, what were those sort of things we tried with the triangle one? . . .
TH: But I was trying to do things like bisect this side.
LS: Yeah, I did that.
TH: It didn't work.

The approach that LS and TH took when working the problem in the laboratory was the same approach that they had taken when working the problem in their own dormitory rooms.

A Brief Analysis of Protocol 10.1

This section offers a brief characterization of LS and TH's work on Problem 1.1. The four categories of analysis in the framework are applied in sequence.

RESOURCES

The impression given by Protocol 10.1—an impression strongly supported by the videotape from which it was transcribed—is that LS and TH had very weak backgrounds in geometry. It appears that they had only dim memories of relevant geometric facts and procedures, and that they barely possessed the domain-specific knowledge required to solve the given problem. In fact, their knowledge base was more than adequate to solve it.

Let us begin with the issue of propositional knowledge. After their initial attempts had failed, many students discussed "fitting the circle in"—an indication that they did not really grasp the rules of the game. Items 76–78 of Protocol 10.1 indicate, however, that LS and TH fully understood and accepted the ground rules for construction problems.

TH: [Looking at a failed attempt] OK, so the radius has got to be smaller because it's going outside of this line. So it's got to be a little smaller and the center has got to be up and over, like here—
LS: But how do we—
TH: But I don't know how to do that, without doing it until it comes out right.

After their work on Problem 1.1, LS and TH were asked to work Problems 1.4 and 1.5. They solved both of those problems, which provide the information required to solve Problem 1.1, in a total of less than 5 minutes. Asked how the solution of Problem 1.5 might help to solve the original problem, TH immediately replied: "When you draw the angle bisector, then the center of the circle is on it." She clearly understood how to use the information provided by the proof problems. Moreover, the students had no difficulty performing any of the relevant constructions. In sum, the students possessed all of the resources necessary for solving the problem.

HEURISTICS

Heuristics à la Pólya were not invoked in this protocol. (This should not necessarily be taken as a negative comment.)

CONTROL

The episode analysis in Figure 10.1 (A and B) indicates that LS and TH's executive behavior was better than average. There was a great deal of local assessment during their solution attempt, and suggestions were rarely made without their correctness or utility being challenged. The exploration episodes in this protocol were coherent, and they were appropriately curtailed when the students discovered that those explorations were not leading to progress. Moreover, the justifications that the students generated for bringing episodes to a close were often based on reason as well as empirical evidence. LS and TH did not fail to solve Problem 1.1 because of bad control. They failed because they ran out of ideas to try.

BELIEF

The approach taken by LS and TH to Problem 1.1 was empirical. With one minor exception—that the endpoint of the diameter was not specified in Items 5–9, so that they could not test their diameter hypothesis by construction—Protocol 10.1 evolved in accord with the model of empiricism presented in Chapter 5. Roughly two-thirds of the allotted time was spent with straightedge and compass in hand. Carefully performed constructions were used to generate hypotheses (e.g., Item 10) and to test them (e.g., Items 44–74). There were times in the solution, in particular, in Items 38 and 80, when legitimate mathematical argumentation was used to provide the final rejection of a hypothetical solution. Close examination reveals, however, that such argumentation served as confirmation of a rejec-

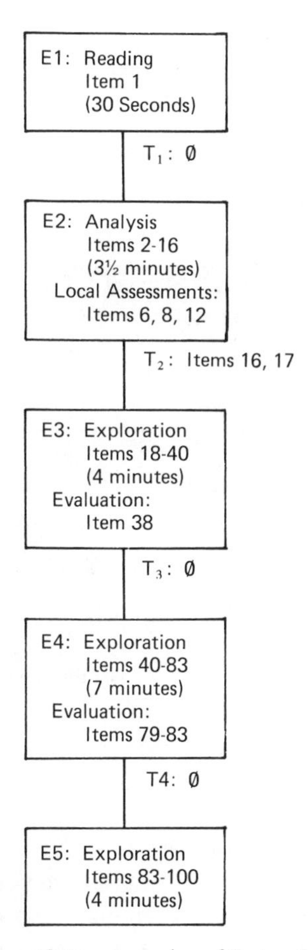

Figure 10.1A A parsing of Protocol 10.1.

tion that had already been made on empirical grounds (in Items 35 and 74–75, respectively). A further discussion of the students' empiricism is found in the sequel.

A Brief Analysis of Protocol 10.2

Protocol 10.2 (Appendix 10.2) was recorded at the conclusion of the 1980–1981 problem-solving course. Geometric constructions were one of the topics discussed in the course. During the semester the students read Chapter 1 of Pólya's (1981) *Mathematical Discovery* and worked roughly a dozen of the construction problems in it. Solutions to many of those prob-

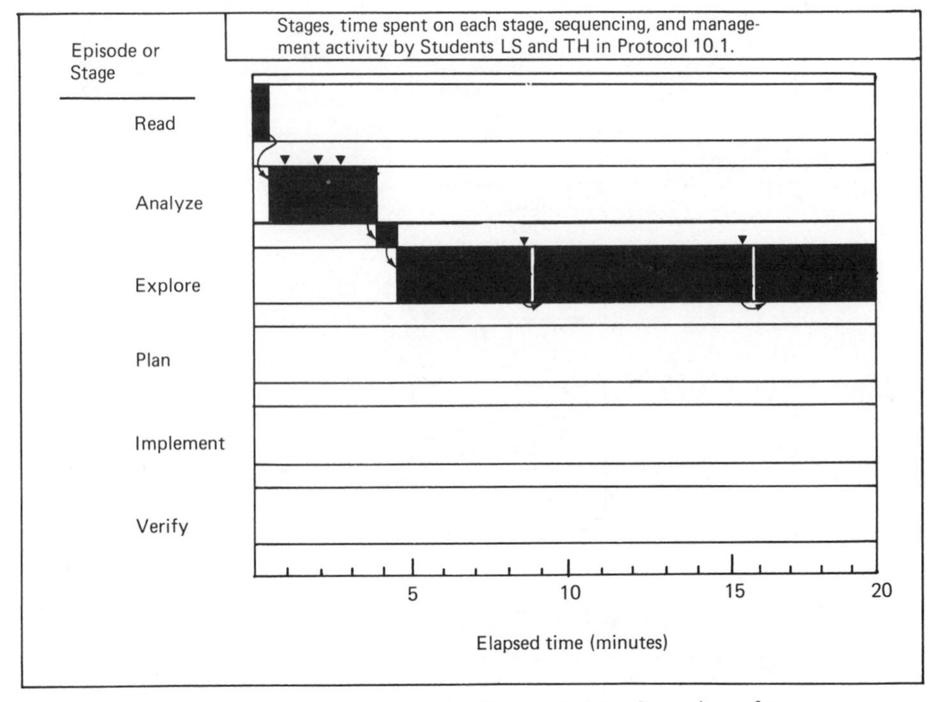

Figure 10.1B A time-line representation of Protocol 10.1. Overt signs of management activity are denoted by inverted triangles.

lems were discussed in class. Proof was often discussed in the course, but in the usual way: "Yes it seems that way, but how do you know it will always be true?" The explicit relationship between mathematical argumentation and mathematical discovery was not discussed. (The issue had not yet been formulated in those terms.)

Aspects of the problem-solving course have been described elsewhere in this book, and this is not the place to provide a detailed characterization of classroom procedures. Additional information can be found in my (1983) *Problem Solving in the Mathematics Curriculum.* In addition to giving a general description of the course, the book provides on pages 42–50 an extended description of a classroom session devoted to the solution of a geometric construction problem. With this as background, we turn to the analysis of Protocol 10.2.

RESOURCES

During the second problem session, LS and TH recalled more relevant information, with more assurance, than they did in the first. More importantly, relevant facts were called into play at appropriate times. As a case in point consider the fact that a tangent to a circle and the radius that meets it at the point of tangency intersect at right angles. In Protocol 10.1, LS and TH almost tripped over that fact by accident (see Items 38–44)—although they did recognize its importance, and use it, once they observed it. In Protocol 10.2 the relevant knowledge was accessed, without difficulty, when it was needed (Items 67–70). Similarly, the students were better able to perform the relevant constructions after the course than before it, and they were more confident about their ability to perform those constructions.

It should be noted, however, that the problem in Protocol 10.2 was a good deal harder than the problem in Protocol 10.1. Moreover, as the discussion above indicated, LS and TH's somewhat shaky resources were a discomfiting but not seriously disabling factor when they worked Problem 1.1 in Protocol 10.1.

HEURISTICS

There is a telling difference in the students' performance at the heuristic level. Students LS and TH's ability to implement a range of heuristic strategies clearly contributed to their success in Protocol 10.2. Such heuristic usage was evident, for example, in the way that they "took the problem as solved," drawing a picture of the goal state in order to determine what properties it had (Items 14 ff.):

LS: Umm, should we try and draw it, maybe, how it would be to see what the relationship of *C* is to the two circles, since that's not drawn. . . . Just draw [i.e., sketch] it—you don't have to use the compass.

LS and TH also examined extreme cases (Items 34–46). They considered obtaining only the partial fulfillment of the conditions as a stepping stone to a solution (Item 52) and used other heuristic strategies as well. One or more of these heuristics, properly used, might have led to success in Protocol 10.1.

CONTROL

The students' control behavior in Protocol 10.2 was quite good, as it was in Protocol 10.1. The episode analysis of their work is given in Figure 10.2. In Protocol 10.2, LS and TH monitored and assessed both the state of their knowledge and the state of their solution with some regularity (Items 10, 23–25, 57, 71), and they avoided the kinds of wild goose chases that often guaranteed failure for less sophisticated students. In fact, control behaviors were a positive force in the evolution of their solution. The two students were well aware of the passage of time as they worked. They imposed strict limits on the amount of time they allowed themselves to spend with straight-

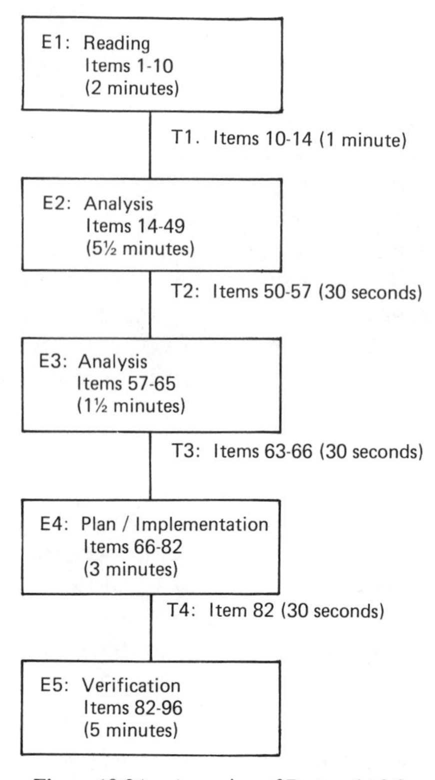

Figure 10.2A A parsing of Protocol 10.2.

edge and compass in hand; see, for example, Items 20–25 and 74–78. As another example, the decision to "bisect" a line segment (Item 63) using a ruler,

LS: All right—wait, we're not allowed to use a ruler, but—yeah, divide it in half.

was a control decision. The students knew that they were capable of bisecting the segment with straightedge and compass if need be, but using the ruler saved time.

BELIEF

LS and TH's behavior in the second problem session indicates that a marked shift had taken place regarding the roles that they assigned accurate constructions and mathematical argumentation in the solution of geometric

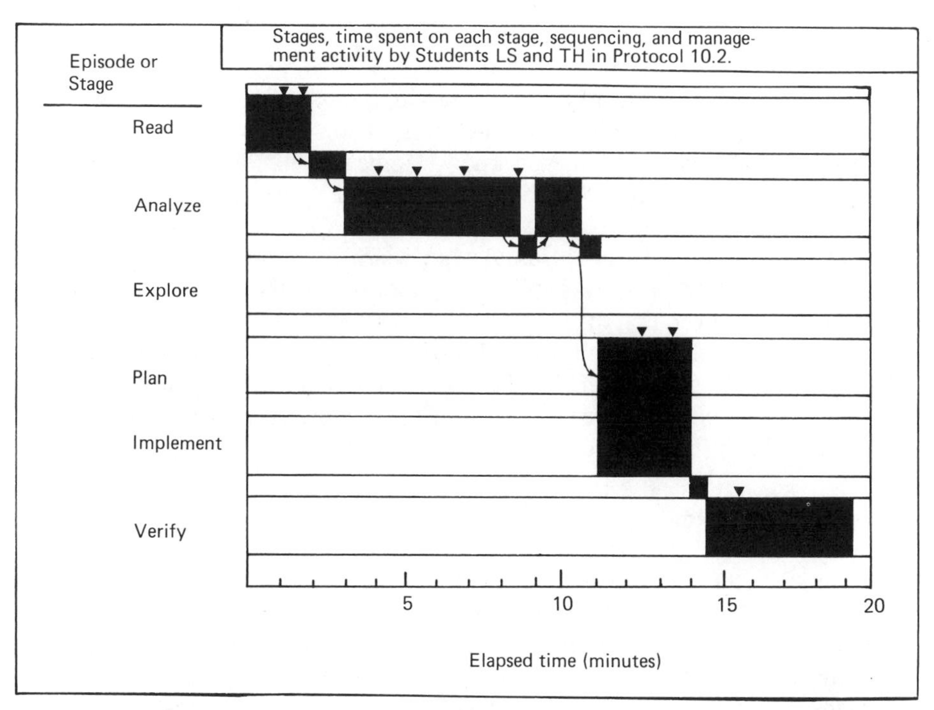

Figure 10.2B A time-line representation of Protocol 10.2. Overt signs of management activity are denoted by inverted triangles.

construction problems. While the two students still used good sketches as a source of ideas, they no longer depended on carefully done constructions as the sole source of hypotheses; while they still tested the plausibility of hypotheses by construction, they no longer depended on carefully done constructions as the sole means of "proof."

The shift in LS and TH's perspective regarding empiricism as a source of inspiration was apparent early in Protocol 10.2. As LS noted in Item 20, "you don't have to use the compass" to see whether an idea makes sense. Nonetheless (Item 23), the compass did help "to see what it would look like more accurately." This more balanced view of empiricism held throughout the problem session; LS and TH were content to argue from rough sketches when those sketches sufficed to make an argument. Thus all of Items 28 – 50 were based on very rough sketches. Yet LS and TH also exploited carefully implemented constructions when there was something to be gained from a higher level of precision. For example, in Item 60 they profited from the accurate sketch that they had made in Item 23. A quick measurement on that sketch substantiated (but did not prove!) their intuition that the midpoint of the line segment between the two given points A and B would play an important role in their construction.

The segment of Protocol 10.2 starting with Item 71 indicates the students' altered view of proof. The comment by LS, "I don't know why this works, I mean, I just seem to see it, you know," indicates that she was no longer content to accept without further examination a construction that simply looked right. Student TH's response, "I think we can do it with similar triangles and things, so let's just make sure it works," shows that preliminary testing by construction was still very important to the students. It also indicates, however, that the accuracy of a construction was no longer considered to be proof of its validity. "Proof by construction" was put to rest (forever, one hopes) in Item 78. Under time pressure, TH suggested that LS abandon the careful construction and "just draw [i.e., sketch] it—it'll work." Then they proceeded to prove that it does.

A Brief Discussion

Between the two problem sessions there were changes in LS and TH's behavior along all four dimensions of the analytical framework. It is clear that improved resources, better heuristic usage, and efficient control all contributed to the students' success in Protocol 10.2 A good case can be made, however, that LS and TH's success in the second problem session rested on changes in their beliefs regarding the roles that accurate constructions and mathematical arguments play in the solution of geometric construction problems.

When LS and TH recorded Protocol 10.1, they had at their disposal adequate subject matter knowledge to solve Problem 1.1. Some of their resources were a bit on the shaky side, but that was a minor factor in determining their success or failure. The issue of importance is that the skills they could have used to solve the problem were not called upon. As long as the students had an empiricist perspective regarding geometrical construction problems—that is, as long as they considered performing accurate constructions to be the primary and perhaps the sole means of gaining information about the solutions to such problems—that perspective dominated their solution attempts and determined which aspects of the knowledge that they might have accessed would actually be used. LS and TH did not use much of their geometric knowledge in Protocol 10.1 because it did not occur to them that it would be useful to do so.

The shift in belief systems between the two problem sessions allowed for significant differences in resource allocation during the second problem session. "Making accurate constructions" no longer had the same status for LS and TH when they recorded Protocol 10.2. Construction still served as a major source of information, but it was no longer the almost exclusive focus of attention. It was considered a valuable method that might be employed, if and when it seemed appropriate to do so. This being the case, control decisions could be more effectively made; the problem solvers could choose among a range of methods that might be of assistance. This being the case, more of the facts, procedures, and strategies potentially at LS and TH's disposal could be accessed and utilized. Thus the shift in belief systems was much more than a mere change in perspective that accompanied improved performance. It was a significant change that enabled LS and TH to use much more of what they knew. In doing so it may well have provided the key to their success.

The Strength of Empiricism: More Data

As indicated by the discussions of belief in Chapters 1 and 5, the kind of empiricism reflected in LS and TH's attempt to solve Problem 1.1 is nearly universal. This section and the next explore some of the causes of that empiricism. Brief selections are given from two research projects in progress: problem-solving interviews that continue the line of research established in Chapter 5, and studies of instruction currently taking place in high schools. The analyses are preliminary, the conclusions tentative.

Some anecdotal evidence introduces the first line of the research, which is the focus of this section. In early 1983 the undergraduate cognitive science society at the University of Rochester asked me to give a talk about "mathematical thinking." The audience for the talk consisted of about 20 cognitive science majors. These students, mostly juniors and seniors, were among the best students at the university. They had put together their own independent interdisciplinary majors, which were heavily based in the sciences. Their course selections generally included advanced offerings in psychology, biology, and computer science; many of these courses had substantial prerequisites in mathematics, physics, and chemistry.

To set the stage for the discussion of some videotapes, the students were given a series of problems to work. The first two problems were Problems 1.4 and 1.5, which the group worked as a whole. These were solved quickly, and their solutions were left on the blackboard. Two other, unrelated problems (from naive physics) were discussed. Then Problem 1.1 was posed. Students came to the board to make suggestions. The following suggestions were made, in order:

1. The center of the circle is the midpoint of the line segment between the point P and its "opposite," P'.
2. The center of the circle is the "midpoint" of the arc segment passing through P, with vertex V, that lies between the two given lines.
3. The center of the circle is the midpoint of the line segment perpendicular to P that lies between the two given lines.

4. The center of the circle lies at the intersection of the perpendicular to P and the bisector of the angle formed by the two given lines.

When they were asked which of the constructions were correct, the students argued on purely empirical grounds for 10 minutes about the merits of the competing hypotheses.* They engaged in these arguments despite the fact that the information that they themselves had produced, and that served to resolve the issue, was still written on the blackboard. (Note that that information rules out the first three hypotheses.)

In a variation on the experimental protocol described in Chapter 5, a series of interviews with students from local high schools has been conducted. Students in half of the interviews were asked to solve the four problem solved by DW and SP in Chapter 5, in the same order: construction problem, angle bisector problem, and proof problems. In the remaining half of the interviews, the problems were given in the following order:

Problem A The circle in Figure 10.3 is tangent to the given lines at the points P and Q. Using the techniques of high school geometry, prove that the length of the line segment PV is equal to the length of the line segment PQ.

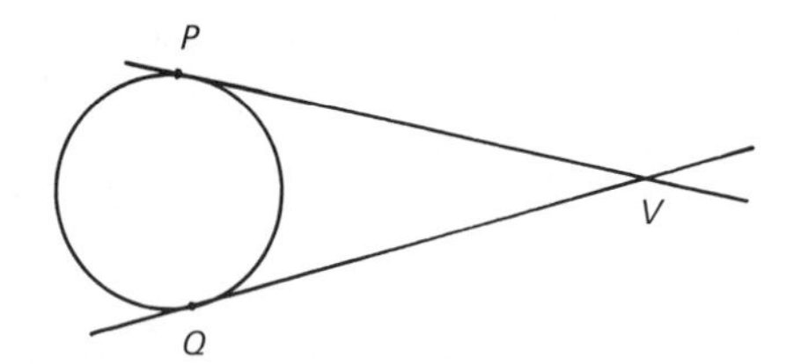

Figure 10.3.

Problem B The circle in Figure 10.4 is tangent to the given lines at the points P and Q; C is the center of the circle. Using the techniques of high school geometry, prove that the line segment CV bisects the angle PVQ.

Problem C Can you remember the procedure for bisecting an angle using straightedge and compass? If so, (1) use the procedure to bisect [a given angle]; (2) explain why this procedure works.

Problem D The same as Problem 1.1.

* To be completely accurate, the one student who proposed hypothesis 4 above, and who could have defended his choice on rational grounds, abstained from the discussion. He sat back, somewhat bemused, and watched the others argue.

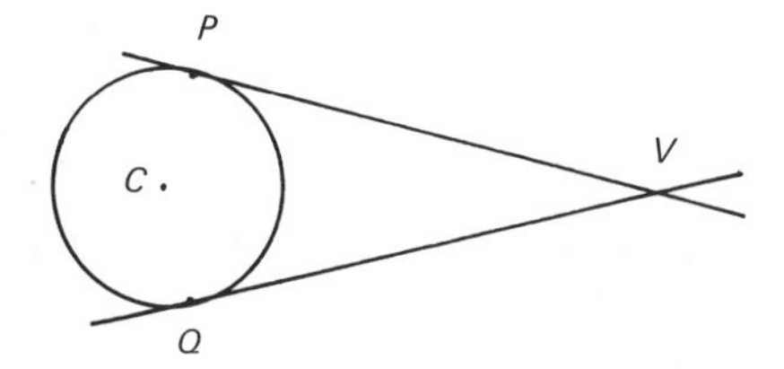

Figure 10.4

In a substantial proportion of the problem sessions (about one in five), the students solved Problems A and B without undue difficulty—and then were at a loss when asked to solve Problem D. In one case the diagram for the construction problem was redrawn directly beneath the completed diagram that a student had used for both of the proof problems. The student was asked if he could see any relationship between the construction problem and the proof problems he had just solved. His response was, "Of course, when you've solved the construction problem the picture will look like the pictures in the other two. But that's all." In a number of other cases the students, like the cognitive science students described above, conjectured and tested hypotheses that flatly contradicted the information they had just derived.

Two points should be stressed here. First, these results are not an aberration. Behavior on the part of students from a local inner-city school, from a highly rated suburban public school, and from a selective regional private school was remarkably consistent. This behavior did not differ from that of undergraduates from Hamilton College and the University of Rochester. Those students, who had come from school districts all over the country, were either voluntarily enrolled in problem-solving courses or had volunteered to be videotaped. They were the successes of our system.

Second, the students' failure to apply the results of the proof problems to the construction problem did not, for the most part, result from failures of understanding. Some of the high school students, including the one quoted above, did not grasp the generic nature of proof arguments. Yet most of the high school students, and virtually all of the college students, did understand the significance of the proof problems. Asked specifically about the implications of Problems A and B, the students described them correctly. Given this understanding, their behavior in generating and testing hypotheses contradicting what they had just proven is all the more striking.

The Origins of Empiricism

This section describes preliminary analyses of data obtained in an observational study conducted during the 1983–1984 school year. Tenth-grade geometry courses were monitored at four high schools in and near Rochester, New York. Observations were made periodically at three of the schools in order to make certain that nothing substantially out of the ordinary took place at the fourth school, where most of the observations took place.*

The bulk of our observations were made in a highly regarded suburban high school. A large percentage of its students go on to college, and the students consistently do well on the Regents examinations. The entire mathematics department at the school was cooperative. One teacher in particular offered us complete access to his classroom and to his students. He gave us permission to videotape any of his classes without prior notice and helped us to make appointments to interview his students. One of this teacher's Math 10 geometry sections became the focus of our study. Class sessions were observed at least once each week, and the complete classroom units on locus theorems and geometric constructions were videotaped for extensive analyses.

The teacher was competent and professional. He respected his students, and they him. His classes were run in a low-key manner. Students paid attention to what was taking place, and discipline was never an issue. The classes were carefully prepared. The teacher had thought about his subject matter presentation and could justify the approaches he took on the grounds of what he thought best for his students. All in all, this was a good class with a good teacher. The teaching practices in this class were typical of those we observed elsewhere.

Classroom observations were supplemented by interviews with students and by the use of questionnaires. In analyzing the data we encountered four recurring dichotomies. Those dichotomies were as follows: empiricism versus deduction, meaning versus form, problems versus exercises, and passive versus active mathematics. The first is discussed at length, the others briefly.

FIRST DICHOTOMY: EMPIRICISM VERSUS DEDUCTION

The segments of dialogue discussed below were recorded during the instructional units on locus theorems and constructions, which occupied about 3 weeks of class time. These units were taught near the end of the year,

* New York has a state-mandated curriculum that assures a large degree of homogeneity in instruction. For that reason all of the courses we observed were similar in content, and the observations recorded here can be extrapolated with confidence to schools around the state. That is, they can be extrapolated to those schools teaching Math 10 Regents geometry. The official New York State Regents curriculum is in transition. A new "integrated sequence" of courses, in which aspects of geometry are taught in grades 9, 10, and 11, is being phased in for statewide implementation in the 1986–1987 academic year.

shortly before the statewide Regents examinations. A great deal of classroom time is devoted to training for these examinations. In addition to counting heavily toward individual students' grades, students' scores on these tests are seen as a reflection of the quality of their teachers, their schools, and their school districts. In consequence, the way any topic is treated on the examinations tends to influence the way that the topic is taught in the classroom. Constructions are treated as follows. There are roughly a dozen specifically designated "required constructions," one or two of which—each worth 2 points out of 100—will appear on the examination. Work on those problems is graded both for correctness (the right sequence of compass markings) and precision. As a result, classroom practice is nearly universal. Students are expected to memorize the constructions and to practice them until they can perform them rapidly without error.

It goes without saying that students are intended to understand the constructions, and that when constructions are first introduced they are usually accompanied by arguments that justify them. However, the focus of instruction quickly shifts from understanding to performance. With that shift comes an emphasis on memorization. To make memorization easier, constructions are presented as step-by-step procedures. To lessen the likelihood that the constructions will be forgotten, constructions are taught shortly before the Regents examinations. To help students reach the desired level of proficiency, a large amount of classroom time is devoted to practice.

During the practice sessions the discussions take on a highly empirical flavor. Selections from classroom discussions of constructions are briefly described below to indicate how strong this empiricism can become. One particularly striking session is discussed at length.

The first day of the unit the teacher made it clear that constructions would be dealt with empirically. Students who came to class without straightedge and compass were sent out to their lockers to get them, and instruction did not begin until all the students had the proper tools in hand. The topic of constructions was then introduced, and the first few constructions (copying a line segment, copying an angle, drawing a line parallel to a given line through a given point) were demonstrated. During the presentation the vast majority of class time was spent on matters of procedure. There were detailed specifications of the sequence of steps required for each construction, and precise demonstrations of how each construction was to be carried out. Since this was the first day on the topic, the students were rather slow at the constructions. They were told that they would have to get faster but that they would not find it difficult: "Mainly with constructions it is all going home and practicing."

During the unit, the importance of memorization was often stressed, as was the degree of speed and precision that the students would need to develop. For example, the teacher made the following comments shortly before the school-wide Math 10 examination on locus theorems and constructions.

> You will have 25 questions. [The length of the examination was 54 minutes. This allowed 2 minutes and 10 seconds per problem.] *You'll have to know all your constructions cold so you don't spend a lot of time thinking about them.* This is where practice at home comes in. . . .
>
> I have to see the arcs. Now, what I will not take will be a lot of trial and error on the constructions. So first you can find some scrap paper or something if you're not sure of it. I have to be able to follow your constructions. If you have 20 different arcs, then I can't follow it. . . .
>
> As long as I can see all the marks so that I can follow your construction, and the construction is correct, then I will not take off if it's [off by] just a minimal distance. But if it's off by more than that, yes.

Amidst this emphasis on memorization and practice, the students eventually lost sight of the rational reasons underlying the constructions. Once the "justifications" for the constructions had been given, they were no longer mentioned; the focus of the classroom sessions was on performing the constructions with accuracy. Comments like the following were made more than once:

> In your constructions, now, make sure that when you draw your straight lines you draw your lines right through the centers of your intersections. . . . If you're off by a little bit, that's what's happening—your lines are not passing through the centers of the intersections. So you want to make sure when you draw the line from the vertex to the point of intersection, you line up the very center of the intersection. If you're slightly off, that's what's happening. You're either below the point of intersection or above the point of intersection.

Though it was not intended, the repetition of such comments and the emphasis on precision gave students the impression that accuracy is the primary criterion used to judge the quality (and perhaps even the correctness) of a construction. Sometimes this message was delivered in indirect ways, as happened one day at the beginning of class. Standard practice was to have students put solutions to homework problems on the blackboard. As usual, six students were chosen, and they went to the front of the room. There was, however, only one large compass for work at the blackboard. The students realized this and said they could not do the constructions. The teacher sent all of the students but one back to their seats. Since rough sketches indicating the sequence of appropriate steps simply would not suffice, the others would have to wait their turns.

At other times the need for accuracy was the subject for humor. For example, the teacher made the following comments while demonstrating the construction for inscribing a circle in a triangle:

[He has just found the intersection of two angle bisectors.] [Now I] have to construct my perpendicular. So from *O* to *AC* you make the two arcs on your line. From those two arcs you mark up two more. You draw your perpendicular, and I've found my point. . . . Ok, now I have the length of the radius. So now using *OD* as your radius, construct the circle. *And if we're lucky we'll get one that will be tangent to every side. If we're not so lucky then we say it was compass error.* [emphasis added]

It was well understood that little compass error would be tolerated on the examinations, and the students knew, of course, that the construction was *supposed* to work. Nonetheless, the reference to "being lucky" undermined the correctness of the construction and reinforced the empirical thrust of the practice sessions. At times the empirical message was even more direct. The following comments were made as the class reviewed the construction for dividing a line segment into equal subsegments:

If you complete these angles and extend the sides far enough so they intersect the other side you should have parallel lines. *And then just to check yourself you measure each line segment and they should be the same length.* [emphasis added]

There is nothing wrong with this suggestion on its own terms. Having students verify the accuracy of their empirical work is sound pedagogical practice, quite appropriate when the students are preparing for an examination. Unfortunately, such advice contributes to the development of students' empiricism. In the context of "practice makes perfect" empirical work, it strengthens the students' growing impression that measurement is the final arbiter of correctness and that a construction is correct when it "looks right."

One of the most interesting classroom discussions took place during the class session in which the following three required constructions were introduced:

Problem C9 Given the arc of a circle, locate the center of the circle;

Problem C10 Given a triangle, circumscribe a circle around the triangle;

Problem C11 Given a triangle, inscribe a circle in the triangle.

For this course the text's treatment of Problems C10 and C11 (Jurgensen, Brown, & King, 1980, p. 295) is given in Figure 10.5.

Construction 10 Given a triangle, circumscribe a circle about the triangle.

Given: $\triangle ABC$

Construct: A circle passing through A, B, and C

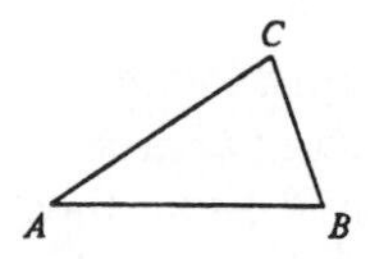

Procedure:

1. Construct the perpendicular bisectors of any two sides of $\triangle ABC$. Label the point of intersection O.
2. Using O as center and OA as radius, draw a circle.

Circle O passes through A, B, and C.

Justification: See Theorem 4-17 on page 143.

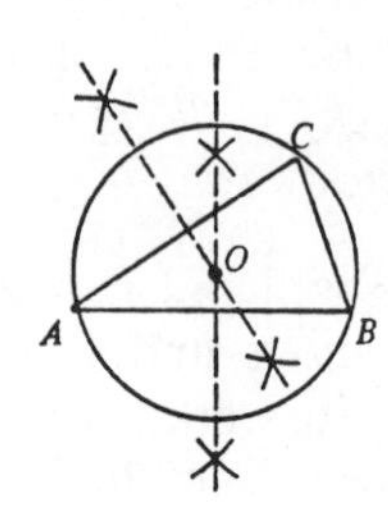

Construction 11 Given a triangle, inscribe a circle in the triangle.

Given: $\triangle ABC$

Construct: A circle tangent to $\overline{AB}$, $\overline{BC}$, and $\overline{AC}$

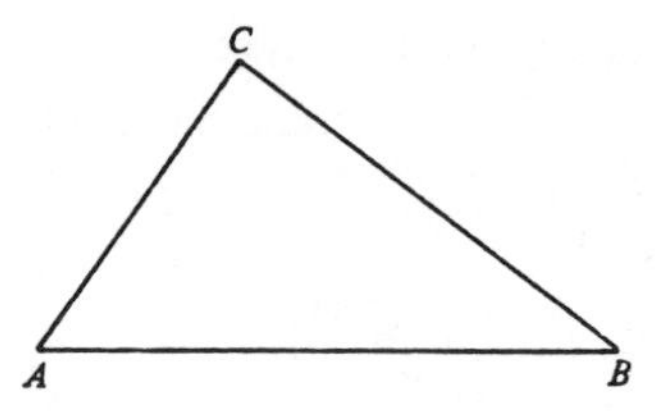

Procedure:

1. Construct the bisectors of $\angle A$ and $\angle B$. Label the point of intersection I.
2. Construct a perpendicular from I to $\overline{AB}$. It intersects $\overline{AB}$ at a point R.
3. Using I as center and IR as radius, draw a circle.

Circle I is tangent to $\overline{AB}$, $\overline{BC}$, and $\overline{AC}$.

Justification: See Theorem 4-16 on page 142.

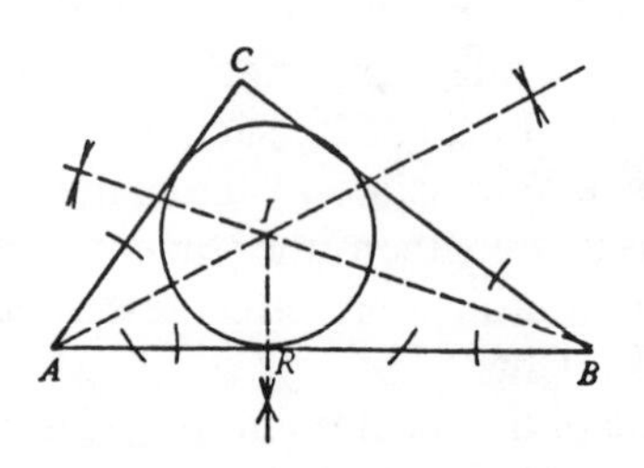

Figure 10.5 Treatment of Problems C10 and C11. (From Jurgensen, Brown, & King, 1980, p. 295.)

The homework assignment the previous day included reading about and practicing the three constructions. The class had been told that the constructions would be discussed in class if the students had problems with them. This was the teacher's standard classroom practice. He fully expected to discuss each construction, but the possibility that a particular problem or construction might not be discussed "if there are no problems" served to pressure students to read the text before class.

The discussion of Problem C9 proceeded as one might expect. The teacher reminded the class that in any circle, the perpendicular bisector of any chord is a diameter. Hence if one draws two chords inside the given arc and constructs their perpendicular bisectors, these two lines intersect at the center of the desired circle. A good deal of time was spent demonstrating the step-by-step implementation of the construction. As the construction was demonstrated, the teacher gave the class pragmatic advice about performing complex multistep constructions such as this one and the ones to follow:

[The teacher draws in two chords.] OK, so if you bisect those two chords, the point of intersection will give you the center.
When you're doing your constructions, finish off one construction before you start the other one. Bisect the one chord [he makes the first set of compass markings], complete the line and so on before you go to the other one. Otherwise you get arcs all over the place, and you get them all mixed up. OK, there's one [he draws in the first diameter]. Then you bisect the other one. . . .

Once Problem C9 has been solved, C10 should be trivial. The sides of the given triangle will be chords of the circumscribed circle, and the problem is to find the center of that circle—exactly the same problem as in C9, with the same solution. In fact, both Problems C10 and C11 provide nice opportunities to illustrate the role that deduction can play in constructions. (Recall, e.g., the expert's solution to Problem C11 given in Chapter 5.) The following transcript gives the full classroom discussion of Problem C10 and the beginning of the discussion of Problem C11.

[The instructor finishes Problem C9.] OK, the next one. To circumscribe a circle about a triangle *ABC*. Any questions on that one? Everybody get that one? No. OK.

In this one we have to construct a circle circumscribed by the triangle *ABC*. If it's circumscribed, that means that the circle has to be where, inside or outside?

[A student replies] Outside.

The circle has to be outside. All right, now, I always have problems remembering what I have to do with this one. I know I have to do one of two things. I know that I either have to bisect the angle or bisect the sides. So what I do is I sketch it and then see what happens. [Draws Figure 10.6] All right, now, for example, if I bisect the angle I know that the distance from the point of intersection to the vertex has to be the same, ok? [sketches Figure 10.7] So, is the distance from here to here [the length of the segment *AD*] the same as from there to there [the length of the segment *BD*]? No. So that

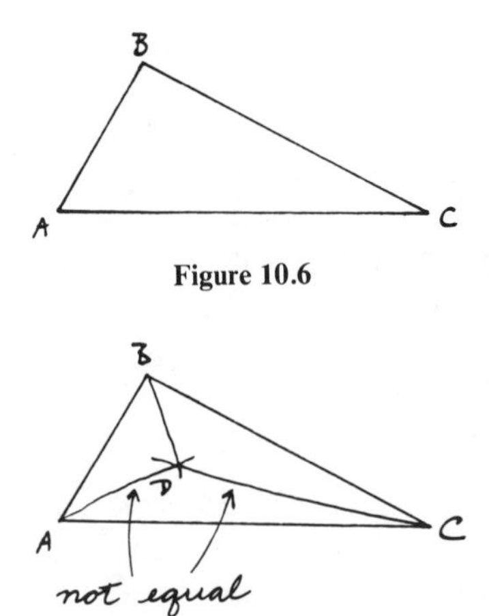

Figure 10.6

Figure 10.7 Line segments *AD* and *DC* are not the same length, therefore D cannot be the center of the circumscribed circle.

means that I can't bisect the angle because it will not give me the radius that I need. So that means then that I have to bisect the sides. Ok, for this one, then, step number one: Bisect any two sides.

Now the other thing I'd better point out is this. Suppose that you start doing this. Some of you like to draw nice-looking triangles. [draws Figure 10.8A] Now that looks to be what kind of a triangle? [A student replies:] Isosceles.

An isosceles, and if I drew it nice enough—equilateral, right? What do you think is going to happen if I bisect the angles? [Draws Figure 10.8B, points to the segments *AD* and *BD*.]

[A student replies:] They look the same.

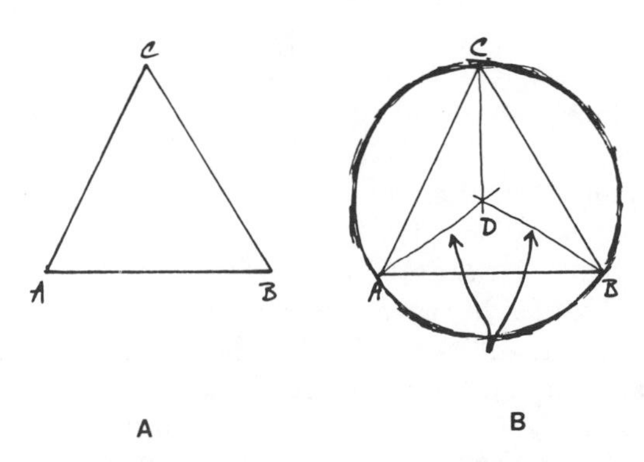

Figure 10.8 In part B, line segments *AD* and *DB* look the same; the point *D* can serve as the center of the circumscribed circle.

The distances will be the same, all right? So, therefore, you have to make sure that, if you're going to see which one's going to work, you'd better pick a triangle that's not equilateral. OK, because otherwise the bisectors won't work. Although when you do the construction you'll find out soon enough that it's not going to work anyway.

So [step number one; see above] bisect any two sides. Find the point of intersection. Again, do one side first and then go on to the other one. OK, so you have the point of intersection of the perpendicular bisectors. Now, step number two: Using O as the center and OV or OC or OA as the radius, construct a circle. Using O as the center, and the distance to any vertex as your radius, you construct the circle. You know that the circle should pass through every vertex if it's to be circumscribed above the triangle, so you can use the distance from O to B as the radius. Then again it will be obvious to you right away if you made a mistake because that circle is going to come out either way outside or way inside if you're way off. Any questions on that one?

There is a brief discussion of an upcoming test, after which the class returns to the discussion of required constructions.

OK, the next one. Construct a circle inscribed in a given triangle. Now to be inscribed that means that the circle must be tangent to every side. In the other case [points to the second construction], we bisected the sides. What do you think we're going to do in this one?

[A student replies:] Bisect the angles.

Bisect the angles. OK, you have a triangle ABC. So you go ahead and bisect any two angles. . . .

There is no need to elaborate on what took place in this session, which speaks eloquently for itself. Rather, it is important to stress two points in concluding this discussion.

First, the approach to the subject matter taken here can be seen as entirely reasonable given the short-term priorities of instruction. As noted above, construction problems are worth only a few points on the statewide examination. Required constructions, like the proofs of required theorems, are universally considered objects appropriate for memorization. Students will enter the examination having tried to memorize all of the required constructions. It is safe to assume that they will recall that the two "circle and triangle" constructions call for finding the points of intersection of important line segments—either angle bisectors or perpendicular bisectors. The short-term problem, then, is to remember which set of important line segments goes with which construction. The empirical crutch suggested by this teacher provides a quick way of determining the answer.

Second, the classroom sessions described above reflect an approach to constructions that is far more the rule than the exception. That approach contributes directly to the kind of empiricism documented earlier. By the time that the students encounter geometric constructions at the end of the school year, they have already had extensive experience with mathematical deduction—but usually in the form of "proof." It was suggested in Chapter 5 that students come to see such argumentation as explanations (to them) of results already known (to their teacher and others). That is, the kind of argumentation involved in proof is seen as confirmation of what is already known to be true; it is therefore not perceived as being relevant to discovery. A research project conducted in parallel with the classroom analyses tends to substantiate this suggestion. Questionnaires investigating students' perceptions regarding the nature of mathematics and mathematics teaching were administered to 230 students from four secondary schools in the Rochester area. The questionnaire used a 4-point Likert Scale (1 = "very true," 2 = "sort of true," 3 = "not very true," 4 = "not at all true"). Of the 30 questions dealing with the nature of mathematics in general and geometry in particular, 5 questions had mean responses of either less than 2 or greater than 3. Two of those five questions were the following:

The math that I learn in school is mostly facts and procedures that have to be memorized (mean = 1.75).

When I do a geometry proof I can only verify something a mathematician has already shown to be true (mean = 1.93).

These data indicate that prior experience has not lead students to sense that mathematical deduction might serve as a tool of invention. A third question with extreme scores suggests that the unit on constructions did little to change their impression:

You have to memorize the way to do constructions (mean = 1.87).

In the classroom we observed, the emphasis on performance ultimately resulted in the students losing sight of the importance of the justifications for the constructions. (Note, incidentally, that the commonly used term *justification* can be interpreted by students as suggesting that the constructions are *reasonable,* rather than guaranteeing that they are correct.) The more the students practiced, the more those justifications receded into the background and took on the status of mere suggestions pointing to the choice of operations—operations that would be validated when the indicated constructions met the requisite standards of accuracy. Thus classroom practice increased the distance between deduction and invention; it strengthened the impression that accuracy serves as empirical proof of a construction's correctness. Contrary to the intention underlying their instruction, these students were trained to be empiricists in their geometry classroom. In view of the data regarding the prevalence of empiricism presented in Chapter 5 and the previous section, there is not much reason to suspect that they are atypical.

SECOND DICHOTOMY: MEANING VERSUS FORM

Some apparently innocuous comments that appear in Protocol 10.1 provide an introduction to this second dichotomy. The following exchange (Items 56–57) took place when LS and TH were convinced, albeit briefly, that they had found a way to solve the given construction problem:

TH: So we can bisect this to find the center, right? So call it center *C*. [She pauses.] *Maybe we should have done our steps.* [emphasis added]

LS: [Referring to the way they had proceeded on the conjecture up to that point] That's all being unmathematical, completely disorganized.

These comments capture LS and TH's perception of what mathematics is all about. It is a perception based in the way that they were taught to do mathematics, and it shapes their behavior. TH's comment about "doing our steps" is a reference to the formal procedures required of students in most high school geometry courses. There is a strictly prescribed form for making mathematical arguments. To write a proof, one draws a large T on the page. The left-hand side of the T contains the "statements" column. One makes a numbered list of statements according to the following set of rules. Only one statement is written on each line. The list of statements begins with the information given in the problem, which is followed in sequence by other statements that are known to be true or that can be derived from statements that already appear in the list. It concludes with the statement that one is supposed to prove true. On the right-hand side of the T one finds the "reasons" column. Opposite each of the entries in the statements column, numbered in the identical way, is the formal justification for that statement. One is allowed and even encouraged to use abbreviations when making entries in the reasons column, as long as the abbreviations are in prescribed form. Thus ASA, meaning "two triangles are congruent if two angles of each triangle, and the sides between them, are congruent," may appear on the right-hand side of an argument. So may CPCT, meaning "corresponding parts of congruent triangles are congruent." A typical formal argument is given in Figure 10.9.

Learning to express oneself in the proper formal way was a major focus of instruction in all of the classrooms we observed. (Among other things, proper form is required on the statewide examinations.) Teachers devoted large amounts of class time to specifying the proper procedures ("Make sure you number your statements. Put the givens first. Make sure you write the

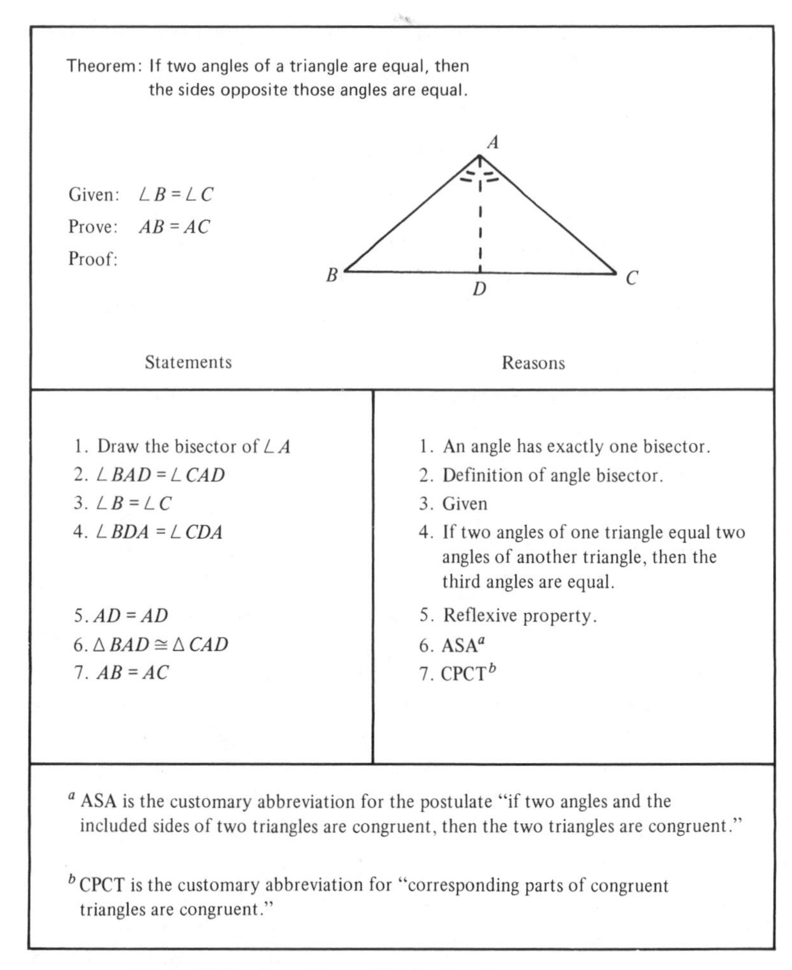

Figure 10.9 An archetypal high school geometry argument.

reasons down. . . ."). When students produced arguments that were mathematically correct but that were not expressed in the right form, those arguments were severely downgraded. In consequence, students learned to focus their attention on matters of form as much as on matters of substance. For some students, in fact, the lesson learned from instruction was that form is far more important than substance. At one particular high school, a number of students managed to solve proof Problems A and B (described in the previous section) in just a few minutes and then spent more than three times as long writing up the answer in the "statements and reasons" format that appears in Figure 10.9. These students had developed the perspective indicated in the comment made above by LS: Failing to write things in the correct form is seen as being "unmathematical." In sum,

As a result of instruction that focuses heavily on writing results in specific ways, and grading procedures that penalize students for not expressing otherwise correct answers in those ways, students can come to believe that "being mathematical" means no less—and no more—than expressing oneself via the prescribed forms. One result of devoting excessive time and attention to the mastery of the forms and procedures of mathematics is that the underlying mathematical substance can be completely obscured. Patterns of productive reasoning, and even the reasons for engaging in such reasoning, can be completely lost.

THIRD DICHOTOMY: PROBLEMS VERSUS EXERCISES

Virtually all of the problems that students are asked to solve during their precollege and lower-division mathematical careers are not really problems at all: They are exercises that can be solved in short order. The geometry classes that we observed provide representative data.

As noted above in the discussion of empiricism versus deduction, the schoolwide examination on the topics of locus problems and constructions contained 25 questions that were to be worked in a total of 54 minutes—an average of 2 minutes and 10 seconds per "problem." This is typical of such examinations and a reflection of standard instructional practice. A sense of that practice can be obtained from the homework assignments.

A large percentage of class time in most of the courses we observed was devoted to having students present the solutions to homework problems at the blackboard. Students worked the problems at the blackboard and were helped by the teacher if help was needed. Homework assignments for those classes generally consisted of 15 or more problems, and an assignment of 25 or 30 problems was not unusual. Thus the expectation was that students would be able to explain the solutions to the assigned problems in just 2 or 3 minutes. This, of course, was while they were learning the material; they would need to be faster on the examination. Rarely did a teacher's presentation of a problem solution last more than 5 minutes, and a 10-minute "problem" was truly extraordinary. Students were never given the impression that one might spend hours (much less days, weeks, or months) working a problem. Rather, the impression was that subject matter in mathematics could be mastered in bite-sized chunks. In the long run this kind of classroom practice has serious consequences.

Students work thousands upon thousands of "problems" in their mathematical careers, all of which could be solved in just a few minutes—if only they knew the right procedures. Rarely if ever do students see problems of any other kind. They are deprived of the opportunity to make slow progress on complicated problems over periods of time, and are thus deprived of the expectation that such progress can be made. They ultimately come to expect that all of the mathematics problems that they encounter will either be solvable within a few minutes or not at all. Confronted with problems that do not yield to their attempts, they act accordingly: After only a few minutes of unsuccessful attempts, they quit.

Students are hardly likely to discover the utility of problem-solving heuristics or the need for good control when working 2-minute exercises. When classroom demonstrations focus almost exclusively on the drill-and-practice mastery of small bits of subject matter, it is unlikely that students will even be aware of the role or importance of such higher-order skills. The fact that most students are ignorant of elementary problem-solving strategies (see Chapter 7) and that they have not developed basic control strategies (see Chapter 9) should come as no great surprise.

It should be stressed that these results apply across the board in mathematics instruction, and not just in the small collection of classes we sampled. For example, Carpenter, Lindquist, Matthews and Silver (1983) describe the results of the Third National Assessment of Educational Progress (NAEP) Mathematics Assessment. The NAEP Assessments have been administered to carefully selected representative national samples of students at ages 9, 13, and 17. Over 45,000 students participated in the third (1982) assessment. "The results supply a valid and reliable measure of American students' achievement in mathematics" (Carpenter *et al.*, 1983, p. 652). The NAEP data on problem solving have been summarized as follows.

> As in previous assessments, a marked discrepancy was noted between performance on routine problems and problems that required some analysis or nonstandard application of knowledge or skills. . . . Data from several of the problems suggest that students do not carefully analyze the problems they are asked to solve. The errors made on several of the problems indicate that students generally try to use all the numbers given in a problem statement in their calculation, without regard for the relationship of either the given numbers or the resulting answers to the problem situation.
>
> Data from this assessment suggest that students may not understand some of the problems they do solve. Most of the routine verbal problems can be solved by mechanically applying a computational algorithm. In such problems, students have no need to understand the problem situation, why the particular computation is appropriate, or whether the answer is reasonable. However, when students are given non-routine problems in which those and other considerations are important, they do not do as well. . . . For example, 13-year-olds were given the following problem:
>
> > An army bus holds 36 soldiers. If 1128 soldiers are being bussed to their training site, how many buses are needed?
>
> Approximately 70 percent of the students performed the correct calculation, but about 29 percent gave the exact quotient (including the remainder), and another 18 percent

> ignored the remainder. These answers reveal a failure to understand the problem situation and the nature of the unknown. Those who gave the exact quotient response ignored the need for a whole number of buses, and those who ignored the remainder failed to provide transportation for all the soldiers. (Carpenter *et al.*, 1983, p. 656)

The origins of this behavior may be seen in the fourth dichotomy.

FOURTH DICHOTOMY: PASSIVE VERSUS ACTIVE MATHEMATICS

Most instruction in mathematics presents students with results which, from the students' perspective, are simply accompanied by their justifications. ("Let me show you why this works," says the teacher.) The rationale for a given procedure generally precedes the bulk of practice on that procedure, and—as we saw in the discussion of the justifications for constructions in the discussion of the first dichotomy—the importance of that rationale can be lost completely during the practice sessions. This is hardly unique to geometry. In algebra, for example, students will be shown the procedure for completing the square at the beginning of the unit on the quadratic formula. After the formula has been derived, it will be used extensively, and the rationale for it recedes into the distance. The few times that students are held responsible for "justifying" such a procedure, it usually suffices to memorize the justification. This happens with the quadratic formula and "required proofs" in high school geometry courses; it happens at the college level when students are asked to reproduce the arguments that justify the rules of differentiation (e.g., the "quotient rule") in their calculus classes.

The geometry classes provide clear evidence of the importance attributed to the rote memorization of procedures. As noted above, students were told to memorize both the required proofs and the required constructions for the Regents examinations. ("You'll have to know all your constructions cold so you don't spend a lot of time thinking about them. This is where practice at home comes in.") Indeed, it was not intended that students spend a lot of time thinking: Procedures were presented in step-by-step fashion, and were intended to be implemented in that way. Recall these excerpts from the presentation of the Problem C10, circumscribing a circle about a triangle:

> OK, for this one, then, *step number one:* Bisect any two sides. . . . Find the point of intersection. Again, do one side first and then go on to the other one. OK, so you have the point of intersection of the two perpendicular bisectors. Now, *step number two,* using O as the center . . . [emphasis added]

This presentation was typical. Students were expected not only to reproduce the correct constructions, but to do so one step at a time, in the right order. For example, the following dialogue took place in the discussion of this problem:

In Figure 10.10 below, construct a line that passes through the given point *P* and is parallel to the base *AB* of the given triangle *ABC*. [This problem] is

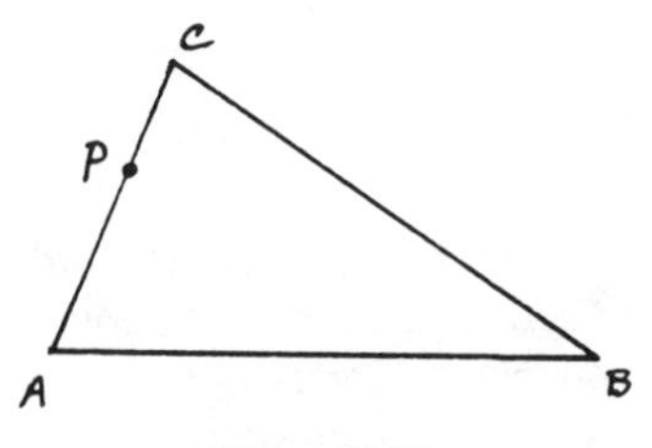

Figure 10.10

saying that you need a line through point *P* parallel to *AB*. So it's the same problem [as the required construction for constructing a line parallel to a given line through a given point, which was problem number 1 on the assignment]. The only difference is that they made a triangle out of this one. So if you have number 1 already worked out, then you have this one already worked out. OK? *In your construction, what is the first step?*

[The student replies:] Draw a transversal.

Draw a transversal. In this problem the transversal's already there. So they save you a step. You already have a point passing through point *P*. That's side *AC*. So if you have a line passing through there, why come up with a new transversal? *In your construction what is step number 2?*

[The student replies:] Copy the angle.

Copy the angle. So you want to copy angle *A* at point *P*. You want a pair of corresponding angles that are equal. OK, so: *Step number one, you mark up the arc at point A and mark up the arc at point P. Step number 2, measure the width of the angle, and you go up and copy the same width. If you connect the two points you will have a line that is parallel to the given side AB.* [emphasis added]

Once again it must be stressed that this teaching practice is quite general. One of the most widely used series of review texts for the Regents examinations is *Barron's Regents Exams and Answers.* Advertisements on the back cover of the 1972 *10th Year Mathematics* text, in large boldface letters, proclaim the following. "Barron's Exams and Answers, written by master teachers, are the best way to review and prepare for all exams. . . . *Students like these books because they offer step-by-step solutions*" (emphasis added). The step-by-step approach is most common. It may produce good short-term results, but the long-term results can be devastating.

Years of being trained to use mathematics that the students do not understand, and years of passively reproducing mathematical arguments handed to them by others, take their toll. Many students come to think of mathematical results as intact, preexisting truths that are passed on "from above." They come to think of mathematics as being beyond the scope of ordinary mortals like themselves. They learn to accept what they are taught at face value without attempting to understand it since such understanding would necessarily be beyond their ability. Moreover, the students come to believe that whatever they forget must be given up as lost forever; not being geniuses, they have no hope of (re-)discovering it on their own. Save for the lucky few who learn (or are taught) that things can be different, these students become the passive consumers of "black box" procedures. Like the students who arrived at the incorrect answers on the NAEP busing problem discussed above, they often apply those procedures without careful analyses of problem situations and without careful attempts to match the operations they choose to the problems they work. Even when they get the right answers, there is some question as to how much of the mathematics they really understand.

Summary

This chapter explored the origins and the strength of students' geometric empiricism. It also touched briefly upon the origins of other beliefs that students have about mathematics—beliefs that, like empiricism, shape their behavior in mathematical situations. The chapter began with a final pair of "before and after" comparisons.

Protocols of the students LS and TH working two geometry problems were examined. The first protocol, recorded before a problem-solving course, reflected the pattern of empiricism documented in Chapter 5. Like most of the students who have studied a year of high school geometry, LS and TH possessed more than enough subject matter knowledge to be able to derive a solution to the construction problem that they were asked to solve. Like almost all of those students, they failed to invoke their deductive knowledge and engaged in a series of trial-and-error attempts (which were unsuccessful). In the second protocol, recorded after the problem-solving course, LS and TH solved a far more difficult problem. A close examination of their performance indicates that the primary determinant of their success was a change in belief; having put "proof by construction" in its place, they were able to take advantage of the cognitive resources potentially at their disposal in ways that would otherwise have been impossible.

The balance of the chapter presented results from studies in progress. The second section described preliminary analyses of problem sets in which students were first given "proof problems" and then asked to solve a construction problem whose answer was guaranteed by the results of the proof problems. In many of the problem sessions, the students solved the proof problems without difficulty and then found themselves at a loss when confronted with the construction problem. In about 20% of the sessions, the students solved the proof problems and went on to try constructions that flatly contradicted the results that they had just proved.

The third section, comprising the bulk of the chapter, described the results of ongoing observational studies in secondary schools. The data suggest that many of the counterproductive behaviors we see in students are learned as unintended by-products of their mathematics instruction. A very strong classroom emphasis on performance—on memorizing constructions and practicing them until they can be performed mechanically with a very high degree of accuracy—ultimately results in the students losing sight of the rational reasons for the correctness of those constructions. So much emphasis is placed on precision, and so many empirical hints are provided for making sure that constructions look good, that students are given the impression that accuracy is the primary determinant of a construction's correctness. The presentation of constructions as step-by-step procedures to be memorized gives students the impression that mathematical knowledge consists of preexisting procedures that are passed on from above, to be used without understanding as "black box" methods. A strong emphasis on writing up mathematical arguments in a rigidly prescribed form, including the practice of severely downgrading students' papers where the work is correct but is not expressed in the proper form, contributes to the impression that form is more important than substance in mathematics—or as a student put it, that "being mathematical" means "doing your steps." When their sole experience with homework assignments consists of working 20 – 30 problems that can all be presented at the blackboard during the next day's class, and their sole experience with tests consists of working 25 "problems" intended to be solved in an average of 2 minutes each, students are led to expect that all the mathematical problems they encounter will be solvable within a few minutes—or not at all. Moreover, 2-minute exercises hardly provide an appropriate experiential base for the development of higher-order problem-solving skills.

Our observations were of good classes taught by good teachers, using standard teaching practice. The difficulties we found are not unique to a local area, as indicated by the striking homogeneity of empiricism found in the undergraduates at Hamilton College and the University of Rochester—successes of our educational system who come from high schools spread over the United States. The difficulties are not unique to geometry, as indicated by data from the National Assessment of Educational Progress. These difficulties point to a major issue in the teaching of mathematics. They indicate that the substance of the subject matter that we teach — the focus of virtually all of our attention in the classroom — determines only part of what our students learn. The ways that we teach mathematics, and the lessons that students abstract from their experiences in doing mathematics, are equally important in shaping their mathematical behavior. Understanding how that behavior is shaped is an important and difficult task.

Postscript

The sometimes harsh realities portrayed in this chapter are not intended to leave a pessimistic impression. As indicated by the pretest-to-posttest comparisons that appeared throughout Part II of this book, there are reasonable empirical grounds for optimism. More importantly, there is reason to believe that at least some small progress is being made in understanding the nature of mathematical thinking.

A huge amount of work lies ahead. We have barely begun to scratch the surface of any of the categories described in Part I: resources, heuristics, control, and belief. Even in the best of laboratory situations, these are complex and elusive. We are only beginning to develop rigorous ways to deal with issues of control; not even that claim can be made with regard to belief. Yet the importance of both phenomena is beginning to emerge clearly, and both are beginning to be better defined. Some of the factors that shape mathematical behavior are also beginning to emerge with some clarity. These three particular areas — metacognition, belief, and "cognition in context" — are all rich and ripe for exploration. Those explorations should make for interesting and exciting times in the years to come.

Appendix 10.1: Protocol 10.1

The following problem was worked the first week of instruction, by students LS and TH (college freshmen who had completed one semester of calculus).

You are given to intersecting straight lines, and a point P marked on one of them, as in the figure below. Show how to construct, using a straightedge and compass, a circle which is tangent to both lines and has the point P as its point of tangency to one of the lines.

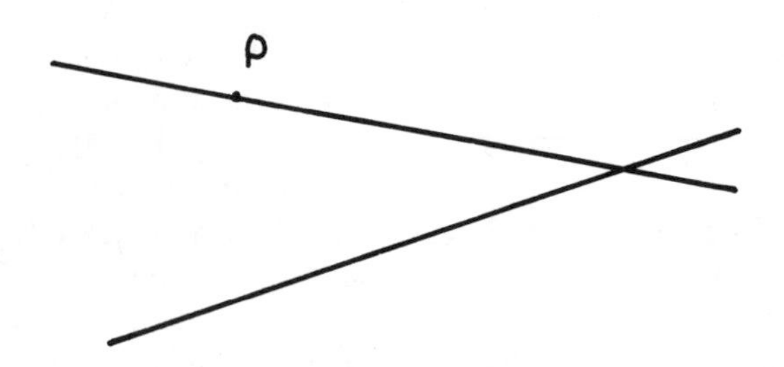

1. TH: [reads the problem] Oh, OK. What you want to do is that [sketches in a circle by hand], basically. OK, how?

2. LS: Now, OK, we have to find the center.

3. TH: Of what?

4. LS: Of the circle. We are trying to find the circle, right? If we did that then we could—oh, and the radius of course.

(ERASES CIRCLE)

5. TH: All right, well, we know the point of tangency on this line is going to be right here [points to *P*]. What we need to find is where the point of tangency is going to be on this other line, I think. So we can find the diameter, in which case we can find the center.

POINTS ABOUT HERE

6. LS: Is that—that's not necessarily true, is it? Is it true that if you have a circle like that [see right], and then that [points with finger] would be the diameter. You know what I mean? Or maybe you couldn't have it that way—

7. TH: The circle has like—no, you don't have a diameter running up through there. No, we have to find the diameter from the point of tangency on this line to the point of tangency on this line, wherever it lies.

8. LS: No, wait; the point of tangency, the point of tangency here, would the line connecting those two points be the diameter? It seems that you

could maybe construct one where it wouldn't always work.

9. TH: Wait, but see, I don't know, we're not drawing it [i.e., sketching it] the right way.

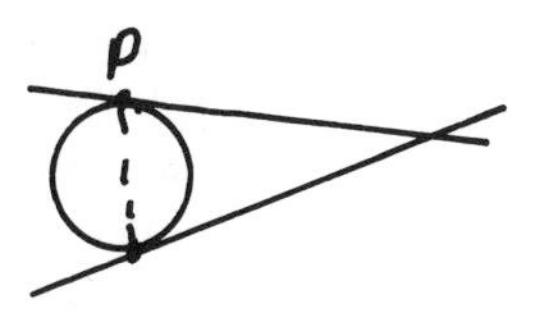

10. LS: Wait, do you want to try drawing it [with the compass] and see—

[Two and a half minutes elapse in empirical work. A reasonably accurate drawing results.]

11. LS: So, maybe it looks like it might be opposite, see?

12. TH: But would that be true for any triangle? Oh, but see—

13. LS: I'm confused. I don't think it would be. Let's say you had your radius over here and you went like that. I don't think that could be—OK, I think there could be, there is a possibility.

14. TH: Remember on the first problem sheet we had to inscribe a circle on a triangle? Could you do that? I couldn't.

15. LS: I couldn't either.

16. TH: We're in pretty sad shape. But just say we draw a triangle even though we don't know how to do it. We will draw a triangle anyway.

17. LS: So how's that going to help?

18. TH: Because we don't have to inscribe it actually. We just have to have something to help us [visualize it]. [draws an apparently arbitrary third line]

19. LS: Although—

20. TH: Does that do anything?

21. LS: Not at this point, I don't think. Maybe further along if we need a radius we could—but I don't think it does anything now.

22. TH: We've gotta do something. With what we have, you just can't do it,

right? We don't have enough lines or whatever there.

23. LS: OK, we need a center and a radius. So how do we locate the center? It has to do with, I think it has something to do with, could we do this?

24. TH: No, maybe you have an equilateral triangle.

25. LS: Wait, let me just try this. [Begins to expand compass.]

26. TH: What are you doing?

27. LS: Don't you want to see if it's true? If you have a center way out there, because it may not connect. Don't you see? [sketch at right]

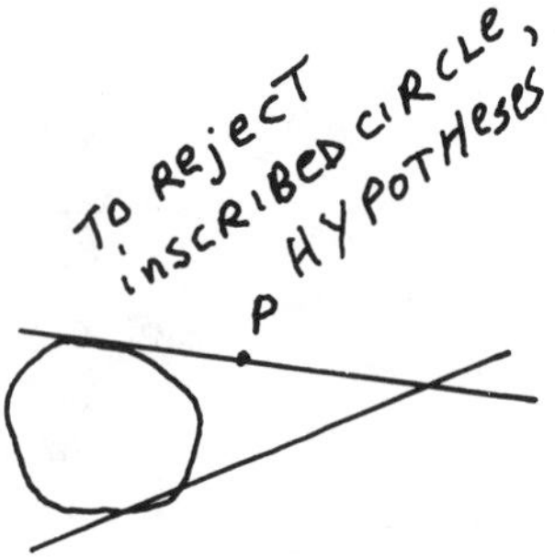

28. TH: I'm pretty sure it won't. I don't think it will.

29. LS: But if it won't make a circle, then that means this circle is ours [points back to earlier sketch]. The one we have to deal with. You know what I mean?

30. TH: I see what you mean. Like, try to draw a circle out here like going through this point. See, it won't. It won't work because in order for it to work [another few minutes with the compass. The dialogue has to do with their attempts to draw a very accurate figure, so that they can draw conclusions from it.]

31. LS: OK, so that's what we're doing, right? We don't need it that big.

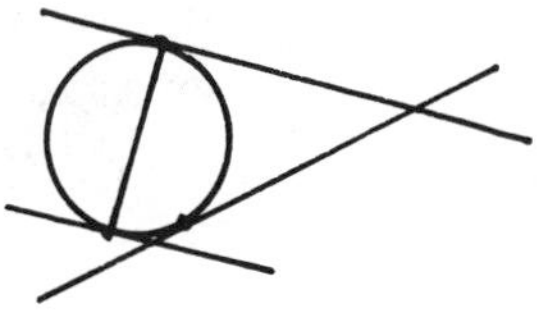

32. LS: Yeah, wait, you couldn't because it is going to go through [the point P]. I think it does have to be, right—

33. TH: If we have these two points, that's definitely our diameter going through it. Now we can draw—

34. LS: But neither is it a tangent.

35. TH: That's just what I was going to say. Can we draw these two lines so that

— see, you can't for in order for this to cut through this, it's too shallow, it's shallow—

36. TH: OK as soon as this—OK, make this a tangent.
37. LS: In order for this to be, do you think it's going to be tangent to—
38. TH: No, because, because we know this one is not going to—I want to see if like we make this a tangent. You see what I mean? But that doesn't look like a diameter either. Well, I don't think that's it. Of course it couldn't be because a diameter is going to be when it's parallel, isn't it?
39. LS: That's the diameter.
40. TH: OK. That's not going to help us [laughs].
41. LS: You figured that out.
42. TH: Right.
43. LS: Can we construct one parallel to it? [looks at original diagram.] But then we still don't know the center. [pause] Could we just draw a perpendicular?
44. TH: Yeah, that's what I was just going to say. If we draw a perpendicular line to this and just call that the diameter, it will work from there. And then it should touch if it's perpendicular. It should be tangent at one point, shouldn't it?
45. LS: Right!
46. TH: Shouldn't it?
47. LS: Yes!
48. TH: Won't it?
49. LS: Yes!
50. TH: OK, draw a perpendicular; oh, good.

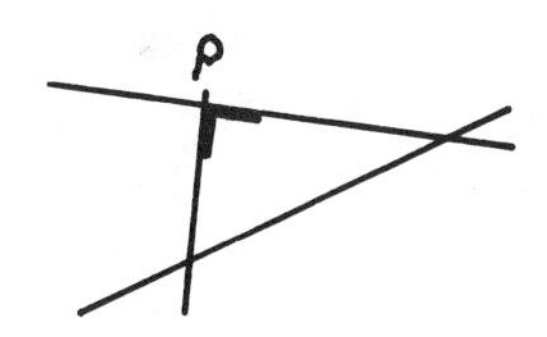

51. LS: Does one know how to do that with a compass? Do you?
52. TH: This is a right angle, so—[uses the corner of the ruler].

53. LS: OK, that's perpendicular, OK. Doesn't look it, but it is.
54. TH: That's our diameter.
55. LS: So if we say this is the point of tangency—
56. TH: So we can bisect this to find the center, right? So call it center *C*. [she pauses.] Maybe we should have done our steps.
57. LS: That's all being unmathematical, completely disorganized.
58. TH: OK, back to the drawing board.
59. LS: I don't know how.
60. TH: Me either.
61. LS: OK, if we just use the ruler with the little numbers on it here.
62. TH: Or isn't that legal?
63. LS: Sure it's legal [does by hand]. Now we have the radius, now we just draw it.
64. TH: Uh, oh, do we know, we have to see if this is going to work. I know! Ugg.
65. LS: My guess is, I think it's not. But we'll try.
66. TH: I would think, though, it would have to, though, wouldn't it?
67: LS: No.

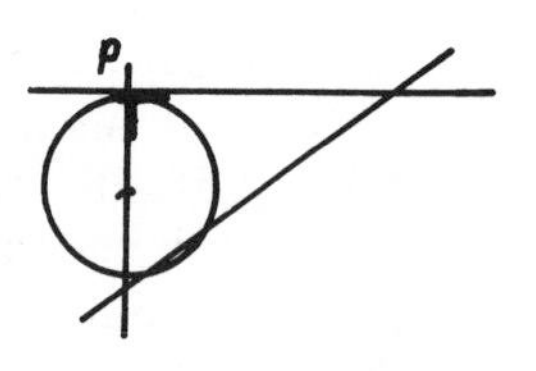

68. TH: The radius is shorter as—
69. LS: I don't know. Well, let's see what happens when it goes through there.
70. TH: Somehow it doesn't look perpendicular, though, doesn't it?
71. LS: See, this line isn't straight relative to the page, which is why it doesn't look perpendicular.
72. TH: Oh, right, but—
73. LS: It looks good. Now we can tell something.
74. TH: Maybe, I think this tells us the point of tangency has to be way more [points to right]. I think.

[Three minutes of constructions]

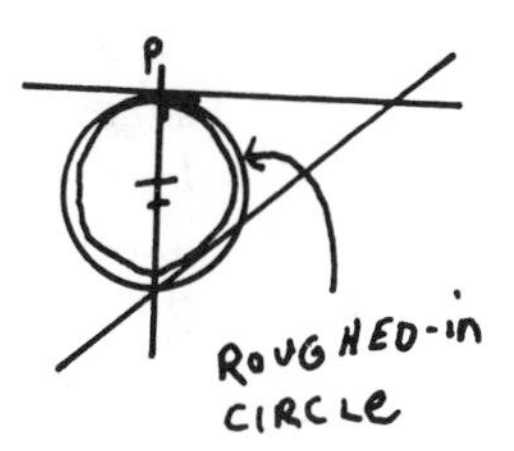

75. LS: What circle was this one? Yup, that was a right angle. Oh, darn it.
76. TH: OK, so the radius has got to be smaller because it's going outside of this line. So it's got to be a little smaller and the center has got to be up and over, like here—
77. LS: But how do we—
78. TH: But I don't know how to do that, without doing it until it comes out right.
79. LS: Yeah.

[pause and evaluation of prior failure]

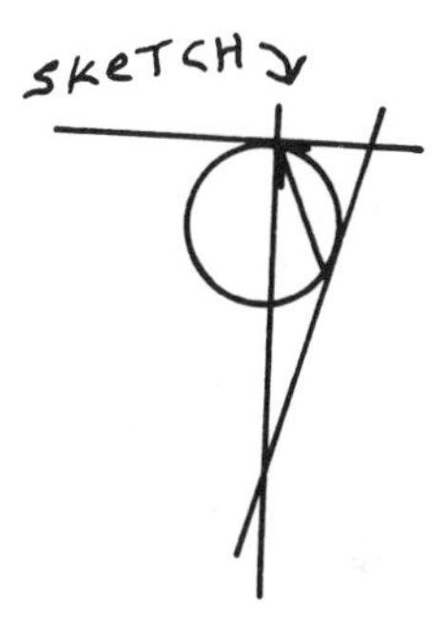

80. TH: That was dumb. By doing that we were saying that no matter what this line looked like, then it looked like this, if we dropped a perpendicular we could do it and we could get the diameter for that angle and still expect to do it. You know what I mean?
81. LS: Yeah, I don't think it will work for any angle though.
82. TH: I know, that's what I mean.
83. LS: Yeah, well, we goofed again.

[pause]

84. TH: Well, the only thing I can think of to do is what we did in class the other—well, what we were supposed to do, you know. The triangle thing, trying to inscribe it.
85. LS: Wait, we know—
86. TH: I know, that's the problem. We don't know how to do it.
87. LS: I don't know what to do.
88. TH: All right, we are going to have to try something else.
89. LS: All right, what are we—what were those sort of things we tried with triangle one? Cause maybe we could

— do the same thing with, on a smaller scale.

90. TH: I got absolutely nowhere.
91. LS: Yeah.
92. TH: But I was trying to do things like bisect this side.

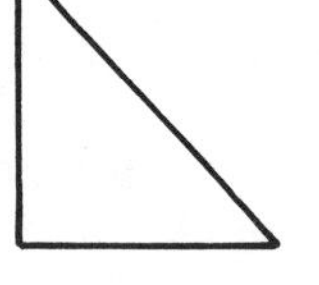

93. LS: Yeah, I did that.
94. TH: It didn't work.
95. LS: Yeah, let's see what we have here. We want to inscribe a circle in this right triangle.
96. TH: Why do you want to do a right triangle?
97. LS: I don't know. It just is one. Oh, I blew it now, no. The ends don't matter because we're, you see, we want to inscribe it. We're putting in the extra conditions, because it doesn't have to touch this line. It doesn't have to—oh, I don't know.
98. TH: I don't think that will get us anywhere.
99. AHS: OK, guys—
100. Both: We give up.

Appendix 10.2: Protocol 10.2

The following problem was worked after the problem-solving course.

The common internal tangent to two circles is the line that is tangent to both but has one circle on each "side" of it, as in the picture to the right.

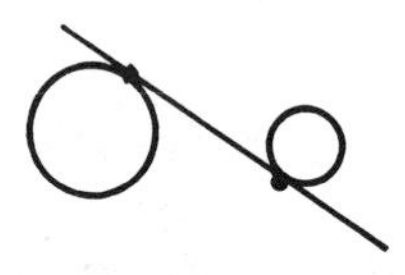

You are given three points A, B, and C as below. Using straightedge and compass, you wish to construct two circles *which have the same radius,* with centers A and B, respectively, such that the common internal tangent to both circles passes through the point C. How do you do it? Justify.

1. TH: [reads problem]
2. LS: Wait, I have to read this. Ummm.
3. TH: What we want basically is this, circles and a line something like this that is

A B

C

going to pass through here [makes sketch].

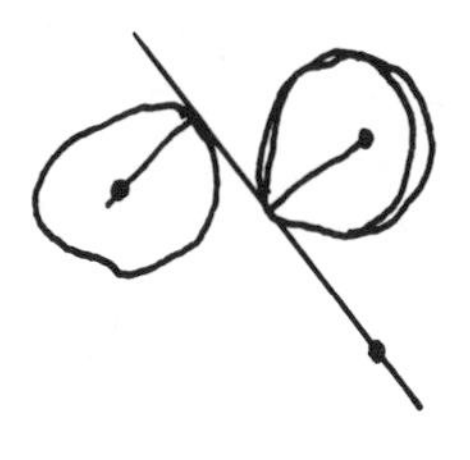

4.	LS:	Right. Ummm.
5.	TH:	Like that.
6.	LS:	Except they have — where is it — have the same radius —
7.	TH:	Uh huh.
8.	LS:	— so it isn't going to look like that.
9.	TH:	Right.
10.	LS:	But, OK. Wait, I've got to think for a second.

[erasing to draw again]

11.	LS:	OK, wouldn't it — no, maybe not.
12.	TH:	What?
13.	LS:	No, that was dumb. Let me think. [pause]
14.	LS:	Umm, should we try and draw it, maybe, how it would be to see what the relationship of *C* is to the two circles, since that's not drawn.
15.	TH:	Right.
16.	LS:	You know how I am with compasses — go ahead.
17.	TH:	Well, how big am I supposed to draw it? [draws with a compass]
18.	LS:	I've made this too big because they're going to overlap one another with that radius.
19.	TH:	Yeah.
20.	LS:	Just draw [i.e., sketch] it — you don't have to use the compass. Just draw it — just draw — no, no, no.

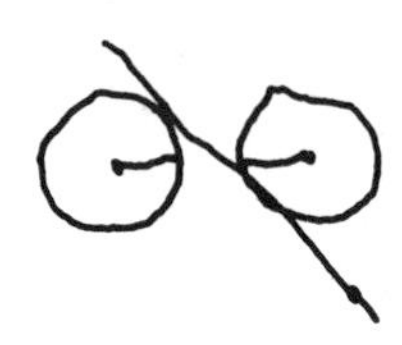

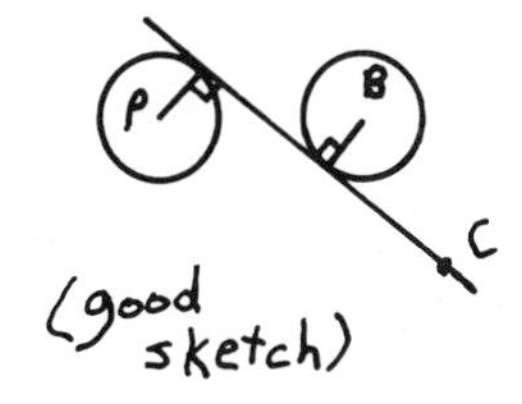

21. TH: OK, and I'll make my circles better. [unclear] OK. [unclear]
22. TH: What are you going to do?
23. LS: I just want to see what it would look like more accurately. [draws with compass]
24. TH: Why?
25. LS: Just so I could see [unclear], but you can think out loud if you have an idea. OK. Can you think of anything? [finishes sketch]
26. TH: Umm. These two radii are the same, right?
27. LS: Yep. Except it doesn't look the same, does it?
28. TH: That's the way you put your centers in the center.
29. LS: [unclear]
30. TH: [unclear] OK. These two centers have to—do you know what I mean?
31. LS: No. Wait, what am I looking for now?
32. TH: [rereads problem] Why don't we first just try to—
33. LS: If we can find [unclear] [pencil placed at center point]

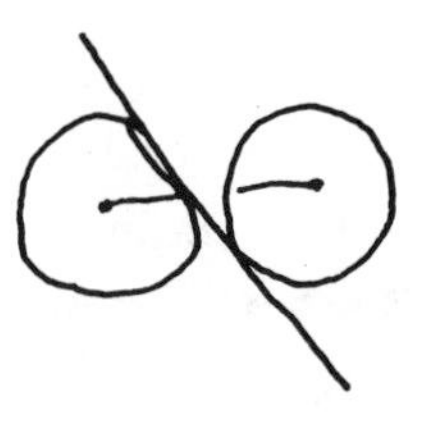

34. TH: All right—if you just have the two centers and you go over—say the radius—the radius will have to be half way in between the centers. All right, and then—
35. LS: Say—wait—wha-wha-wha-what?
36. TH: If we just try to draw the two circles and the tangent line without worrying about point *C* for right now.
37. LS: Right.
38. TH: OK. Since they have to be of equal radius—the radius will be halfway between the two centers?* It's like the tangent line would be like this.

* She meant to say that the *length* of the radius in this extreme case was half the distance between the centers of the two circles.

H: That's just what I was going to measure.
S: Ummm. Because if we did that, we were given points A, B, and C.
H: Yes [looks at her sketch], that crosses it, too. That's exactly what we're going to do.

THEY MEASURE THIS
THE GOOD SKETCH

S: All right—wait, we're not allowed to use a ruler, but—yeah, divide it in half.
H: Yeah, bisect.
S: Why don't you actually do it—
H: Let's try it on here since we're not sure. [Begins new sketch]
S: Wait, I think it was the other line. [unclear] Just connect point B. We're going to have to drop a perpendicular from B to the line.
H: What are you doing that for?
S: Because this is perpendicular and that's what the radius would be, a perpendicular and from A coming to the line also.
H: Right.
S: OK. I don't know why this works, I mean, I just seem to see it, you know.
H: I think we can do it with similar triangles and things so let's just make sure it works [unclear].

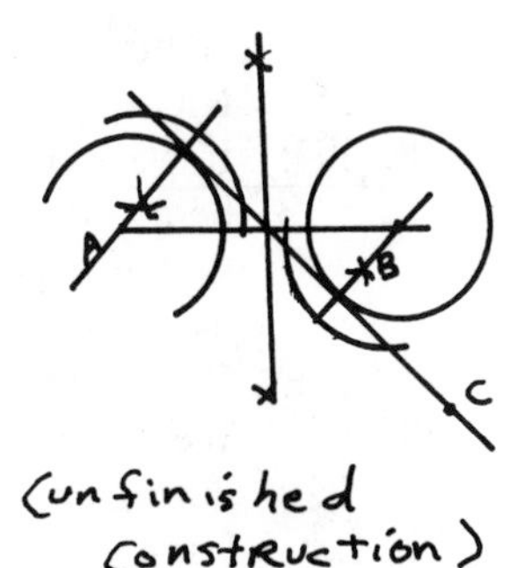

S: We can do it here, too—this isn't a very nice compass.
H: We're running out of time. [whispering] Draw faster, draw faster.
S: I can't—this is hard.
H: Draw faster anyway.
S: I didn't construct it right.
H: Well, just draw it—it'll work.
S: Oh, wait, maybe I did actually. OK, that's the radius then.
TH: Right.
S: Perpendicular. Then we just have to draw—I think that's just the right thing.
TH: That'll do it, that'll do it—wait, we've got to draw—OK, we did it. We've got

39. LS: I don't get this about the radius being halfway between two centers.
40. TH: Me neither.
41. LS: I don't get what you mean. How's the radius halfway—I don't get what you mean.
42. TH: If it was like this and the tangent line would just be [unclear]
43. LS: OK, yeah.
44. TH: OK? These two have to be the same length.
45. LS: Right.
46. TH: And the thing that is going to determine how long they are is the angle on this line. What I mean, like, if they are exactly—halfway in between the two centers then the line is vertical.
47. LS: Right.
48. TH: If we make it somehow shorter right here and here, the circles would be like this, and the tangent would be on a slant like this.
49. LS: OK. Ummm.
50. TH: We have to figure out how they go through point *C*. So—
51. LS: I don't know either.
52. TH: Can we just start with *C* and draw a line through it somewhere and then make the circles tangent to it?
53. LS: No.
54. TH: Or—
55. LS: No, we're given the centers.
56. TH: We're also given *C*.
57. LS: Uh huh. But just drawing the line can't guarantee you could end it with something like this if you just drew the line here. Ummm. Isn't there another way we can characterize the line? Find the locus.
58. TH: Ummm.
59. LS: This might not work for all of them, but, look here, doesn't this look like—that's just like the center?

to show why. We have to show that these—the reason that these are halfway in between these two points is because—angle side—we have to show that—what this side—

83. LS: Like we have an angle.

84. TH: But what are we trying to show—we want to show why this is in between A and B.

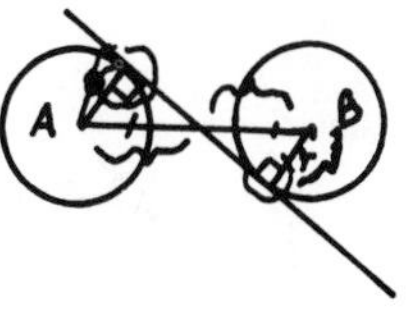

85. LS: Right.

86. TH: So we want to show that this is equal to this—that they—

87. Both: —are congruent.

88. TH: OK, we have that. We have—

89. LS: —an angle and a side. How do we know—

90. TH: And we need to show that this side is compared to that side. And—

91. LS: [to A] Must we prove why something works or just show you the construction?

92. AHS: If you can justify it, I would be happy.

93. LS: OK, let's try to justify it.

94. TH: Now the angle—

95. LS: Well, we know, I mean, r is equal to r so it is just like—

96. TH: We have these angles, so this angle equals this one.

After a few minutes, and with some slight confusion, they prove that their construction has the desired properties.

References

Aiken, L. R. Attitudes toward mathematics. *Review of Educational Research,* 1970, *40,* pp. 551–596.

Aiken, L. R. Update on attitudes and other affective variables in learning mathematics. *Review of Educational Research,* 1976, *46,* pp. 293–311.

Association for Supervision and Curriculum Development. *Mathematics Education Research: Implications for the 80's.* Alexandria, VA: ASCD, 1981.

Atkinson, John W., & Joel O. Raynor. *Motivation and Achievement.* Washington, DC: Winston, 1974.

Ball, S. (Ed.). *Motivation in Education.* New York: Academic Press, 1977.

Begle, E. G. *Critical Variables in Mathematics Education.* Washington, DC: Mathematical Association of America and National Council of Teachers of Mathematics, 1979.

Beth, Evert W., & Jean Piaget. *Mathematics, Epistemology, and Psychology* (W. Mays, Trans.), New York: Gordon and Breach, 1966.

Bloom, Benjamin S. *Taxonomy of Educational Objectives. Handbook I: Cognitive Domain.* New York: David McKay, 1956.

Bobrow, Daniel, & Allen Collins. *Representation and Understanding.* New York: Academic Press, 1975.

Brown, Ann L. Knowing when, where, and how to remember: A problem of metacognition. In R. Glaser (Ed.), *Advances in Instructional Psychology* (Vol. 1). Hillsdale, NJ: Lawrence Erlbaum Associates, 1978.

Brown, Ann L. Metacognition, executive control, self-regulation, and other even more mysterious mechanisms. In F. E. Weinert & R. H. Kluwe (Eds.), *Learning by Thinking.* West Germany, Kuhlhammer, 1984.

Brown, Ann L., John Bransford, Robert Ferrara, & Joseph Campione. Learning, remembering, and understanding. In. P. H. Mussen (Ed.), *Handbook of Child Psychology* (Vol. 3). New York: Wiley, 1983.

Brown, J. S., & R. R. Burton. Diagnostic models for procedural bugs in basic mathematical skills. *Cognitive Science,* 1978, *2,* pp. 155–192.

Brown, J. S., & K. vanLehn. Repair theory: A generative theory of bugs in procedural skills. *Cognitive Science,* 1980, *4,* pp. 379–426.

Brown, John Seely, Richard R. Burton, & Johann de Kleer. Pedagogical, natural language, and knowledge engineering techniques in SOPHIE I, II, and III. In D. Sleeman & J. S. Brown (Eds.), *Intelligent Tutoring Systems.* London: Academic Press, 1982.

Burton, R. R. Diagnosing bugs in a simple procedural skill. In D. Sleeman & J. S. Brown (Eds.), *Intelligent Tutoring Systems.* London: Academic Press, 1982.

Buxton, Laurie. *Do You Panic about Maths? Coping with Maths Anxiety.* London: Heneimann Educational Books, 1981.

Caramazza, A., M. McCloskey, & B. Green. Naive beliefs in "sophisticated" subjects: Misconceptions and trajectories of objects. *Cognition,* 1981, *9,* pp. 117–123.

Carpenter, Thomas P., Mary M. Lindquist, Westina Matthews, & Edward A. Silver. Results of the third NAEP mathematics assessment: Secondary school. *Mathematics Teacher,* 1983, *76*(9), pp. 652–659.

Carroll, Lewis. *The Annotated Alice.* New York: Bramhall House, 1960.

Chartoff, B. T. An exploratory investigation utilizing a multidimensional scaling procedure to discover classification criteria for algebra word problems used by students in grades 7–13. (Doctoral dissertation, Northwestern University, 1976.) *Dissertation Abstracts International,* 1977, *37,* 7006A.

Chase, W. G., & H. A. Simon. Perception in chess. *Cognitive Psychology,* 1973, *4,* pp. 55–81.

Chi, M., P. Feltovich, & R. Glaser. Categorization and representation of physics problems by experts & novices. *Cognitive Science,* 1981, *5,* pp. 121–152.

Clancey, William J. Tutoring rules for guiding a case method dialogue. In D. Sleeman & J. S. Brown (Eds.), *Intelligent Tutoring Systems.* London: Academic Press, 1982.

Clement, John. Algebra word problem solutions: Thought processes underlying a common misconception. *Journal for Research in Mathematics Education,* 1982, *13,* pp. 16–30.

Clement, John. A conceptual model discussed by Galileo and used intuitively by physics students. In Dedre Gentner & Albert Stevens (Eds.), *Mental Models.* Hillsdale, NJ: Erlbaum, 1983.

Clement, John, Jack Lochhead, & G. S. Monk. Translation difficulties in learning mathematics. *American Mathematical Monthly,* 1981, *88,* pp. 286–290.

Cole, M., J. Gay, J. Glick, & D. Sharp. *The Cultural Context of Learning and Thinking.* New York: Basic Books, 1971.

Dahl, O., E. Dijkstra, & C. Hoare. *Structured Programming.* New York: Academic Press, 1972.

de Groot, Adriaan D. *Thought and Choice in Chess.* The Hague: Mouton, 1965.

de Groot, Adriaan D. Perception and memory versus thought: Some old ideas and recent findings. In Benjamin Kleinmuntz (Ed.), *Problem Solving: Research, Method, and Theory.* New York: Wiley, 1966.

Descartes, Rene. *Rules for the Direction of the Mind* and *Discourse on Method* (E. S. Haldane & G. R. I. Ross, Trans.). In Great Books of the Western World (Vol. 31). Chicago: Encyclopedia Brittanica, Inc., 1952.

Dijkstra, E. *A Discipline of Programming.* Englewood Cliffs, NJ: Prentice-Hall, 1976.

diSessa, Andrea A. Unlearning aristotelian physics: A study of knowledge-based learning. *Cognitive Science,* 1982, *6,* pp. 37–75.

diSessa, Andrea A. Phenomenology and the evolution of intuition. In D. Gentner & A. Stevens (Eds.), *Mental Models.* Hillsdale, NJ: Erlbaum, 1983.

Doise, Willem, Gabriel Mugny, & Anne-Nelly Perret-Clermont. Social interaction and the development of cognitive operations. *European Journal of Social Psychology,* 1975, *5*(3), pp. 367–383.

Duncker, K. *On Problem Solving.* Psychological Monographs 58, No. 5. Washington, DC: American Psychological Association, 1945. [Whole #270]

Einhorn, J. J., & R. M. Hogarth. Behavioral decision theory: Processes of judgment and choice. *Annual Review of Psychology,* 1981, *32,* pp. 53–88.

Ericsson, K. A., & H. A. Simon. Verbal reports as data. *Psychological Review,* 1980, *87*(3), pp. 215–251.

Ericsson, K. A., & H. A. Simon. Sources of evidence on cognition: A historical overview. In T. Merluzzi, C. Glass, & M. Genest (Eds.), *Cognitive Assessment.* New York: Guilford Press, 1981.

Ericsson, K. A., & H. A. Simon. *Protocol Analysis.* Cambridge, MA: MIT Press, 1984.

Ernst, G., & A. Newell. *GPS: A Case Study in Generality and Problem Solving.* New York: Academic Press, 1969.

Fahlman, S. A planning system for robot construction tasks. *Artificial Intelligence,* 1974, *5,* pp. 1–49.

Federation of Behavioral, Cognitive, and Psychological Sciences. *Research on cognition, and behavior relevant to education in mathematics, science and technology.* (Report compiled by James Greeno, Robert Glaser, Allen Newell.) A report submitted by the Federation to the National Science Board Commission on Precollege Education in Mathematics, Science, and Technology, April, 1983.

Fennema, Elizabeth, & Merlyn Behr. Individual differences and the learning of mathematics. In R. J. Shumway (Ed.), *Research in Mathematics Education.* Reston, VA: National Council of Teachers of Mathematics, 1980.

Fikes, R. Knowledge representation in automatic planning systems. In A. Jones (Ed.), *Perspectives on Computer Science.* New York: Academic Press, 1977.

Flavell, John. Metacognitive aspects of problem solving. In Lauren Resnick (Ed.), *The Nature of Intelligence.* Hillsdale, NJ: Erlbaum, 1976.

Flavell, John, & Henry Wellman. Metamemory. In R. V. Kail & J. W. Hagen (Eds.), *Perspectives on the Development of Memory and Cognition.* Hillsdale, NJ: Erlbaum, 1977.

Freudenthal, Hans. *Mathematics as an Educational Task.* Dordrecht, Holland: D. Reidel, 1973.

Gardner, A. Search. In A. Barr & E. A. Feigenbaum (Eds.), *The Handbook of Artificial Intelligence* (Vols. 1 and 2). Los Altos, CA: William Kaufman, 1981, 1982. (Also Stanford University Computer Science Department Report No. STAN-CS-79-742, June 1979.)

Ginsburg, H., N. Kossan, R. Schwartz, & D. Swanson. Protocol methods in research on mathematical thinking. In H. Ginsburg (Ed.), *The Development of Mathematical Thinking.* New York: Academic Press, 1983.

Goldberg, D. J. The effects of training in heuristics methods on the ability to write proofs in number theory. (Unpublished doctoral dissertation, Columbia University, 1974.) *Dissertation Abstracts International,* 1974, 4989B. (University Microfilms, 75-7, 836.)

Goldin, G., & E. McClintock (Eds.). *Task Variables in Mathematical Problem Solving.* Columbus, OH: ERIC, 1980.

Greeno, James G., & H. A. Simon. Problem solving and reasoning. (draft, February 1984). In R. C. Atkinson, R. Herrnstein, G. Lindzey, & R. D. Luce (Eds.), *Stevens' Handbook of Experimental Psychology.* New York: John Wiley and Sons, to appear.

Hadamard, J. W. *Essay on the Psychology of Invention in the Mathematical Field.* New York: Dover, 1954.

Harvey, J., & T. Romberg. *Problem-Solving Studies in Mathematics.* Madison, WI: Wisconsin R & D Center, 1980.

Hayes, J. R., & H. A. Simon. Understanding written problem instructions. In L. W. Gregg (Ed.), *Knowledge and Cognition.* Hillsdale, NJ: Erlbaum, 1974.

Hayes-Roth, B., & F. Hayes-Roth. A cognitive model of planning. *Cognitive Science,* 1979, *3,* 275–310.

Heller, J. L., & J. G. Greeno. Information processing analyses of mathematical problem solving. In R. Lesh, D. Mierkiewicz, and M. Kantowski (Eds.), *Applied Mathematical Problem Solving.* Columbus, OH: ERIC, 1979. (ERIC Document Reproduction Service No. ED 180 816).

Hinsley, Dan A., John R. Hayes, & Hebert A. Simon. From words to equations: Meaning and representation in algebra word problems. In P. A. Carpenter & M. A. Just (Eds.), *Cognitive Processes in Comprehension.* Hillsdale, NJ: Erlbaum, 1977.

Hoffer, Alan. Geometry is more than proof. *Mathematics Teacher,* January 1981, pp. 11–17.

Hoffer, Alan. Van Hiele-based research. In R. Lesh & M. Landau (Eds.), *Acquitision of Mathematics Concepts and Processes.* New York: Academic Press, 1983.

James, William. *Principles of Psychology* (Vol. 1). New York: Holt, 1890.

Janvier, Claude. The interpretation of complex Cartesian graphs representing situations: Studies and teaching experiments. Unpublished doctoral dissertation, the University of Nottingham, England, 1978.

Johnson, S. C. Hierarchical clustering schemes. *Psychometrika,* 1967, *32,* pp. 241–254.

Jurgensen, Ray C., Richard G. Brown, & Alice M. King. *Geometry* (new ed.). Boston: Houghton-Mifflin, 1980.

Kahneman, Daniel, & Amos Tversky. On the study of statistical intuitions. In D. Kahneman, P. Slovic, & A. Tversky (Eds.), *Judgment under Uncertainty: Heuristics and Biases.* Cambridge: Cambridge University Press, 1982.

Kahneman, Daniel, Paul Slovic, & Amos Tversky. *Judgment under Uncertainty: Heuristics and Biases.* Cambridge: Cambridge University Press, 1982.

Kantowski, M. G. Processes involved in mathematical problem solving. *Journal for Research in Mathematics Education,* 1977, *8,* 163–180.

Kaput, James, J. Mathematics and learning: Roots of epistemological status. In Jack Lochhead & John Clement (Eds.), *Cognitive Process Instruction.* Philadelphia: Franklin Institute Press, 1979.

Kilpatrick, J. Analyzing the solution of word problems in mathematics: An exploratory study. (Unpublished doctoral dissertation, Stanford University, 1967.) *Dissertation Abstracts International,* 1968, *28,* 4380A. (University Microfilms, 68-5, 442.)

Kilpatrick, J. Problem solving and creative behavior in mathematics. In J. W. Wilson & L. R. Carry (Eds.), *Studies in Mathematics* (Vol. 19). Stanford, CA: School Mathematics Study Group, Stanford University, 1969. (A briefer version, Problem Solving in Mathematics, appeared in *Review of Educational Research,* 1970, *39,* pp. 523–534.)

Kilpatrick J. & I. Wirszup (Eds. and Trans.). *Soviet Studies in the Psychology of Learning and Teaching Mathematics* (14 Vols). Chicago: University Press, 1969–1975.

Kitcher, Philip. *The Nature of Mathematical Knowledge.* New York: Oxford University Press, 1983.

Krutetskii, V. A. *The Psychology of Mathematical Abilities in School Children.* (Joan Teller, Trans., Jeremy Kilpatrick & Izaak Wirszup, Eds.). Chicago: University of Chicago Press, 1976.

Kulm, Gerald. Research on mathematics attitude. In R. J. Shumway (Ed.), *Research in Mathematics Education.* Reston, VA: National Council of Teachers of Mathematics, 1980.

Kulm, G., P. Campbell, M. Frank, & G. Talsma. *Analysis and synthesis of mathematical problem-solving processes.* Paper presented at the annual meeting of the National Council of Teachers of Mathematics, St. Louis, 1981, April.

Lachman, Janet, Roy Lachman, & Carroll Thronesbery. Metamemory through the adult life span. *Developmental Psychology,* 1979, *15*(5), pp. 543–551.

Lakatos, I. *Proofs and Refutations* (rev. ed.). Cambridge: Cambridge University Press, 1977.

Larkin, J. Teaching problem solving in physics: The psychological laboratory and the practical classroom. In F. Reif & D. Tuma (Eds.), *Problem Solving in Education: Issues in Teaching and Research.* Hillsdale, NJ: Erlbaum, 1980.

Larkin, J., J. McDermott, D. Simon, & H. A. Simon. Expert and novice performance in solving physics problems. *Science,* 1980, *208,* pp. 1335–1342.

Lave, J. What's special about experiments as contexts for thinking? *The Quarterly Newsletter of the Laboratory of Comparative Human Cognition,* 1980, *2,* pp. 86–91.

Leder, Gilah. Bright girls, mathematics, and fear of success. *Educational Studies in Mathematics,* 1980, *11,* pp. 411–422.

Lefcourt, Herbert M. *Locus of Control.* Hillsdale, NJ: Erlbaum, 1982.

Lesh, R. Modeling students' modeling behaviors. In *Proceedings of 4th Annual PME meeting,* Athens, GA, October 1982.

Lesh, R. *Metacognition in mathematical problem solving.* Evolving manuscript. Available from author. WICAT Corporation, Orem, UT, 1983a.

Lesh, R. *Modeling middle school students' modeling behaviors in applied mathematical problem solving.* Paper delivered at AERA, Montreal, April, 1983b.

Lester, F. *Mathematical Problem Solving: Issues in Research,* Philadelphia: Franklin Institute Press, 1982.

Lindquist, M. M. (Ed.). *Selected Issues in Mathematics Education.* Reston, VA: National Council of Teachers of Mathematics, 1980.

Lipson, S. H. The effects of teaching heuristics methods to student teachers in mathematics. (Doctoral dissertation, Columbia University, 1972.) *Dissertation Abstracts International,* 1972, *33,* 2221A. (University Microfilms No. 72-20,334.)

Lochhead, Jack. Research on students' scientific misconceptions: Some implications for teaching. Paper delivered at the annual AERA meetings, Montreal, April, 1983.

Loomer, Norman. A multidimensional exploratory investigation of small group-heuristic and expository learning in calculus. In J. G. Harvey & T. A. Romberg (Eds.), *Problem-Solving Studies in Mathematics.* Madison, WI: Wisconsin R&D Center Monograph Series, 1980.

Lucas, J. F. An exploratory study of the diagnostic teaching of heuristic problem-solving strategies in calculus. (Doctoral dissertation, University of Wisconsin, 1972.) *Dissertation Abstracts International* 1972, 6825-A. (University Microfilms No. 72-15,368.)

Lucas, J. F. The Teaching of Heuristic Problem-Solving Strategies in Elementary Calculus. *Journal for Research in Mathematics Education,* 1974, *5,* pp. 36–46.

Lucas, J. F. An exploratory study on the diagnostic teaching of heuristic problem-solving strategies in calculus. In J. G. Harvey & T. A. Romberg (Eds.), *Problem-Solving Studies in Mathematics.* Madison, WI: Wisconsin R&D Center Monograph Series, 1980.

Lucas, J. F., N. Branca, D. Goldberg, M. G. Kantowski, H. Kellogg, & J. P. Smith. A Process-Sequence Coding System for Behavioral Analysis of Mathematical Problem Solving. In G. Goldin and E. McClintock (Eds.), *Task Variables in Mathematical Problem Solving.* Columbus, OH: ERIC, 1980.

McCloskey, M. Intuitive Physics. *Scientific American,* April 1983a, pp. 122–130.

McCloskey, M. Naive theories of motion. In Dedre Gentner & Albert Stevens (Eds.), *Mental Models.* Hillsdale, NJ: Erlbaum, 1983b.

McCloskey, M., A. Caramazza, & B. Green. Curvilinear motion in the absence of external forces: Naive beliefs about the motions of objects. *Science,* 1980, *210,* pp. 1139–1141.

McDermott, L. *Identifying and overcoming students' conceptual difficulties in physics.* Paper delivered at the annual AERA meetings, Montreal, 1983, April.

Malone, J., G. Douglas, B. Kissane, & R. Mortlock. Measuring problem-solving ability. In *Problem Solving in School Mathematics.* Reston, VA: National Council of Teachers of Mathematics, 1980.

Markman, Ellen M. Realizing that you don't understand: A preliminary investigation. *Child Development,* 1977, *48,* pp. 986-992.

Matz, M. Towards a process model for high school algebra errors. In D. Sleeman & J. S. Brown (Eds.), *Intelligent Tutoring Systems.* London: Academic Press, 1982.

Messick, Samuel (Ed.). *Individuality in Learning.* San Francisco: Jossey-Bass, 1976.

Miller, M. A structured planning and debugging environment for elementary programming. In D. Sleeman & J. S. Brown (Eds.), *Intelligent Tutoring Systems.* London: Academic Press, 1982.

Minsky, Marvin. A framework for representing knowledge. In P. Winston (Ed.), *The Psychology of Computer Vision.* New York: McGraw-Hill, 1975.

Monsell, Stephen. Representations, processes, memory mechanisms: The basic components of cognition. *Journal of the American Society for Information Science, 32*(5), September 1981, pp. 378-390.

Mugny, Gabriel, & Willem Doise. Socio-cognitive conflict and structure of individual and collective performances. *European Journal of Social Psychology,* 1978, *8,* pp. 181-192.

National Council of Teachers of Mathematics. *Problem Solving in School Mathematics.* (1980 Yearbook, S. Krulik, Ed.) Reston, VA: NCTM, 1980.

Neisser, Ulric. General, academic, and artificial intelligence. In. L. Resnick (Ed.), *The Nature of Intelligence.* Hillsdale, NJ: Erlbaum, 1976.

Newell, Allen. *On the analysis of human problem-solving protocols.* Paper given at an international symposium on mathematical and computational methods in the social sciences, Rome, 1966, July 4-9.

Newell, Allen, & Herbert A. Simon. *Human Problem Solving.* Englewood Cliffs, NJ: Prentice-Hall, 1972.

Nickerson, Raymond S. Understanding understanding. Draft manuscript, March, 1982. Available from author at Bolt Beranek and Newman, Cambridge, MA.

Nilsson, Nils J. *Principles of Artificial Intelligence.* Palo Alto, CA: Tioga Publishing Co., 1980.

Nisbett, R. E., & T. Wilson. Telling more than we know: Verbal reports on mental processes. *Psychological Review,* 1977, *84,* pp. 231-260.

Norman, D. Twelve issues for cognitive science. In D. A. Norman (Ed.), *Perspectives on Cognitive Science.* Norwood, NJ: Ablex, 1980. Also Hillsdale, NJ: Lawrence Erlbaum Associates, 1980.

Osborne, A. R. *Models for Learning Mathematics: Papers from a Research Workshop.* Columbus, OH: ERIC/SMEAC, 1976.

Papert, Seymour. *Mindstorms.* New York: Basic Books, 1980.

Perkins, David. *Difficulties in everyday reasoning and their change with education.* Final report to the Spencer Foundation. Cambridge, MA: Harvard University, November 1982.

Perkins, David, Richard Allen, & James Hafner. Difficulties in everyday reasoning. In William Maxwell (Ed.), *Thinking: The Frontier Expands.* Philadelphia: Franklin Institute Press, 1983.

Petitto, Andrea. Collaboration in problem solving. Manuscript available from author, University of Rochester, 1985.

Piaget, Jean. *Genetic Epistemology* (Eleanor Duckworth, Trans.). New York: W.W. Norton, 1970.

Poincaré, H. *The Foundations of Science* (G. H. Halstead, Trans.). New York: Science Press, 1913.

Polanyi, Michael. *Personal Knowledge.* Chicago: University of Chicago Press, 1962.

Pólya, G. *How to Solve It.* Princeton: Princeton University Press, 1945.

Pólya, G. *Mathematics and Plausible Reasoning* (2 Vols.). Princeton: University Press, 1954.
Pólya, G. *How to Solve It* (2nd ed.). New York: Doubleday, 1973.
Pólya, G. *Mathematical Discovery* (Vols. 1 and 2). New York: Wiley, 1962 (Vol. 1) and 1965 (Vol. 2) Combined paperback edition, New York: Wiley, 1980.
Resnick, Lauren B., & Wendy W. Ford. *The Psychology of Mathematics for Instruction.* Hillsdale, NJ: Lawrence Erlbaum Associates, 1981.
Ringel, Barbara, & Carla Springer. On knowing how well one is remembering: The persistence of strategy use during transfer. *Journal of Experimental Child Psychology,* 1980, *29,* pp. 322–333.
Rissland, Edwina. Artificial intelligence and the learning of mathematics: A tutorial summary. In E. A. Silver (ed.), *Learning and Teaching Mathematical Problem-Solving; Multiple Research Perspectives.* Hillsdale, NJ: Erlbaum, 1985.
Rogoff, Barbara, & Jean Lave (Eds.), *Everyday Cognition: Its Development in Social Context.* Cambridge: Harvard University Press, 1984.
Rosnick, Peter, & John Clement. Learning without understanding: The effect of tutoring strategies on algebra misconceptions. *Journal of Mathematical Behavior,* 1980, *3*(1), pp. 3–27.
Rubinstein, M. *Patterns of Problem Solving.* Englewood Cliffs, NJ: Prentice-Hall, 1975.
Rumelhart, David E., & Donald A. Norman. Representation in memory. CHIP paper 116, University of California, San Diego, June, 1983. (To appear as a chapter in the revised version of Stevens' *Handbook of Experimental Psychology.*)
Sacerdoti, E. *A Structure for Plans and Behavior.* New York: American Elsevier, 1977.
Schaie, K. W. Quasi-experimental research designs in the psychology of aging. In J. E. Birren & K. W. Schaie (Eds.), *Handbook of the Psychology of Aging.* New York: Van Nostrand Reinhold, 1977.
Schank, R., & R. Abelson. *Scripts, Plans, Goals, and Understanding.* Hillsdale, NJ: Lawrence Erlbaum Associates, 1977.
Schoenfeld, A. H. *Integration: Getting It All Together.* Undergraduate Modules and Application Project Modules 203, 204, 205. Newton, MA: Educational Development Corporation/Undergraduate Modules and Applications Project, 1977.
Schoenfeld, A. H. Presenting a strategy for indefinite integration. *American Mathematical Monthly,* 1978, *85*(8), pp. 673–678.
Schoenfeld, A. H. Can heuristics be taught? In *Cognitive Process Instruction.* Philadelphia, PA: Franklin Institute Press, 1979a.
Schoenfeld, A. H. Explicit heuristic training as a variable in problem-solving performance. *Journal for Research in Mathematics Education,* 1979b, *10*(3), pp. 173–187.
Schoenfeld, A. H. Teaching problem-solving in college mathematics: The elements of a theory and a report on the teaching of general mathematical problem-solving skills. In R. Lesh, D. Mierkiewicz, & M. Kantowski (Eds.), *Applied Mathematical Problem Solving.* Columbus, OH: ERIC, 1979c.
Schoenfeld, A. H. Heuristics in the classroom. In *Problem Solving in School Mathematics.* Reston, VA: National Council of Teachers of Mathematics, 1980a.
Schoenfeld, A. H. Teaching problem-solving skills. *American Mathematical Monthly,* 1980b, *87*(10), pp. 794–805.
Schoenfeld, A. H. Measures of problem solving performance and of problem solving instruction. *Journal for Research in Mathematics Education,* 1982a, *13*(1), pp. 31–49.
Schoenfeld, A. H., & D. Herrmann. Problem perception and knowledge structure in expert and novice mathematical problem solvers. *Journal of Experimental Psychology: Learning, Memory, and Cognition,* 1982b, *8*(5), pp. 484–494.
Schoenfeld, A. H. Some thoughts on problem solving research and mathematics education.

In F. Lester & J. Garofalo (Eds.), *Mathematical Problem Solving: Issues in Research.* Philadelphia: Franklin Institute Press, 1982c.

Schoenfeld, A. H. Beyond the purely cognitive: Belief systems, social cognitions, and metacognitions as driving forces in intellectual performance. *Cognitive Science* 1983a, *7,* pp. 329–363.

Schoenfeld, A. H. Episodes and executive decisions in mathematics problem solving. In R. Lesh & M. Landau (Eds.), *Acquisition of Mathematical Concepts and Processes.* New York: Academic Press, 1983b.

Schoenfeld, A. H. *Problem solving in the mathematics curriculum: A report, recommendations, and an annotated bibliography.* (M.A.A. Notes #1). Washington, DC: Mathematical Association of America, 1983c.

Schoenfeld, A. H. Artificial intelligence and mathematics education, or, AI and ME. In E. A. Silver (Ed.), *Learning and Teaching Mathematical Problem Solving: Multiple Research Perspectives.* Hillsdale, NJ: Erlbaum, 1985.

Schoenfeld, A. H. Making sense of "out loud" problem-solving protocols. *Journal of Mathematical Behavior,* in press (a).

Schoenfeld, A. H. Psychology and mathematical method: A capsule history and a modern view. In H. Beilin (Ed.), *Recent Trends in the Psychology of Mathematics Instruction.* New York: Sage Press, in press (b).

Selfridge, O. Pandemonium: A paradigm for learning. *Symposium on the Mechanization of Thought.* London, Her Majesty's Stationery Office, 1959.

Shavelson, R. J. Some aspects of the correspondence between content structure and cognitive structure in physics instruction. *Journal of Educational Psychology,* 1972, *63*(3), pp. 225–234.

Shavelson, R. J. Methods for examining representations of a subject-matter structure in a student's memory. *Journal of Research in Science Teaching,* 1974, *11*(3), pp. 231–249.

Shavelson, R. J., & G. C. Stanton. Construct validation: Methodology and application three measures of cognitive structure. *Journal of Educational Measurement,* 1975, *12*(2), pp. 67–85.

Silver, Edward A. Student perceptions of relatedness among mathematical verbal problems. *Journal for Research in Mathematics Education,* 1979, *10*(3), pp. 195–210.

Silver, Edward A. Knowledge organization and mathematical problem solving. In F. Lester (Ed.), *Mathematical Problem Solving: Issues in Research.* Philadelphia: Franklin Institute Press, 1982.

Silver, Edward A. *Thinking about problem solving: Toward an understanding of metacognitive aspects of mathematical problem solving.* Paper presented at the Conference on Thinking, Suva, Fiji, 1982b, January.

Silver, Edward A., N. Branca, & V. Adams. Metacognition: The missing link in problem solving? In R. Karplus (Ed.), *Proceedings of the Fourth International Conference for the Psychology of Mathematics Education,* Berkeley, CA, 1980.

Simon, D. P., & H. A. Simon. Individual differences in solving physics problems. In R. Siegler (Ed.), *Children's Thinking: What Develops?* Hillsdale, NJ: Erlbaum, 1978.

Simon, Herbert A. Information processing models of cognition. *Annual Review of Psychology,* 1979, *30,* pp. 363–396.

Simon, Herbert A. Problem solving and education. In D. Tuma & F. Reif (Eds.), *Problem Solving and Education: Issues in Teaching and Research.* Hillsdale, NJ: Erlbaum, 1980.

Skinner, B. F. *About Behaviorism.* New York: Knopf, 1974.

Slagle, J. R. A heuristic program that solves symbolic integration problems in freshman calculus. In E. Feigenbaum & J. Feldman (Eds.), *Computers and Thought.* New York: McGraw-Hill, 1963.

Sleeman, D., & J. S. Brown (Eds.). *Intelligent Tutoring Systems.* London: Academic Press, 1982.

Smith, J. P. The effect of general versus specific heuristics in mathematical problem-solving tasks. (Unpublished doctoral dissertation, Columbia University, 1973.) *Dissertation Abstracts International,* 1973, *34,* 2400A. (University Microfilms 73-26, 637.)

Soviet Studies in the Psychology of Learning and Teaching Mathematics (14 Vols.) (Isaac Wirzsup & Jeremy Kilpatrick, Trans. and Eds.) Stanford, CA: SMSG, 1969–1975. (Later reissued by the NCTM.)

Stevens, A., A. Collins, & S. Goldin. Misconceptions in students' understanding. In D. Sleeman & J. S. Brown (Eds.), *Intelligent Tutoring Systems.* London: Academic Press, 1982.

Stewart, P. *Jacobellis v. Ohio.* Decision of United States Supreme Court, 1964.

Stipek, Deborah J., & John R. Weisz. Perceived personal control and academic achievement. *Review of Educational Research,* 1981, *51*(1), pp. 101–137.

Suinn, R. M., Edie, C. A., Nicoletti, J., & Spinelli, P. R. The MARS, a measure of mathematics anxiety: Psychometric data. *Journal of Clinical Psychology* 1972, *28,* pp. 373–375.

Sussman, G. *A computational model of skill acquisition.* Massachussetts Institute of Technology Artificial Intelligence Laboratory Technical Report 297, 1973.

Thomas, George B. *Calculus and Analytic Geometry* (3rd ed.). Reading, MA: Addison-Wesley, 1960.

Tobias, Sheila. *Overcoming Math Anxiety.* New York: Norton, 1978.

Tresemer, D. The cumulative record on research on "Fear of Success." *Sex Roles,* 1976, *2,* pp. 217–236.

Tversky, Amos, & Daniel Kahneman. The framing of decisions and psychology of choice. *Science,* 1981, *211* (4481), 30, pp. 453–458.

van Hiele, P. M. *De Problematiek van het inzicht.* Unpublished doctoral dissertation, University of Utrecht, The Netherlands, 1957.

van Hiele, P. M. How can one account for the mental levels of thinking in math class? *Educational Studies in Mathematics,* 1976, *7,* pp. 157–159.

van Hiele, P. M., & D. van Hiele-Geldof. A method of initiation into geometry. In H. Freudenthal (Ed.), *Report on Methods of Initiation into Geometry.* Groningen: J. B. Wolters, 1958.

van Hiele-Geldof, D. *De didaktiek van de meetkunde in de eerste klas van het V.H.M.O.* Unpublished doctoral dissertation, University of Utrecht, The Netherlands, 1957.

VanLehn, K. *Bugs are not enough: Empirical studies of procedural flaws, impasses, and repairs in procedural skills.* Palo Alto, CA: Xerox C.I.S. – 11, 1981.

VanLehn, K. On the representation of procedures in repair theory. In H. P. Ginsburg (Ed.), *The Development of Mathematical Thinking.* New York: Academic Press, 1983.

VanLehn, K., & J. S. Brown. Planning nets: A representation for formalizing analogies and semantic models of procedural skills. In R. E. Snow, P. A. Federico, & W. E. Montague (Eds.), *Aptitude, Learning and Instruction: Cognitive Process Analyses.* Hillsdale, NJ: Erlbaum, 1980.

Voltaire, Emile. Candide ou l'optimisme (1759). In H. Benac (Ed.), *Voltaire: Romans et Contes.* Paris: Editions Garnier Freres, 1960.

Vygotsky, Lem Semyonovich. *Thought and Language.* (Eugenia Hanfmann and Gertrude Vakar, Eds. and Trans.). Cambridge, MA: MIT Press and John Wiley and Sons, 1962.

Vygotsky, Lem Semyonovich. *Mind in Society.* (Michael Cole, Vera John-Steiner, Silvia Scribner, and Ellen Souberman, Trans.). Cambridge, MA: Harvard University Press, 1978.

Wearne, Diana C. Development of a test of mathematical problem solving which yields a comprehension, application, and problem-solving score. In J. G. Harvey & T. A. Romberg (Eds.), *Problem-Solving Studies in Mathematics.* Madison, WI: Wisconsin R&D Center Monograph Series, 1980.

Weiner, B. *Achievement Motivation and Attribution Theory.* Morristown, NJ: General Learning Press, 1974.

Wertheimer, M. *Productive Thinking.* New York: Harper & Row, 1959.

White, B. Y. *Designing computer games to facilitate learning* (MIT AI TR-619). Cambridge, MA: MIT Artificial Intelligence Laboratory, February 1981.

Wickelgren, W. *How to Solve Problems.* San Francisco: Freeman, 1974.

Wilson, J. W. Generality of heuristics as an instructional variable. (Unpublished doctoral dissertation, Stanford University, 1967.) *Dissertation Abstracts International,* 1967, *28,* 2575A. (University Microfilms 67-17, 526.)

Winston, P. H. *Artificial Intelligence.* Reading, MA: Addison-Wesley, 1977.

Wirszup, Izaak. Breakthroughs in the psychology of learning and teaching geometry. In J. L. Martin (Ed.), *Space and Geometry: Papers from a Research Workshop.* Columbus, OH: ERIC/SMEAC, 1976.

Witkin, Herman A., & Donald R. Goodenough. *Cognitive Styles: Essence and Origins.* New York: International Universities Press, 1981.

Woods, D. R. Articles and Ideas. *P.S. News 29* (Nov.–Dec. 1983). Department of Chemical Engineering, McMaster University.

Zalewski, Donald L. A study of problem-solving performance measures. In J. G. Harvey & T. A. Romberg (Eds.), *Problem-Solving Studies in Mathematics.* Madison, WI: Wisconsin R&D Center Monograph Series, 1980.

Author Index

Subject Index